Sustainable Development and Application of Renewable Chemicals from Biomass and Waste II

Sustainable Development and Application of Renewable Chemicals from Biomass and Waste II

Mohamad Nasir Mohamad Ibrahim

MDPI
Basel • Beijing • Wuhan • Barcelona • Belgrade • Novi Sad • Cluj • Manchester

Editor
Mohamad Nasir Mohamad Ibrahim
School of Chemical Sciences
Universiti Sains Malaysia
Minden
Malaysia

Editorial Office
MDPI AG
Grosspeteranlage 5
4052 Basel, Switzerland

This is a reprint of articles from the Special Issue published online in the open access journal *Molecules* (ISSN 1420-3049) (available at: www.mdpi.com/journal/molecules/special_issues/KW126S22XJ).

For citation purposes, cite each article independently as indicated on the article page online and as indicated below:

Lastname, A.A.; Lastname, B.B. Article Title. *Journal Name* **Year**, *Volume Number*, Page Range.

ISBN 978-3-7258-2174-7 (Hbk)
ISBN 978-3-7258-2173-0 (PDF)
doi.org/10.3390/books978-3-7258-2173-0

Contents

About the Editor . . . . **vii**

Preface . . . . **ix**

Olga Gómez-de-Miranda-Jiménez-de-Aberasturi, Javier Calvo, Ingemar Svensson, Noelia Blanco, Leire Lorenzo and Raquel Rodriguez
Novel Determination of Functional Groups in Partially Acrylated Epoxidized Soybean Oil
Reprinted from: *Molecules* **2024**, *29*, 4582, doi:10.3390/molecules29194582 . . . . **1**

Kapil Khandelwal and Ajay K. Dalai
Prediction of Individual Gas Yields of Supercritical Water Gasification of Lignocellulosic Biomass by Machine Learning Models
Reprinted from: *Molecules* **2024**, *29*, 2337, doi:10.3390/molecules29102337 . . . . **19**

Chidiebere Millicent Igwebuike, Sary Awad and Yves Andrès
Renewable Energy Potential: Second-Generation Biomass as Feedstock for Bioethanol Production
Reprinted from: *Molecules* **2024**, *29*, 1619, doi:10.3390/molecules29071619 . . . . **46**

Mariacristina Compagnone, José Joaquín González-Cortés, María Pilar Yeste, Domingo Cantero and Martín Ramírez
Sustainable Recovery of Platinum Group Metals from Spent Automotive Three-Way Catalysts through a Biogenic Thiosulfate-Copper-Ammonia System
Reprinted from: *Molecules* **2023**, *28*, 8078, doi:10.3390/molecules28248078 . . . . **73**

Myeongjong Lee, Hyeongtak Ko and Seacheon Oh
Pyrolysis of Solid Recovered Fuel Using Fixed and Fluidized Bed Reactors
Reprinted from: *Molecules* **2023**, *28*, 7815, doi:10.3390/molecules28237815 . . . . **90**

David O. Usino, Päivi Ylitervo and Tobias Richards
Primary Products from Fast Co-Pyrolysis of Palm Kernel Shell and Sawdust
Reprinted from: *Molecules* **2023**, *28*, 6809, doi:10.3390/molecules28196809 . . . . **105**

João Moreira Neto, Josiel Martins Costa, Antonio Bonomi and Aline Carvalho Costa
A Novel Kinetic Modeling of Enzymatic Hydrolysis of Sugarcane Bagasse Pretreated by Hydrothermal and Organosolv Processes
Reprinted from: *Molecules* **2023**, *28*, 5617, doi:10.3390/molecules28145617 . . . . **120**

Yuchen Bai, Mingke Tian, Zhiwei Dai and Xuebing Zhao
Improving the Cellulose Enzymatic Digestibility of Sugarcane Bagasse by Atmospheric Acetic Acid Pretreatment and Peracetic Acid Post-Treatment
Reprinted from: *Molecules* **2023**, *28*, 4689, doi:10.3390/molecules28124689 . . . . **135**

Ittichai Kanchanakul, Thongchai Rohitatisha Srinophakun, Sanchai Kuboon, Hiroaki Kaneko, Wasawat Kraithong and Masahiro Miyauchi et al.
Development of Photothermal Catalyst from Biomass Ash (Bagasse) for Hydrogen Production via Dry Reforming of Methane (DRM): An Experimental Study
Reprinted from: *Molecules* **2023**, *28*, 4578, doi:10.3390/molecules28124578 . . . . **152**

Asim Ali Yaqoob, Nabil Al-Zaqri, Muhammad Alamzeb, Fida Hussain, Sang-Eun Oh and Khalid Umar
Bioenergy Generation and Phenol Degradation through Microbial Fuel Cells Energized by Domestic Organic Waste
Reprinted from: *Molecules* **2023**, *28*, 4349, doi:10.3390/molecules28114349 . . . . **162**

Kateřina Hájková, Michaela Filipi, Roman Fojtík and Ali Dorieh
Application of Alkali Lignin and Spruce Sawdust for the Effective Removal of Reactive Dyes from Model Wastewater
Reprinted from: *Molecules* **2023**, *28*, 4114, doi:10.3390/molecules28104114 . . . . **179**

Joaquín Hernández-Fernández, Heidis Cano and Ana Fonseca Reyes
Valoration of the Synthetic Antioxidant Tris-(Diterbutyl-Phenol)-Phosphite (Irgafos P-168) from Industrial Wastewater and Application in Polypropylene Matrices to Minimize Its Thermal Degradation
Reprinted from: *Molecules* **2023**, *28*, 3163, doi:10.3390/molecules28073163 . . . . **190**

About the Editor

Mohamad Nasir Mohamad Ibrahim

Mohamad Nasir Mohamad Ibrahim obtained his B.Sc. (1994), M.Sc. (1997), and PhD (1999) from Missouri S&T (formerly known as the University of Missouri-Rolla, USA). He currently serves as a Professor in the School of Chemical Sciences, Universiti Sains Malaysia (USM). He had published more than 200 journal articles, 17 book chapters, and 12 academic books throughout his 24-year career at USM, including "Graphene: A Versatile Advanced Material". Currently, his Scopus h-index is 47, with 7,750 citations, and he has been granted 13 patents for his R&D products/processes, where 5 of which are at an international level. More than thirty research grants (one international grant and eight industry grants) were secured and utilized to support his team's research activities. He has received 14 international awards for his research outputs. He has supervised more than 30 graduate students. His main research areas include Lignocellulosic Materials, Microbial Fuel Cells, Nanomaterials, Biocomposites, and Petroleum Chemistry. Recently, he has been busy working in the microbial fuel cells field, especially in developing a novel electrode, and he has published several papers in well-reputed journals such as *Chemical Engineering Journal, Journal of Cleaner Production*, etc. Currently, he is the pioneer of the MFCs research topic in the School of Chemical Sciences, USM. He enjoys sharing his industrial experience, where he spent two and half years working as an R&D Manager at the NSE Resources Corporation Sdn Bhd., in his classes.

Preface

The growing demand for sustainable industrial solutions has led to significant research advancements in various fields. This reprint presents a collection of journal articles focused on environmental sustainability, including wastewater treatment, renewable energy, biomass valorization, and industrial waste recycling. These studies reflect ongoing efforts to minimize industrial impacts on the environment while promoting efficient resource use.

The first article discusses the removal and recovery of Irgafos P-168, a synthetic antioxidant, from petrochemical wastewater. Using advanced extraction methods, this research mitigates environmental harm and demonstrates the potential to reuse recovered substances, extending product life.

The second article highlights the use of lignin and spruce sawdust as natural sorbents for dye removal from wastewater. This study emphasizes repurposing industrial by-products for environmental remediation, offering an eco-friendly solution to wastewater treatment.

Continuing the focus on green technologies, the third article explores microbial fuel cells (MFCs) for phenol degradation from wastewater while generating bioenergy. This dual-purpose approach paves the way for sustainable innovations in waste degradation and renewable energy.

In hydrogen production, the fourth article introduces an innovative approach that uses bagasse ash as a catalytic support for dry reforming of methane (DRM). By employing waste-derived catalysts, this study provides a sustainable alternative to traditional fossil fuel-based methods.

The reprint also addresses biofuel production, with articles five and six exploring pretreatment and enzymatic hydrolysis of sugarcane bagasse. These studies offer critical insights into improving cellulose digestibility and biofuel output, contributing to advancements in renewable energy.

Other articles cover topics such as the co-pyrolysis of biomass, plastic waste pyrolysis for recycling, bioleaching for platinum group metal recovery, and machine learning applications in gasification optimization. The diversity of subjects showcases the breadth of research aimed at solving environmental challenges while improving industrial efficiency.

Each article contributes to the vision of a sustainable future, where waste becomes a resource, and environmental harm is mitigated through innovation. The findings provide practical solutions for industries seeking to adopt more eco-friendly practices. We hope that this collection inspires further research and implementation of sustainable technologies across sectors.

By bringing together these pioneering studies, this reprint offers a comprehensive look at progress in environmental research and industrial sustainability, serving as a resource for scientists, engineers, and decision-makers working towards achieving a sustainable future.

Mohamad Nasir Mohamad Ibrahim
Editor

molecules

MDPI

Article

Novel Determination of Functional Groups in Partially Acrylated Epoxidized Soybean Oil

Olga Gómez-de-Miranda-Jiménez-de-Aberasturi [1,*], Javier Calvo [2], Ingemar Svensson [1], Noelia Blanco [1], Leire Lorenzo [1] and Raquel Rodriguez [1]

[1] Tecnalia R&I, Basque Research and Technology Alliance (BRTA), Parque Tecnológico de Alava 01510, Leonardo da Vinci 11, 01510 Vitoria-Gasteiz, Spain; ingemar.svensson@tecnalia.com (I.S.); noelia.blanco@tecnalia.com (N.B.); leire.lorenzo@tecnalia.com (L.L.); raquel.rodriguez@tecnalia.com (R.R.)

[2] Center for Cooperative Research in Biomaterials (CIC BiomaGUNE), Basque Research and Technology Alliance (BRTA), Paseo de Miramon 194, 20014 Donostia-San Sebastián, Spain; jcalvo@cicbiomagune.es

* Correspondence: olga.gomez@tecnalia.com

Abstract: The acrylation degree of vegetable oils plays a relevant role in determining the mechanical properties of the resulting polymers. Both epoxide and acrylate functionalities participate in polymerization reactions, producing various types of chemical bonds in the polymer network, which contribute to specific properties such as molecular size distribution, crosslinking degree, and glass transition temperature (Tg). The accurate identification of epoxide and acrylated groups in triglyceride molecules helps to predict their behavior during the polymerization process. A methodology based on analytical spectrometric techniques, such as direct infusion, mass spectrometry with electrospray ionization, and ultra-high-performance liquid chromatography, is used in combination with FTIR and ^{1}H NMR to characterize the epoxy and acrylic functionalities in the fatty chains with different numbers of carbon atoms of partially acrylated triglycerides obtained by a non-catalytic reaction.

Keywords: soybean oil; epoxidized intermediate; acrylated monomer; mass spectrometry; electrospray ionization; ultra-high-performance liquid chromatography

check for updates

Citation: Gómez-de-Miranda-Jiménez-de-Aberasturi, O.; Calvo, J.; Svensson, I.; Blanco, N.; Lorenzo, L.; Rodriguez, R. Novel Determination of Functional Groups in Partially Acrylated Epoxidized Soybean Oil. *Molecules* **2024**, *29*, 4582. https://doi.org/10.3390/molecules29194582

Academic Editors: Gavino Sanna and Irene Panderi

Received: 8 July 2024
Revised: 6 September 2024
Accepted: 20 September 2024
Published: 26 September 2024

1. Introduction

There is a global shift from using petroleum-based products to bio-based ones that are obtained from renewable resources like biomass. This is composed from organic matter such as vegetable or animal oils, garbage, waste, lignin, and cellulose [1]. Utilizing bio-based products from biomass, such as bio-epoxides from seed oils, bio-polyols, and bio-based acrylates or polyesters, is crucial for introducing biogenic carbon into the product life cycle, ultimately leading to a reduction in the carbon footprint by sequestering atmospheric carbon dioxide through photosynthesis [2]. Using biomass-derived raw materials for various polymer applications, such as coatings, adhesives, elastomers, and composites, is the most direct and effective approach. Examples include components made from seed oils or vegetable oils, bio-based polymers, UV-cured systems, polymers from recycled residues, and byproducts from secondary bio-based streams. The commitment to developing more sustainable products and processes constitutes a necessity for continuous innovation for the future industrial manufacturers [3]. In this context, soybean oil (SO) is a great renewable resource that is abundant, versatile, and low cost. It was employed to obtain acrylated monomers that have the necessary reactivity to synthesize new bio-based polymers as a sustainable alternative to traditional petroleum-based platform molecules.

Acrylated oils constitute a very promising alternative to be employed in diverse coating applications. The lower viscosity of these materials compared to conventional epoxy acrylates eliminates the need for their reduction with a reactive diluent [4]. They are primarily flexible and possess an aliphatic acrylic backbone, which can significantly improve the UV coating flexibility and adhesion of the polymer [4,5]. Special interest has been paid to UV-curable coatings, where acrylated oils react by means of UV-initiated free radical polymerizations in the presence of specific photoinitiators. As an example, new photocurable biocompatible liquid resins have been developed for 3D stereolithography-based bioprinting [6,7]. Acrylated coatings have also been developed to reduce the moisture sensitivity and permeability of bio-based films [8].

The acrylation degree of the oil is an important factor that is very relevant for the mechanical properties of the polymer. In this way, vegetable oils are not always fully acrylated to obtain monomers with specific characteristics. Boucher et al. [9] epoxidized and partially acrylated linseed oil and copolymerized it with (3,4-dihydroxyphenetyl)-acrylamide and N,N-dimethyl acrylamide to develop a coating for corrosion protection. A series of partially acrylated vegetable oils with different functional groups was synthesized and further polymerized with styrene [10]. Several methyl esters of C16 to C24 derived from camelina oil and linseed oil were epoxidized, fully or partially acrylated, and then polymerized in emulsion with various amounts of bio-based derivatives (5–30 wt.% in monomer mixture) to obtain polymeric latexes for coating formulations [11].

Typically, acrylated epoxidized oils are obtained by adding acrylic acid (AA) to epoxidized vegetable oils (ESO if they are obtained from SO), which are more reactive than the original raw materials. The commercial production of epoxidized oils is carried out with homogeneous acidic catalysts, carboxylic acids (principally acetic or formic acids), and an aqueous hydrogen peroxide solution, usually in concentrations between 30 and 50% *w*/*w* [12].

The acrylation process is usually carried out with an excess of AA, hydroquinone as a free radical inhibitor, and different catalysts. 1,4-diazobicyclooctane with temperatures up to 95 °C and times up to 11 h has been used [13]. Acrylated epoxidized soybean oil (AESO) was also synthesized using triphenylphosphine oxide as the catalyst and a temperature of 120 °C for 6 h, achieving a conversion of 95% [5]. Triethylamine was also considered as an efficient catalyst for the acrylation of epoxidized oils. In this case, a temperature of 80 °C and 8 h were employed to achieve an average conversion of the epoxidized groups of 70% [14]. Although they are common catalysts, they need to be handled with extra precautions due to their high oral and inhalation toxicity [15]. A solution of 40–60% chromium (III) 2-ethylhexanoate in a mixture of di (heptyl, nonyl, undecyl) phthalates) was demonstrated to be one of the most effective catalysts in the acrylation of vegetable oils. Complete acrylation of camelina oil with AA was obtained at 80 °C with 2% of this catalyst in 12 h [16].

Catalytic acrylation reactions offer several advantages, such as higher reaction rates at lower temperatures, but they also show some limitations. Once the acrylation process is finished, it is essential to purify the product to remove any traces of remaining AA and catalyst, which may negatively affect the product's further applications. Spent catalyst residues must be properly handled.

Meléndez et al. [17] studied the synthesis of an oligo-acrylated product from soya oil and acrylic acid without using any catalyst, which was obtained in 86% yield after 12 h at 120 °C. The acrylated monomers studied in this paper were obtained without the use of any catalyst to avoid the generation of any residual stream. The reactant ratio and operational conditions were selected to obtain a product with a high acrylation degree with lower reaction times to avoid the formation of undesired acrylic oil auto-polymerization.

On the other hand, the application of different analytical technologies is crucial for a deep analysis of the molecular structure of acrylated molecules. The most commonly used methods to characterize acrylated monomers include FTIR spectroscopy, which qualitatively describes the different functional groups present in triglyceride molecules, and ^{1}HNMR to calculate the average inclusion degree of AA in the epoxidized fatty chains. Chemical titrations such as the iodine value (IV) and the oxirane content (COOe) are used to determine the double bond conversion in the epoxidation reaction and the remaining epoxide presence after the acrylation reaction, respectively [18]. A better understanding of partially acrylated structures using mass spectrophotometric techniques would permit a visual identification of the different functionalities present in fatty chains of various lengths and original unsaturation degree. Triglycerides with no functionalities, those with a single functionality, and molecules with more than one acrylate group could be distinguished, which is interesting for controlling polymerization reactions.

Kuki et al. [19] employed matrix-assisted laser desorption ionization and electrospray ionization mass spectrometry (MALDI-MS and ESI-MS) for the characterization of epoxidized soybean and linseed oils. These techniques allowed for the identification of different epoxidized triglyceride (TG) mass spectral peak series and the number of epoxide groups in the products without any complicated and time-consuming sample preparation.

In this study, different spectroscopy techniques, such as direct infusion, mass spectrometry with electrospray ionization, and ultra-high-performance liquid chromatography were combined with FTIR and ^{1}HNMR to define a methodology for the characterization of the epoxy and acrylic functionalities in fatty chains with different numbers of carbon atoms.

2. Results and Discussion

First, SO was epoxidized with acetic acid and hydrogen peroxide (50% *w*/*w*) using H_2SO_4 (96%) as the acid catalyst. The purification process of the intermediates did not involve the use of any hazardous organic solvents, since impurities, unreacted materials, and catalyst residues were removed from the epoxidized intermediate through 4 washes with a warm brine solution (60 °C).

The epoxidized samples presented a COOe of 6.6 g O/100 g, and unsaturation degree of 8 g I_2/100 g. These values were determined by measuring the oxirane content and the IV of ESO. According to this, the oxirane yield calculated for the ESO intermediate was 88%.

The FTIR spectra of the samples demonstrated the development of distinctive peaks for the epoxidized products. A peak at 830 cm^{-1}, characteristic of the epoxide group, was observed, which evidenced that the reaction took place. Other representative peaks related to the fatty unsaturations =C–H at 3000 cm^{-1} and C=C at 1651 cm^{-1} appeared in the spectrum of SO but disappeared in the ESO spectrum, indicative that the internal double bonds of SO reacted with the peracid.

Table 1 displays the most relevant peaks found in the spectra of SO and ESO. It is noted that no peaks were present at wavelengths between 3300 and 3500 cm^{-1}, which are typical of -OH groups. This confirmed that no undesired reactions for producing glycols occurred. (Figure 1).

It is well known that the oxirane rings of ESO consist of tensioned structures that easily react with nucleophiles such as alcohols, amines, or protic acids. These molecules interact with the electrophilic carbons of the epoxide rings, polarizing them, breaking the initial bonds, and creating new ones as ester and hydroxyl functionalities (Scheme 1).

Table 1. FTIR characteristic peaks of SO and ESO.

Wavenumber (cm^{-1})	Functional Group
3000	=C–H stretching
2925	CH_2 stretching
2860	CH_3 stretching
1740	C=O asymmetric stretching
1650	C=C stretching
1460	CH_2 scissors
1230	C=C–C–O stretching
1160	C–O asymmetric axial stretching
823	C–O–C asymmetric epoxide stretching
723	CH_2 Rocking

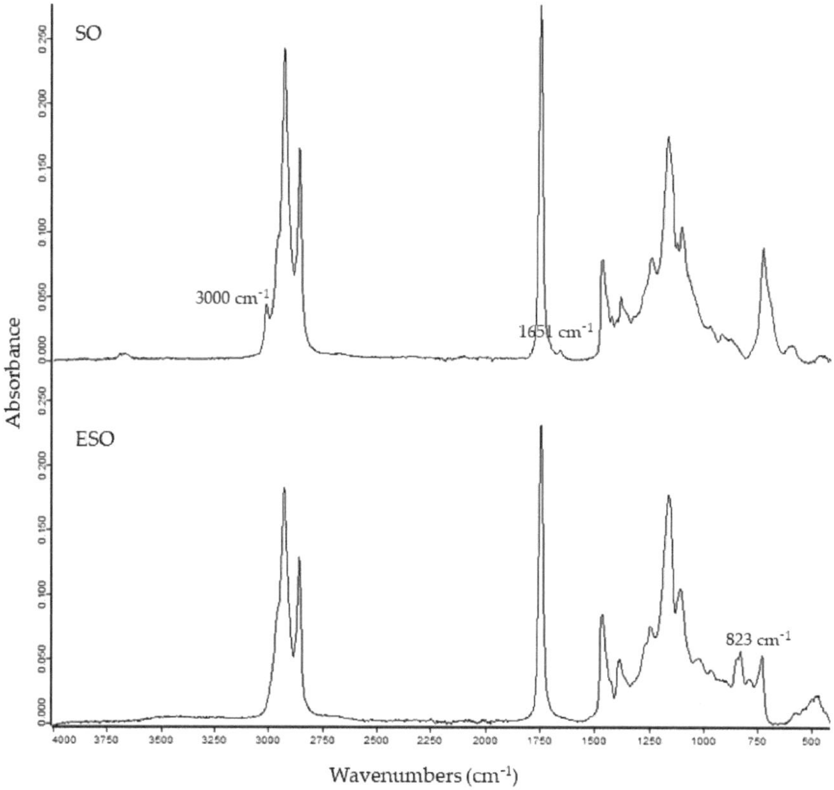

Figure 1. FTIR spectra of SO and ESO.

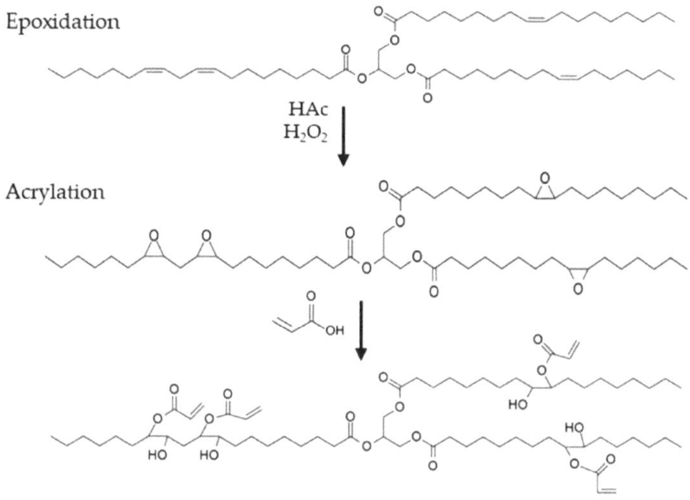

Scheme 1. Acrylation of SO in two steps through ESO intermediates.

In the second step, the epoxidized intermediates were reacted with AA to produce acrylated monomers. A slight excess of acid was used (RM AA/epoxide groups 1.1:1) along with hydroquinone (0.2% *w*/*w*) to prevent the auto-polymerization of AA.

Additionally, the acrylation process was carried out without a catalyst to avoid the generation of effluents containing spent catalyst, which are difficult to recover. AESO was obtained by controlling both the reagent ratio and temperature. Previous references described that the synthesis of an acrylated oligomer could be performed at 120 °C without generating toxic effluents containing spent catalyst [17]. Here, the formation of AESO was maintained for 4 h and monitored using FTIR spectroscopy to track the evolution of the different molecules that were being synthesized. The remaining epoxide groups were determined by the oxirane content measurement, which gave a value of 0.81 g O/100 g.

It was observed that longer reaction times led to undesired auto-polymerization of AA and/or AESO molecules, evidenced by a decrease in the acrylic peak length (Figure 2). These could limit the overall product yield despite full conversion. When the oxirane ring is opened with AA, it produces an AESO molecule with a secondary hydroxyl group. However, this secondary hydroxyl group can further esterify with another acrylic acid and then produce water. Alternatively, water can react with other epoxide rings, yielding diols as a byproduct [17].

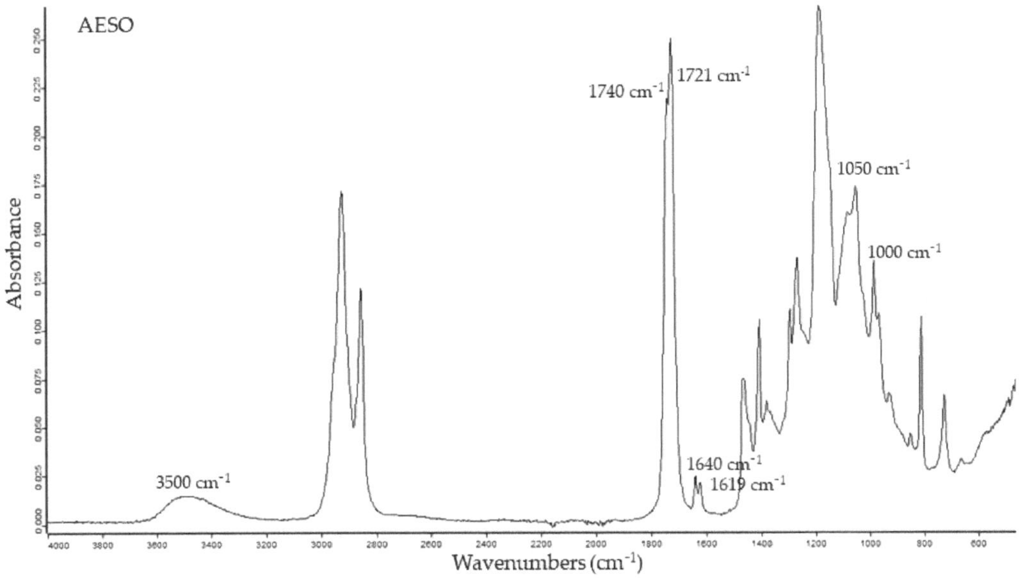

Figure 2. FTIR spectra of thermally obtained AESO without catalyst.

The disappearance of peaks at 823–830 cm^{-1} corresponding to epoxy groups indicated high reaction conversion. The remaining oxirane functionalities in the product could not be observed in the spectrum with enough precision. Additionally, new peaks attributed to acrylate groups (CH2=CH–COO–) appeared at 1619 cm^{-1}. The signals at 3400–3500 cm^{-1} corresponding to the formation of hydroxyl groups confirmed the opening of oxirane rings to form acrylate groups and alcohols. Other notable peaks included the C=O stretching vibration of acrylate ester groups at 1721 cm^{-1} close to the triglyceride ester peak at 1740 cm^{-1}. Other relevant peaks were C–O stretching at 1000 cm^{-1}, =C–H bending (twisting) at 1050 cm^{-1}, and the trans and cis C=C stretching vibration at 1640 cm^{-1}, all of which demonstrated the attachment of acrylate groups to fatty acid chains. The principal FTIR signals are listed in Table 2.

Table 2. Summary of principal new peaks that appeared in the AESO molecules' spectra.

Wavenumber (cm^{-1})	Functional Group
3400–3500	–OH stretching
1721	C=O acrylate stretching
1635	CH_2=CH– acrylate stretching
1619	CH_2=CH–COO– tension
1405	CH_2=CH- acrylate flexion
1050	=C–H bending (twisting)
1000	–C–O-C stretching
810	–HC=CH– out of plane bending vibration

Figure 3 shows the ^1H NMR spectrum of thermically acrylated epoxidized soybean oil. The two sets of peaks from 4.0 to 4.4 ppm are produced by the four methylene hydrogen atoms attached to the glycerol center. The peak at 2.3 ppm is produced by the six methylene hydrogen atoms alpha to the carbonyl groups. The peak areas of the four methylene hydrogens in glycerol were used as the internal standard to determine the number of acrylates per triacylglycerol by comparing with the peak areas from the three acrylate protons (5.7–6.6 ppm). Considering an equivalent methodology to the ones described by Zhang et al. [20] and Su et al. [21], the acrylated molecule/TG ratio was determined to give a value of 2.1 using the following formula:

$$n = \frac{\text{Area (vinyl protons : 5.7–6.6 ppm)}}{\text{Area (glycerol methylens : 4.4–4.4 ppm)}} \cdot \frac{4}{3} \quad (1)$$

The authors defined an initial number of double bonds per SO molecule as 4.08. Therefore, the conversion of double bonds to acrylates was calculated to be 51%.

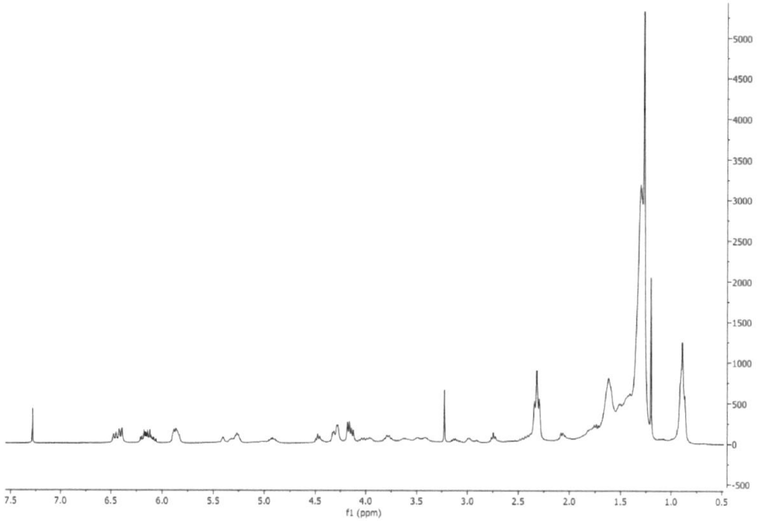

Figure 3. ^1H NMR spectra of the thermally synthesized AESO.

The signals of the protons [–CHOCH–] corresponding to the remaining epoxide groups in the acrylate monomers at 2.9 ppm could not be distinguished with high precision. Additional techniques were necessary for a more detailed analysis of the different functionalities present in the product.

A specific methodology based on electrospray ionization techniques was developed to clearly distinguish AESO molecules with similar masses but different geometries, which can cause different steric effects and reactivity and affect in a different way further polymerization processes. These technologies are favored over others due to their gentle ionization process, ensuring intact ionization and analysis of samples. They are most suitable for measuring lipid substances and can efficiently identify changes in their different structures, such as unsaturated, epoxidized, or acrylated functionalities. Additionally, ultra-high-performance liquid chromatography and mass spectrometry detection are used before ionization to improve detection. These steps ensure that the ionization of the main substances does not interfere with the ionization and detection of less abundant ones, also known as ion suppression.

Three different types of samples were analyzed by mass spectrometry techniques: (1) unreacted SO; (2) ESO intermediate, and (3) AESO sample obtained by a non-catalytic reaction. Two different approaches were carried out for the characterization of the functional groups present in the molecules.

- Method 1: Direct sample infusion and mass spectrometry detection with an electrospray source and high-resolution time-of-flight analyzer (DI-ESI-TOF-MS).
- Method 2: Ultra-high-performance liquid chromatography and mass spectrometry detection with an electrospray source and high-resolution time-of-flight analyzer (UHPLC/ESI-TOF-MS).

To make the identification of unsaturations, epoxidations, and acrylated groups easier, the diverse TG structures were named as shown in Figure 4.

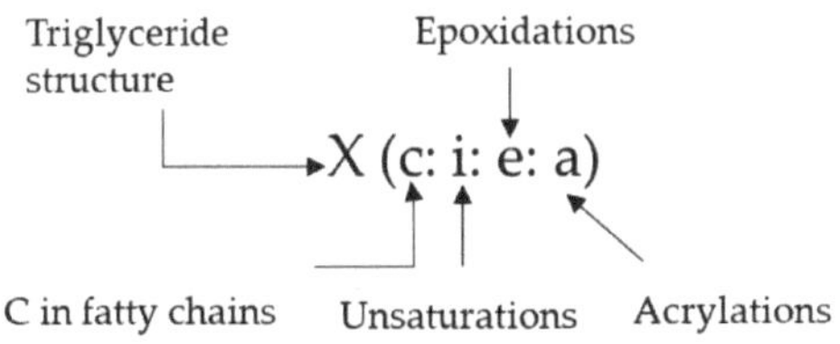

Figure 4. Nomenclature used in mass spectrometry analysis.

The possibilities of the TG distribution analyzed are very broad. Molecules of different fatty chain lengths, unreacted TGs (TGs), fully epoxidized TGs (ETG), partially epoxidized TGs (pETG), completely acrylated TGs (ATG), and partially epoxidized and acrylated TGs (pATG) could be detected. In Figure 5, an example of the nomenclature used for the determination of sample molecules is shown.

It was evidenced that by using DI-ESI-TOF-MS (method 1), no epoxidation or acrylation functionalities appeared in the SO sample (Figure 6a) due to the absence of signals corresponding to these structures in comparison to the corresponding commercial extracts. Different fully epoxidized molecules (even for the less abundant TGs) were found in the ESO intermediates thanks to the high resolution of the technique (Figure 6b).

In the AESO sample synthesized without using any catalyst, epoxidized and acrylated functionalities were found in TGs with different chain lengths. This revealed the uncomplete conversion of the reaction (Figure 7). Both functionalities were more clearly observed with this analytical technique than with others.

Taking into consideration a mass difference $\Delta m/z = 72$ between the epoxy and the acrylate groups, a series of TGs with several acrylation degrees was identified considering the mass differences between the epoxidized/acrylated groups.

Analyzing more in detail the series (54:4), in which the assignment was unambiguous, it was possible to calculate the epoxidation/acrylation ratio, assuming that all triglycerides, having very similar structures, ionize in a similar way in the mass spectrometer.

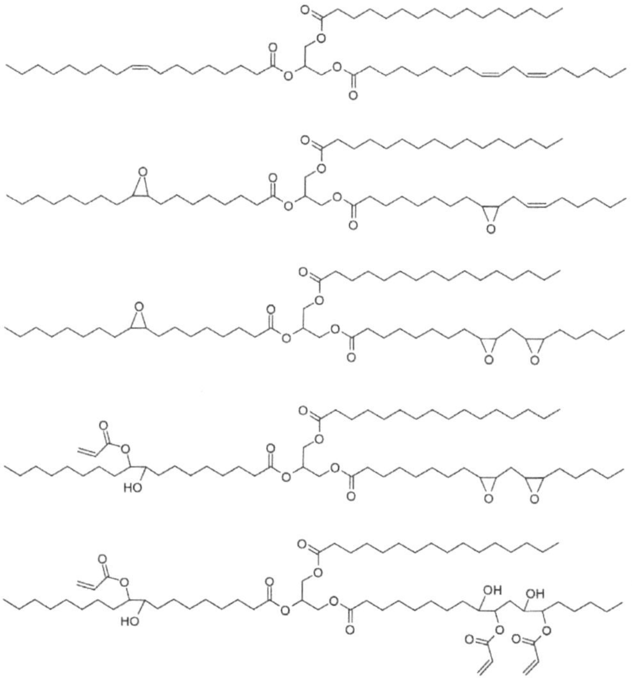

Figure 5. Hypothetical structures of a partially acrylated TG of 52C with 3 original unsaturations and with different degree of epoxidation and acrylation.

If m is defined as the total number of functionalized triglyceride unsaturations, either as epoxy or as acrylate, the total concentration of this group is given by the following formula:

$$C_{\mathrm{func}\ (m,n)} = C_{\mathrm{total}\ (m,n)} \cdot m \tag{2}$$

Similarly, if n is defined as the number of acrylate groups in a sample, the concentration of this group can be calculated as follows:

$$C_{\mathrm{acrylated}\ (m,n)} = C_{\mathrm{total}\ (m,n)} \cdot n \tag{3}$$

And the concentration of the epoxy group in the sample is defined as follows:

$$C_{\mathrm{epoxi}\ (m,n)} = C_{\mathrm{total}\ (m,n)} \cdot (m - n) \tag{4}$$

Thus, the acrylation percentage %AC of a specific triglyceride defined by m functionalized unsaturations and n acrylate ones is determined by the following equation:

$$\%\mathrm{Ac} = 100 \times \frac{\sum_{n=0}^{m} n \cdot C_{\mathrm{total}\ (m,n)}}{\sum_{n=0}^{m} m \cdot C_{\mathrm{total}\ (m,n)}} \tag{5}$$

On the other hand, if we define ionizability obtained by mass spectrometry, i, as the ratio between the intensity (Int) obtained from a signal and its relationship to the concentration of the analyte, assuming that the ionizability of fully epoxidized, partially acrylated, or fully acrylated triglycerides is similar and proportional, the acrylation percentage can be defined as follows:

$$\%\mathrm{Ac} = 100 \times \frac{\sum_{n=0}^{m} n \cdot \mathrm{Int}_{(m,n)}}{m \cdot \sum_{n=0}^{m} \mathrm{Int}_{(m,n)}} \tag{6}$$

It is worth mentioning that making an unequivocal assignment between different lipids is quite difficult since the increase in mass due to one more carbon in a fatty acid and the presence of an epoxy group offer very similar signals (*m*/*z* 13.98, 14.02). However, it is possible to estimate the percentage of ionization based on the assigned signals corresponding to unique structures and assuming that the ionizability of these structures is similar concerning the number of epoxy/acrylate groups. Using the intensities for the different components Int (m,n) obtained in the ETG(54:4:4:0); ATG(54:4:4:4), in relation to the acrylation number (n) and the number of total epoxides groups (m) (Figure 8 and Table 3), it could be confidently approximated that the percentages of acrylation with respect to the epoxidized groups were around 61.1% (ec 7) and 53.7% of the initial double bonds of SO (taking in consideration that the oxirane yield calculated for the ESO intermediate was 88%).

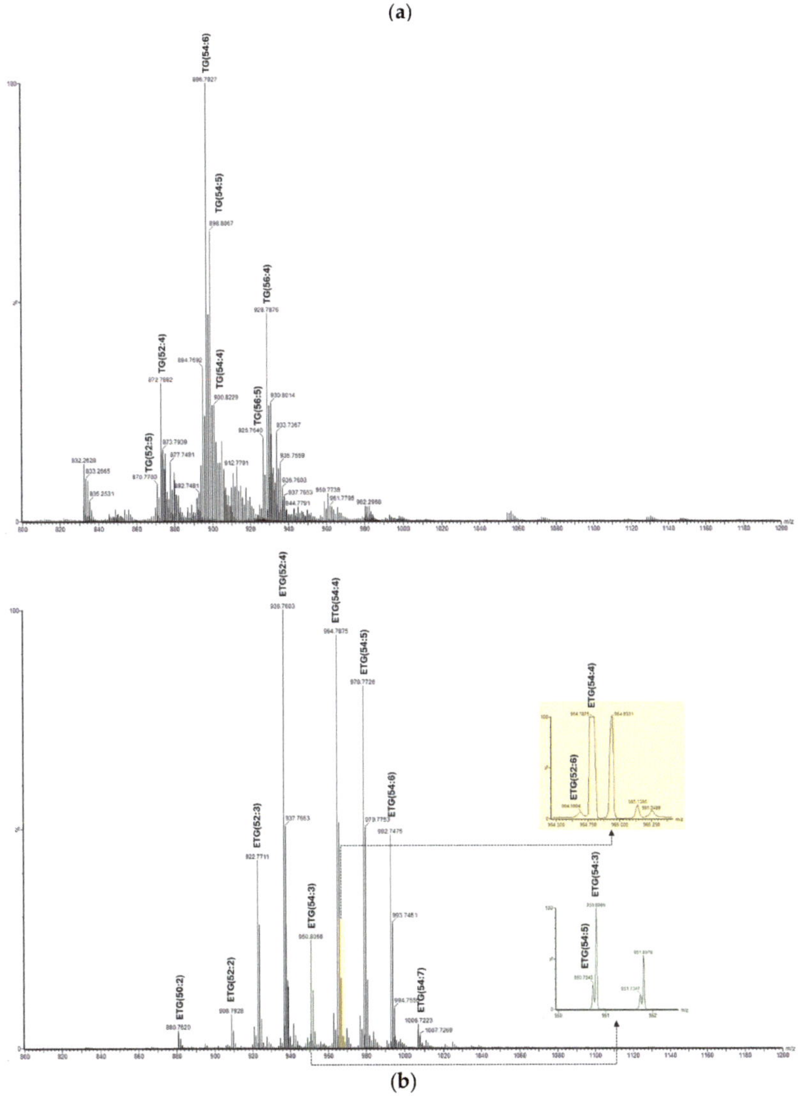

Figure 6. DI-ESI-TOF-MS spectra of SO (**a**) and ESO samples (**b**).

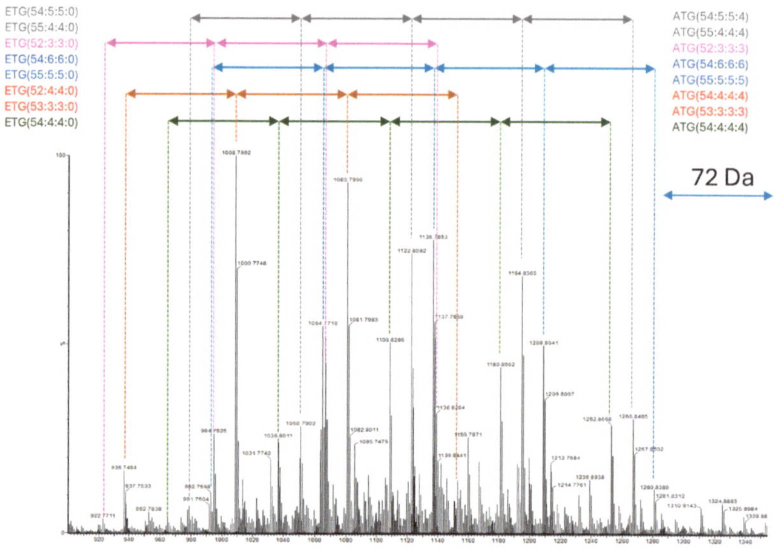

Figure 7. DI-ESI-TOF-MS spectra of thermally synthesized AESO. Identification of different acrylation series: ETG; pETG; ATG.

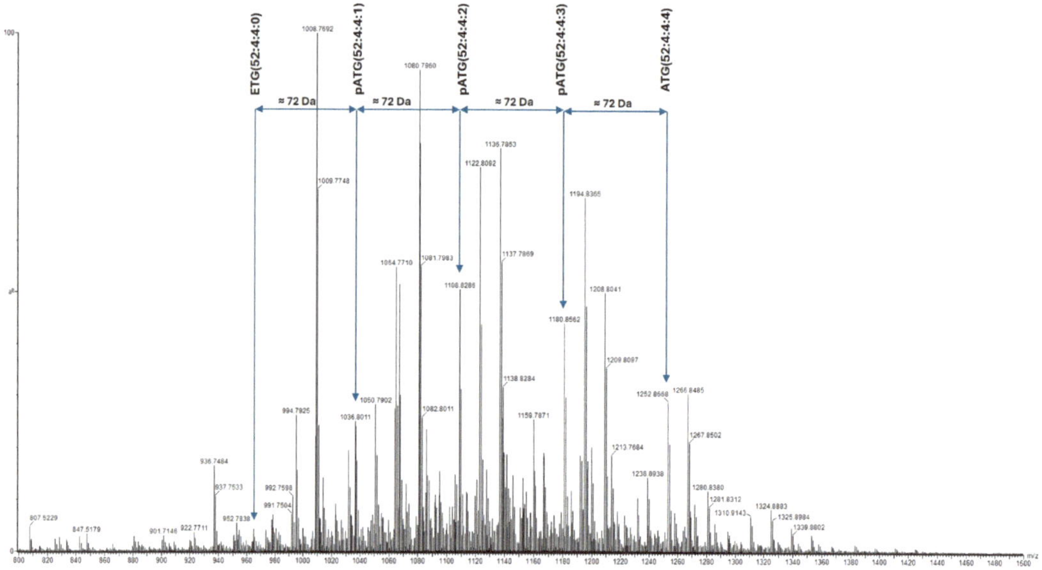

Figure 8. DI-ESI-TOF-MS spectra of AESO synthesized without catalyst—series assignation ETG (54:4:4:0); ATG(54:4:4:4). Total and relative intensities.

Table 3. Assignment of MS signals and their absolute and relative intensities. Series (54:4).

Lipid	Mass Intensity (a.u.)	Relative Mass Intensity (%)	Acrylic Groups
ETG(54:4:4:0)	469,000	2.8	0
pATG(54:4:4:1)	2,740,000	16.4	1
pATG(54:4:4:2)	5,520,000	33.1	2
pATG(54:4:4:3)	4,800,000	28.8	3
ATG(54:4:4:4)	3,140,000	18.8	4

The results obtained fit with the corresponding ones obtained by ^{1}H NMR, taking into consideration that the oxirane yield calculated for the ESO intermediate was 88%.

On the other hand, different assignations were made using UPLC/MS (method 2) for the ESO and AESO molecules. Due to the sample's complexity, only some TGs were unequivocally identified. The chromatograms corresponding to ETG(50:2) and ETG(50:3) showed that these compounds were not detectable by direct infusion (method 1) and were determined using this methodology instead. Some isomers could also be separated based on their polarity (Figure 9).

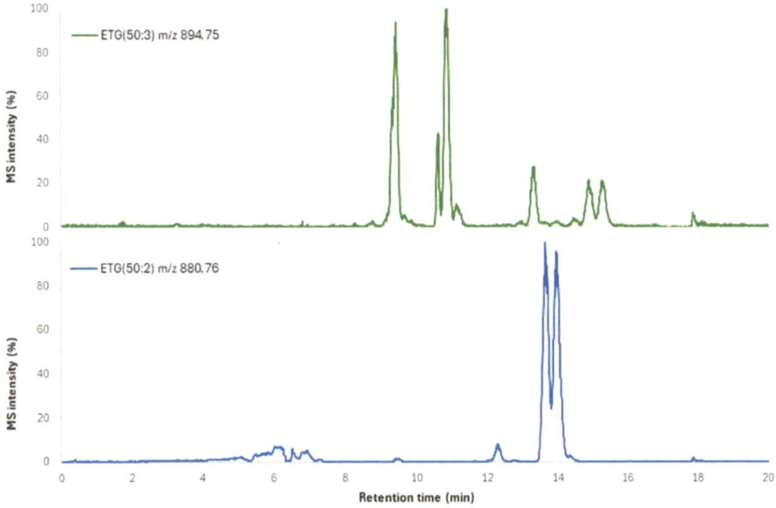

Figure 9. Extracted-ion chromatograms (XIC) at m/z 880.76 and 894.75 corresponding to the structures ETG(50:2) and ETG(50:3), respectively, obtained by UPLC-ESI-TOF-MS.

The series of TGs with 52 and 54 carbons with different degrees of epoxidation were monitored by extracting the ion chromatograms. Thus, various isomers of these ETGs could be separated based on their different polarities (Figure 10). ETG(52:2), ETG(52:3), ETG(52:4), ETG(52:5), and ETG(52:6) were successfully monitored in the following chromatograms (Figure 9). The retention times were according to the higher polarities of more epoxy groups in the TGs.

Similar structures could be found in the AESO sample, but in this case, the signals corresponded to acrylated groups instead of epoxide rings (Figure 11). It is worth mentioning that chromatographic separation allowed for the separation and detection of numerous isobaric structures of different acrylated TGs based on their polarity.

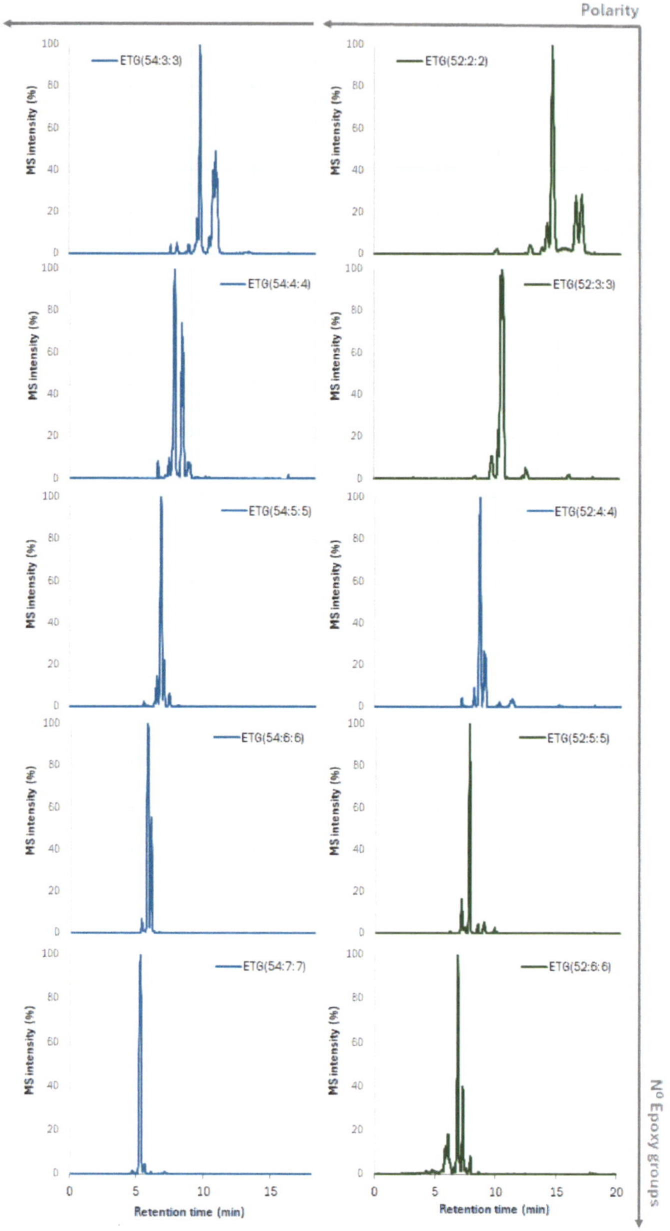

Figure 10. UPLC-ESI-TOF-MS/ESO/Extracted ion-chromatograms at different m/z corresponding to ETG (54:3);ETG (54:7) and ETG (52:2); ETG (52:6) series.

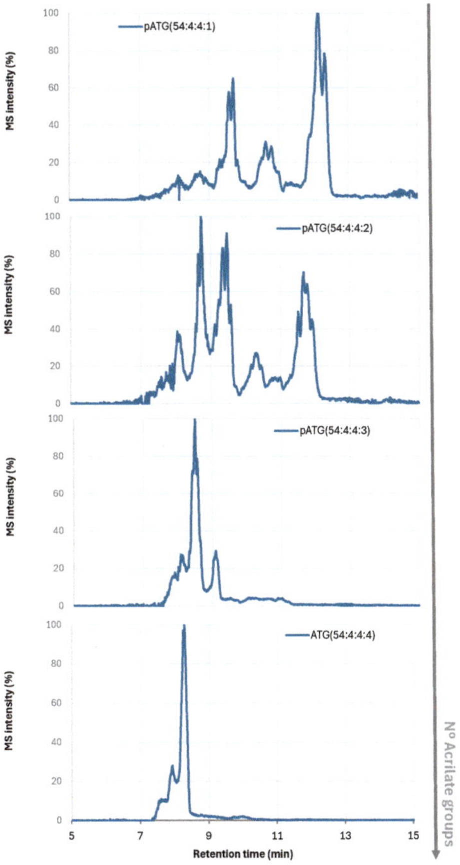

Figure 11. UPLC-ESI-TOF-MS/AESO/Extracted ion-chromatograms at different *m/z* corresponding to pATG(54:4:4:1), pATG(54:4:4:2), pATG(54:4:4:3), and ATG(54:4:4:4).

3. Materials and Methods

3.1. Materials

3.1.1. SO Epoxidation

The following reactants were used: SO (cold-pressed, 5 liters, Mystic Moments (Madar Corporation Ltd), Fordingbridge, UK Batch #4446101, www.mysticmomentsuk.com URL accessed on 4 February 2024). Glacial acetic acid (HAc; HPLC-grade, Scharlau, Barcelona, Spain Batch #23893408), hydrogen peroxide (H_2O_2; 50% *w*/*w*, Scharlau, Batch #24267501), H_2SO_4, (95–98%, Scharlau, Batch #24332404), Na_2SO_4 (extra pure, anhydrous, Thermo Scientific, Madrid, Spain Lot: A0449840), NaCl (99.5%, laboratory reagent-grade, Fisher Chemical, Madrid, Spain, Lot: 2222866), and $NaHCO_3$ (99.7%, Scharlau, Batch #22779201).

3.1.2. Partial Acrylation of the Epoxidized Intermediates

Partially epoxidized samples were synthetized using AA (extra pure, stabilized, Thermo Scientific, Lot: A0459748) and hydroquinone (99.5%, Acros Organics, Madrid Spain Lot: A0346098). Solvents and purification reagents were the same as previously described.

3.2. Methods

3.2.1. Synthesis of the Epoxidized Intermediates

The epoxidation of SO (129.3 g I_2/100 g) was previously described elsewhere [12,22]. Reactions were conducted using SO and H_2SO_4 as the catalyst. Reactions were started by adding 100 g of SO to an isothermal jacketed reactor, which was mechanically stirred and provided with a reflux condenser. The oil was kept stirring at 400 rpm until it reached the desired reaction temperature (65 °C). Then, 0.3 moles of acetic acid per mol of double bonds of SO were added. A solution of 1.25 mol of H_2O_2 per mol of double bonds of the oil and sulfuric acid (2% w/w) was added drop by drop by means of an HPLC pump with a constant flow of 1 mL/min. The reaction mixture was maintained at constant temperature for 3 h.

To purify the reaction product, it was transferred to 250 mL PE centrifuge tubes and then centrifuged at 3000 rpm for 10 min. The oily phase was recovered and washed 4 times with warm brine solution (60 °C). No organic solvent was used to dissolve the epoxidized oil and favorize the water extraction. This positively resulted in saving costs and reduced residue generation.

When neutral pH of the washing water was achieved, the product was dried and clarified by adding Na_2SO_4. Next, it was filtered and further dried in a rotavapor at 45–50 °C.

3.2.2. Production of the Acrylated Epoxidized Monomers

Hydroquinone (70 mg) was dissolved at 80 °C for more than 10 min in a reactor that contained 70 g of the epoxidized oil, then 21.9 mL of AA was added slowly for 1 h. Then, the reaction was heated to 120 °C and kept at this temperature for 3 h and finally cooled down.

The acrylated product was purified with an equal volume of ethyl acetate as the organic phase, which was added to the reactor. Next, it was neutralized with $NaHCO_3$ (1% aq.) until the bubbling of CO_2 stopped. The upper organic phase was separated in 250 mL plastic tubes that were centrifuged at 3000 rpm for 10 min. The white precipitate (presumably sodium polyacrylate) that formed in the interphase was discarded. The washing and centrifugation steps were repeated with $NaHCO_3$ (1%) and NaCl (sat.) until neutrality. The organic solution was dried with Na_2SO_4 salt, filtered, and the solvent was evaporated on a rotavapor at 45–50 °C.

3.3. Characterization

3.3.1. Oxirane Content COOe

It was determined according to the ASTM D1652-04 standard [23]. This process determined the percentage of epoxide groups in an epoxidized intermediate sample and in the final acrylated product through the titration of a standardized 0.1 N perchloric acid solution in glacial acetic acid with a tetraethyl ammonium bromide solution in acetic acid, which includes the sample for analysis.

Likewise, the oxirane yield of the epoxidized intermediate was calculated from the ratio between the experimental oxirane content COO_{exp} and the theoretical oil maximum oxirane content of the oil, according to the following formula [12]:

$$Y_{COO}(\%) = \frac{COO_{exp}}{COO_{max}} \cdot 100 \tag{7}$$

Whereas COO_{exp} was measured empirically, COO_{max} is the theoretical maximum oxirane content of the SO, which was calculated as follows:

$$COO_{max} = \left[\frac{{}^{IV_0}/_{2 \cdot A_i}}{100 + \left(\left({}^{IV_0}/_{2 \cdot A_i} \right) \cdot A_0 \right)} \right] \cdot A_0 \cdot 100 \tag{8}$$

A_i is the atomic weight of iodine, 126.9; and A_O is that of O, 16.0.

3.3.2. Iodine Value (IV)

IV quantifies the unsaturation degree of SO. It was determined using Wij's method [24]. It is based on the iodometric titration of the remaining iodine content after the reaction with an appropriate reagent.

$$IV = \frac{12.69(V_b - V_s)\cdot N}{W} \quad (9)$$

where $(V_b - V_s)$ represents the difference, in mL, of the sodium thiosulfate solution required for the blank and the sample, respectively. N stands for the normality of the sodium thiosulfate solution in eq/L. The value 12.69 serves as the conversion factor from meq of sodium thiosulfate to grams of iodine (considering that the molecular weight of iodine is 126.9 g/mol). Lastly, W represents the weight of the sample in grams.

3.3.3. FTIR

A Bruker Alpha Platinum ATR spectrometer was used, which collects spectra at a resolution of 4 cm^{-1} in absorbance mode within the range of 400–4000 cm^{-1}. For each analysis, a total of 24 scans were carried out, respectively, and they were analyzed using OPUS 6.5 software.

3.3.4. ^{1}H NMR

It was used to estimate the acrylation degree of the monomers, which was obtained by the integration of the peak areas referenced to stable protons considered as reference peaks. A Bruker Avance 400 MHz spectrometer equipped with a QNP z gradient probe was used at room temperature. The signals obtained were represented in parts per million (ppm) in relation to the internal standard tetramethylsilane. The spin multiplicity was expressed by s = singlet, d = doublet, t = triplet, q = quartet, and m = multiplet. The analysis was performed after dissolving the acrylated samples in deuterated chloroform (40 mg/mL).

3.3.5. Mass Spectrometry Analysis

This was carried out to determine the position/distribution of the acrylated groups in the TG molecules. Two different approaches were adopted:

Method 1: Direct sample infusion and mass spectrometry detection with an electrospray source and high-resolution time-of-flight analyzer (DI-ESI-TOF-MS).

Mass analyses were carried out on an ESI-QTOF, Synapt XS (Waters, Mildford, MA, USA). All instrument parameters were optimized to obtain the best signal for TGs and their derivatives. Thus, the capillary and cone voltages were set at 700 and 50 V, the nebulizer and cone gas flow rates were 700 and 150 L/h, and the nebulizer and source temperatures were set at 350 and 150 °C, respectively.

The equipment was calibrated externally and prior to analysis using a dilution of sodium iodide at 1 μg/mL in water/isopropanol (1:1). In addition, during the measurements, a dissolution of leucine enkephalin (200 ng/mL in water/isopropanol 1:1) was introduced in parallel in the source (lock mass) in order to adjust the measurements during the analyses and to achieve accuracies below 5 ppm (high resolution—accurate mass).

All samples were diluted to 100 μg/mL in methanol and 1 μL was injected into the equipment using an FNT-type automatic injector (Acquity, Waters, Mildford, MA, USA) and a mobile phase of 10 mM ammonium formiate in water and ACN.

The acquisition time was 30 sec in the mass ranges m/z 800–1200 and 800–1500 (acrylates). All data obtained were processed using Masslynx v4.2 software (Waters, Mildford, MA, USA) and signal assignment was performed by monitoring mainly the $[M + NH4]^+$ adduct.

Method 2: Ultra-high-performance liquid chromatography and mass spectrometry detection with an electrospray source and high-resolution time-of-flight analyzer (UHPLC/ESI-TOF-MS).

Before detection by mass spectrometry, the samples were separated by liquid chromatography using Acquity Premier-type equipment equipped with a BEH C18-type column of dimensions 100 × 2.1 mm and particle size of 1.7 um (Waters, Milford, MA, USA).

A gradient was employed for the chromatographic separation using a mixture of water with 50 mM ammonium formiate (A) and ACN (B) as the mobile phases, as indicated in Table 4:

Table 4. Mixtures of A and B used for chromatographic separation.

Time (min)	Flow (mL/min)	% A	% B
0	0.3	30	70
1	0.3	30	70
7	0.3	1	99
17	0.3	1	99
17.5	0.3	30	70
20	0.3	30	70

The column temperature was set at 30 °C, and the injection volume was 1 μL. The total analysis time was 20 min. The detection was carried out with using ESI-QTOF Synapt XS equipment (Waters, Milford, MA, USA) with the use of mass spectrometry. All instrumental parameters were optimized to obtain the best signal for TGs and their derivatives. To achieve this, the capillary voltage and cone voltage were firmly set to 700 and 50 V, respectively, while the nebulization and cone gas flow rates were firmly set at 700 and 150 L/h, respectively, and the nebulization and source temperatures were set at 350 °C. The equipment was externally calibrated using a NaI solution with a concentration of 1 μg/mL in a water–isopropanol solution at a ratio of 1:1. During measurements, a leucine enkephalin solution with a concentration of 200 μg/mL in a water–isopropanol solution in a 1:1 ratio was introduced in parallel into the source (lock mass). The acquisition was performed in positive mode within the precise mass ranges of m/z 800–1200 and 800–1500 (acrylates). All data obtained were processed using Masslynx v4.2 software (Waters, Milford, MA, USA), and the signal assignment was performed by monitoring the $[M + NH4]^+$ adduct.

4. Conclusions

A novel methodology based on electrospray ionization techniques (UPLC/ESI-TOF-MS) was developed to accurately analyze the acrylated species formed during the reaction between ESO and AA. These techniques can effectively differentiate between isomers with equivalent mass but different spatial configurations as a function of their polarity and their affinity to the chromatographic column. The characterization and quantification of these structures is a complex task that requires advanced MS/MS characterization to ensure that each isomer is unequivocally assigned. These methods are very useful for controlling the functionalities (epoxide groups and acrylic bonds) of modified triglycerides, which can affect the properties of acrylate polymers, such as the crosslinking degree, Tg, or durability.

Author Contributions: Conceptualization, O.G.-d.-M.-J.-d.-A., I.S. and R.R.; Methodology, O.G.-d.-M.-J.-d.-A., J.C. and L.L; Formal analysis, J.C., N.B. and L.L.; Investigation, O.G.-d.-M.-J.-d.-A., I.S., N.B. and L.L.; Writing—original draft, O.G.-d.-M.-J.-d.-A. and J.C.; Writing—review & editing, O.G.-d.-M.-J.-d.-A., J.C. and I.S.; Supervision, O.G.-d.-M.-J.-d.-A.; Project administration, R.R.; Funding acquisition, O.G.-d.-M.-J.-d.-A. and R.R. All authors have read and agreed to the published version of the manuscript.

Funding: This research was funded by EUROPEAN UNION'S HORIZON EUROPE RESEARCH AND INNOVATION PROGRAMME, through TORNADO project, grant number N° 101091944. The views and opinions expressed are, however, those of the authors only and do not necessarily reflect those of the European Union or HaDEA. Neither the European Union nor the granting authority can be held responsible for them.

Institutional Review Board Statement: Not applicable.

Informed Consent Statement: Not applicable.

Data Availability Statement: The data presented in this study are available upon request from the corresponding authors.

Acknowledgments: The authors acknowledge the CIC Energigune (Albert Einstein 48, Miñano, Spain) for the ^{1}HNMR analysis.

Conflicts of Interest: The authors declare no conflicts of interest.

References

1. Mielenz, J.R. Small-scale approaches for evaluating biomass bioconversion for fuels and chemicals. In *Bioenergy: Biomass to Biofuels and Waste to Energy*; Academic Press: Cambridge, MA, USA, 2015; pp. 545–571. [CrossRef]
2. Thomas, J.; Patil, R. Enabling Green Manufacture of Polymer Products via Vegetable Oil Epoxides. *Ind. Eng. Chem. Res.* **2023**, *62*, 1725–1735. [CrossRef]
3. Thomas, J.; Patil, R.S.; Patil, M.; John, J. Navigating the Labyrinth of Polymer Sustainability in the Context of Carbon Footprint. *Coatings* **2024**, *14*, 774. [CrossRef]
4. Kopyciński, B.; Langer, E. UV-curable coatings—Environment-friendly solution for sustainable protection of metal substrates. *Ochr. Przed Koroz.* **2023**, *1*, 4–13. [CrossRef]
5. Chu, X.M.; Liu, S.J.; Zhao, F.Q. Preparation of acrylated epoxidized soybean oil with excellent properties. *Appl. Mech. Mater.* **2014**, *662*, 7–10. [CrossRef]
6. Miao, S.; Zhu, W.; Castro, N.J.; Nowicki, M.; Zhou, X.; Cui, H.; Fisher, J.P.; Zhang, L.G. 4D printing smart biomedical scaffolds with novel soybean oil epoxidized acrylate. *Sci. Rep.* **2016**, *6*, 27226. [CrossRef]
7. Mendes-Felipe, C.; Isusi, I.; Gómez-Jiménez-Aberasturi, O.; Prieto-Fernandez, S.; Ruiz-Rubio, L.; Sangermano, M.; Vilas-Vilela, J.L. One-Step Method for Direct Acrylation of Vegetable Oils: A Biobased Material for 3D Printing. *Polymers* **2023**, *15*, 3136. [CrossRef]
8. Ge, X.; Yu, L.; Liu, Z.; Liu, H.; Chen, Y.; Chen, L. Developing acrylated epoxidized soybean oil coating for improving moisture sensitivity and permeability of starch-based film. *Int. J. Biol. Macromol.* **2019**, *125*, 370–375. [CrossRef]
9. Boucher, D.; Ladmiral, V.; Negrell, C.; Caussé, N.; Pébère, N. Partially acrylated linseed oil UV-cured coating containing a dihemiacetal ester for the corrosion protection of an aluminium alloy. *Prog. Org. Coat.* **2021**, *158*, 106344. [CrossRef]
10. Fu, L.; Yang, L.; Dai, C.; Zhao, C.; Ma, L. Thermal and mechanical properties of acrylated expoxidized-soybean oil-based thermosets. *J. Appl. Polym. Sci.* **2010**, *117*, 2220–2225. [CrossRef]
11. Kolář, M.; Honzíček, J.; Podzimek, Š.; Knotek, P.; Hájek, M.; Zárybnická, L.; Machotová, J. Derivatives of linseed oil and camelina oil as monomers for emulsion polymerization. *J. Mater. Sci.* **2023**, *58*, 15558–15575. [CrossRef]
12. Gómez-De-Miranda-Jiménez-De-Aberasturi, O.; Perez-Arce, J. Efficient epoxidation of vegetable oils through the employment of acidic ion exchange resins. *Can. J. Chem. Eng.* **2019**, *97*, 1785–1791. [CrossRef]
13. Campanella, A.; Fahimian, M.; Wool, R.P.; Raghavan, J. Synthesis and rheology of chemically modified canola oil. *J. Biobased Mater. Bioenergy* **2009**, *3*, 91–99. [CrossRef]
14. Habib, F.; Bajpai, M. Chemistry synthesis and characterization of acrylated epoxidized soybean oil for UV cured coatings. *Chem. Chem. Technol.* **2011**, *5*, 317–326. [CrossRef]
15. Ho, Y.H.; Parthiban, A.; Thian, M.C.; Ban, Z.H.; Siwayanan, P. Acrylated Biopolymers Derived via Epoxidation and Subsequent Acrylation of Vegetable Oils. *Int. J. Polym. Sci.* **2022**, *2022*, 6210128. [CrossRef]
16. Li, Y.; Wang, D.; Sun, X.S. Epoxidized and Acrylated Epoxidized Camelina Oils for Ultraviolet-Curable Wood Coatings. *J. Am. Oil Chem. Soc.* **2018**, *95*, 1307–1318. [CrossRef]
17. Melendez-Zamudio, M.; Donahue-Boyle, E.; Chen, Y.; Brook, M.A. Acrylated soybean oil: A key intermediate for more sustainable elastomeric materials from silicones. *Green Chem.* **2022**, *25*, 280–287. [CrossRef]
18. Cogliano, T.; Turco, R.; Russo, V.; Di Serio, M.; Tesser, R. 1H NMR-based analytical method: A valid and rapid tool for the epoxidation processes. *Ind. Crop. Prod.* **2022**, *186*, 115258. [CrossRef]
19. Kuki, Á.; Nagy, T.; Hashimov, M.; File, S.; Nagy, M.; Zsuga, M.; Kéki, S. Mass spectrometric characterization of epoxidized vegetable oils. *Polymers* **2019**, *11*, 394. [CrossRef]
20. Zhang, P.; Xin, J.; Zhang, J. Effects of Catalyst Type and Reaction Parameters on One-Step Acrylation of Soybean Oil. *ACS Sustain. Chem. Eng.* **2014**, *2*, 181–187. [CrossRef]
21. Su, Y.; Lin, H.; Zhang, S.; Yang, Z.; Yuan, T. One-Step Synthesis of Novel Renewable Vegetable Oil-Based Acrylate Prepolymers and Their Application in UV-Curable Coatings. *Polymers* **2020**, *12*, 1165. [CrossRef]
22. Vianello, C.; Piccolo, D.; Lorenzetti, A.; Salzano, E.; Maschio, G. Study of Soybean Oil Epoxidation: Effects of Sulfuric Acid and the Mixing Program. *Ind. Eng. Chem. Res.* **2018**, *57*, 11517–11525. [CrossRef]

23. *ASTM-D1652*; Standard Test Method for Epoxy Content of Epoxy Resins. Document Center, Inc.: Oakland, CA, USA, 2011. Available online: https://www.document-center.com/standards/show/ASTM-D1652 (accessed on 10 May 2024).
24. Samanta, A.; Kataria, N.; Dobhal, K.; Joshi, N.C.; Singh, M.; Verma, S.; Suyal, J.; Jakhmola, V. Wijs, Potassium Iodate, and AOCS Official Method to Determine the Iodine Value (IV) of Fat and Oil. *Biomed. Pharmacol. J.* **2023**, *16*, 1201–1210. [CrossRef]

molecules

MDPI

Article

Prediction of Individual Gas Yields of Supercritical Water Gasification of Lignocellulosic Biomass by Machine Learning Models

Kapil Khandelwal and Ajay K. Dalai *

Department of Chemical and Biological Engineering, University of Saskatchewan, Saskatoon, SK S7N 5A9, Canada; kak368@usask.ca

* Correspondence: ajay.dalai@usask.ca; Tel.: +1-(306)-966-4771

Abstract: Supercritical water gasification (SCWG) of lignocellulosic biomass is a promising pathway for the production of hydrogen. However, SCWG is a complex thermochemical process, the modeling of which is challenging via conventional methodologies. Therefore, eight machine learning models (linear regression (LR), Gaussian process regression (GPR), artificial neural network (ANN), support vector machine (SVM), decision tree (DT), random forest (RF), extreme gradient boosting (XGB), and categorical boosting regressor (CatBoost)) with particle swarm optimization (PSO) and a genetic algorithm (GA) optimizer were developed and evaluated for prediction of H_2, CO, CO_2, and CH_4 gas yields from SCWG of lignocellulosic biomass. A total of 12 input features of SCWG process conditions (temperature, time, concentration, pressure) and biomass properties (C, H, N, S, VM, moisture, ash, real feed) were utilized for the prediction of gas yields using 166 data points. Among machine learning models, boosting ensemble tree models such as XGB and CatBoost demonstrated the highest power for the prediction of gas yields. PSO-optimized XGB was the best performing model for H_2 yield with a test R^2 of 0.84 and PSO-optimized CatBoost was best for prediction of yields of CH_4, CO, and CO_2, with test R^2 values of 0.83, 0.94, and 0.92, respectively. The effectiveness of the PSO optimizer in improving the prediction ability of the unoptimized machine learning model was higher compared to the GA optimizer for all gas yields. Feature analysis using Shapley additive explanation (SHAP) based on best performing models showed that (21.93%) temperature, (24.85%) C, (16.93%) ash, and (29.73%) C were the most dominant features for the prediction of H_2, CH_4, CO, and CO_2 gas yields, respectively. Even though temperature was the most dominant feature, the cumulative feature importance of biomass characteristics variables (C, H, N, S, VM, moisture, ash, real feed) as a group was higher than that of the SCWG process condition variables (temperature, time, concentration, pressure) for the prediction of all gas yields. SHAP two-way analysis confirmed the strong interactive behavior of input features on the prediction of gas yields.

Keywords: machine learning; artificial intelligence; biofuel; hydrogen; lignocellulosic biomass; supercritical water gasification

Citation: Khandelwal, K.; Dalai, A.K. Prediction of Individual Gas Yields of Supercritical Water Gasification of Lignocellulosic Biomass by Machine Learning Models. *Molecules* **2024**, *29*, 2337. https://doi.org/10.3390/molecules29102337

Academic Editor: Mohamad Nasir Mohamad Ibrahim

Received: 15 April 2024
Revised: 10 May 2024
Accepted: 11 May 2024
Published: 16 May 2024

1. Introduction

Ever-increasing urbanization, modernization, and industrialization of human society have led to an exponential rise in energy consumption. Worldwide primary energy consumption reached to 604 exajoules in 2022, which is a 2.1% rise from the 2021 level even after the slowdown due to COVID-19 [1]. Nearly 494 exajoules of this energy demand is fulfilled by fossil fuel sources, which amounts to 81.8% of total energy consumption [2]. Consumption of non-renewable fuel sources not only leads to fuel scarcity but also results in environmental issues [3]. There is a pressing need of human society to reduce this dependency on fossil fuels and shift to alternative renewable fuel sources.

Lignocellulosic biomass is easily available and abundant in nature, making it a cost-effective, renewable, and sustainable source of energy generation [4]. Use of lignocellulosic biomass as a fuel source will also democratize the access to energy and improve the socio-economics of countries that do not have reservoirs of fossil fuels [5]. At the recent COP 28 summit, 130 counties participated in the Global Renewables and Energy Efficiency Programs aimed to accelerate the clean energy transition by tripling the worldwide renewable energy generation capacity to 11,000 GW by 2030 and rapidly improving the efficiency [6]. Sixty-six counties also set a target to reduce the emissions by 68% by 2050 compared to 2022 levels [7]. Sustainable production of biofuels from renewable lignocellulosic biomass has the potential to not only aid the achievement of the net zero scenario, but also reduce our dependency on non-renewable fossil fuel sources.

However, raw lignocellulosic biomass contains a high amount of moisture, which requires pre-treatment for production of biofuels by conventional thermochemical processes [8]. This reduces the efficiency of the convectional thermochemical processes for processing of lignocellulosic biomass. Gasification of lignocellulosic biomass in the presence of supercritical water (SCW) can efficiently process high-moisture-containing biomass without needing to pre-dry the feedstock [9]. At temperatures $\geq$ 371 °C and pressures $\geq$ 22.1 MPa, water exists in its supercritical state, and is used as the reaction medium in the supercritical water gasification (SCWG) process [10]. SCWG of lignocellulosic biomass is a promising process for the production of hydrogen. Hydrogen can be used as a clean source of energy for industry and transportation as it produces water as its only combustion product apart from energy generation. It also finds an industrial use as a reducing agent for steel production, in the hydrodesulfurization process in refineries, and for the production of green chemicals, green ammonia, and methanol [11,12].

The SCWG process is a complex thermochemical conversion process involving various competing reaction and complex reaction mechanisms [13]. Furthermore, due to the heterogenous nature of lignocellulosic biomass, the chemical composition of biomass differs between different biomass. Even the composition of biomass may change for the same biomass depending on the source of biomass, difference in crop production, season, and aging of the biomass. Biomass availability is also a serious supply chain issue for biofuel production processes. The availability of biomass varies dramatically depending on season, geographical location, cost, and ease of access. Additionally, industries typically use a mixture of a variety of lignocellulosic biomass, the composition of which varies drastically. Therefore, understanding the interactive influence and effects of the properties of biomass on the gas yields of SCWG is pivotal for its commercialization. These properties of biomass along with SCWG reaction conditions have interactive, complex, and non-linear relationships on the gas yields of the SCWG process.

However, most of the studies have investigated only the effects of SCWG reaction parameters on the SCWG process. Even though the conventional design of experiments using response surface methodology or single univariate methods utilizes experimental datasets, it can only study the effects of a limited number of input features on target features. The number of experimental runs quickly increases with the increase in the number of input features and the level of these features, making the performance of such experiments infeasible. Moreover, univariate methods can only study the effect of one parameter at a time and do not account for interactive behavior, whereas RSM utilizes linear regression techniques which are prone to overfitting of the data. On the other hand, traditional modeling techniques such as thermodynamic simulation, kinetics, and computational fluid dynamics (CFD) are not capable of capturing these non-linear relationships and utilize assumptions for simplifying the complex differential equations.

Since the advent of machine learning modeling, which is capable of efficiently solving complex equations and captures these non-linear relationships using experimental data, it is finding its use in various applications such as fraud detection [14], sentimental analysis [15], and recommender systems [16]. Interpretable machine learning models are also being used for thermochemical processes for conversion of biomass into biofuels [17–19].

Despite the widespread use of machine learning for investigation of complex processes, the literature on the application of machine learning models to the SCWG process is still scarce. These studies have focused mostly on the prediction of hydrogen yields, while other gas yields of SCWG are not considered; these other gases also constitute a significant portion of SCWG gas yield and entail the mechanism of the SCWG process. Furthermore, attention has been paid only towards the prediction power of machine learning models themselves, and very little discussion has been provided about the interpretability of these machine learning models; previous research also lacks the detailed reasoning of the observed results of the effects of input features on gas yields. Moreover, extra parameters such as reactor type, biomass type such as lignocellulosic or non-lignocellulosic, and nature of feedstock such as real feed or model compounds are not considered or accounted for.

To address these knowledge gaps, in this study, eight machine learning models, namely, linear regression (LR), Gaussian process regression (GPR), artificial neural network (ANN), support vector machine (SVM), decision tree (DT), random forest (RF), extreme gradient boosting (XGB), and categorical boosting regressor (CatBoost) with particle swarm optimization (PSO) and a genetic algorithm (GA) optimizer were developed for hyperparameter tuning of machine learning models for prediction of H_2, CH_4 CO, and CO_2 gas yields of various lignocellulosic biomasses from the SCWG process. Reaction temperature (temperature), reaction time (time), feedstock concentration (concentration), and reaction pressure (pressure) were used as input features for SCWG reaction conditions. For biomass characteristics, ultimate analysis (carbon content (C), hydrogen content (H), nitrogen content (N), and sulfur content (S)), proximate analysis (volatile matter (VM), moisture content (moisture), ash content (ash)), and feed type (real feed) were used. In total, a dataset of 166 datapoints for the prediction of the H_2 yield and 118 datapoints each for predictions of CO, CO_2, and CH_4 gas yields, with no missing values, was developed using literature studies, which utilized similar reactor setups and minimized variation in unaccounted-for parameters. Shapley additive explanation (SHAP) was used for interpretable machine learning to determine the most dominant features in the prediction of gas yields. Furthermore, one-way and two-way SHAP analyses were used to investigate the effects of input features and their interactions on the prediction of gas yields. A detailed discussion on the results of these analyses in the perspective of SCWG is provided with detailed reasoning.

2. Methods and Materials

2.1. Data Collection, Preprocessing, and Exploratory Data Analysis (EDA)

Data for the development of machine learning models were reviewed from research articles on SCWG of biomass. Collected data were screened to only include lignocellulosic biomass and batch reactors while filtering out the data points for non-lignocellulosic biomass, and continuous and semi-batch reactors. Studies of only batch reactors were included as most of the studies investigating SCWG of lignocellulosic biomass utilize the batch reactor. This is due to the feed pumping limitation of lignocellulosic biomass slurry in continuous reactors, especially at small research scale reactors. Furthermore, continuous reactors suffer from fouling and scaling with local heat spots, deposition of feedstock, and non-uniform heat transfer, especially for solid biomass. Moreover, the reaction mechanism of degradation of biomass in continuous reactors differs greatly with the degradation mechanism in batch reactors. Hence, to limit and eliminate the effects of the reactor on gas yields to ensure consistency, only batch reactor studies of SCWG of lignocellulosic biomass were considered. The dataset consisted of 28 types of different lignocellulosic biomass collected from 16 research articles, which utilized a stainless steel (SS) 316 batch reactor having nearly similar dimensions and reactor setups, which also minimized the effects of unaccounted-for variables such as reactor material, heating rate, and reactor dimensions. These lignocellulosic biomasses comprise cellulose, xylose, lignin, kraft lignin, soybean straw, flax straw, canola straw, rice straw, cotton stalk, wheat straw, canola hull, canola meal, pinewood, orange peel, aloe vera rind, banana peel, coconut shell, lemon peel, pineapple peel, sugarcane bagasse, timothy grass, horse manure, pinecone, canola hull fuel pellet, canola meal fuel pellet, oat hull fuel pellet, barley straw fuel pellet, and partially burnt wood.

Individual gas yields (H_2, CO, CO_2, and CH_4) of SCWG of lignocellulosic biomass in batch reactors were predicted. In total, 166 datapoints were used for the prediction of hydrogen yield and 118 datapoints each were used for the predictions of CO, CO_2, and CH_4 gas yields, as some of the studies only reported hydrogen yield. SCWG of lignocellulosic biomass is primarily dependent on SCWG process conditions and biomass characteristics. Hence, SCWG reaction parameters and biomass characteristics were used as input variables. Reaction temperature (temperature), feedstock concentration (concentration), reaction time (time), and reaction pressure (pressure) were used as SCWG process condition variables. For biomass characteristics, proximate analysis, ultimate analysis, and type of lignocellulosic biomass were used as input features. Carbon content (C), hydrogen content (H), nitrogen content (N), sulfur content (S) for ultimate analysis; while, ash content (ash), volatile matter (VM), and moisture content (moisture) were used for proximate analysis. A categorical feature (real feed) was used to represent the real feedstock or model compound. A value of 1 was assigned to 'Real Feed' for a real lignocellulosic biomass and a value of 0 was assigned for model compounds of lignocellulosic biomass. Fixed carbon (FC) and oxygen content (O) were filtered out to avoid collinearity, as these are indirectly calculated by subtracting other components from 100% of proximate and ultimate analyses of biomass, respectively.

Exploratory data analysis (EDA) of the developed dataset was conducted for detailed analysis, exploration of the dataset, and identifying the preliminary relationship between input variables and target variables. The distribution of the input dataset is represented via a box plot to analyze the spread of data and identify the outliers (Figure 1). The data were preprocessed to remove outliers and the final dataset consisted of 166 datapoints for hydrogen yield and 118 datapoints each for CO, CO_2, and CH_4 gas yields. The relationships between the input variables and gas yields were analyzed using the Pearson correlation coefficient (PCC) matrix (Figure 2). The values of PCC for any two variables vary from +1 to −1. A positive correlation between two features is represented by positive sign and +1 shows a strong positive relation, which indicates that a change in the value of one feature will increase the value of the other feature positively with the same proportion. A negative sign of the PCC value shows a negative relation between two features, meaning an increase in the value of one feature will cause a decrement in the value of the other feature. A PCC value of 0 shows that the two features are not related to each other and a change in the value of either feature will not affect the other feature. PCC values among input features or output features also represent the correlation between a pair of two input variables or a pair of two output variables.

2.2. Machine Learning Model Development and Feature Analysis

Supervised machine learning models such as regression, artificial neural networks, support vector machines, decision trees, and ensemble trees are commonly used machine learning models for biomass conversion processes [17–19]. Supervised machine learning utilizes probabilistic and statistical approaches for model building, especially for a structured dataset of a complex process [20]. The performance of these models varies depending on the nature of the relationship between inputs and outputs, number and quality of datapoints, and the complexity of the process. Thus, eight different types of machine learning models (LR, GPR, ANN, SVM, DT, RF, XGB, and CatBoost) were screened for the prediction of gas yields of SCWG.

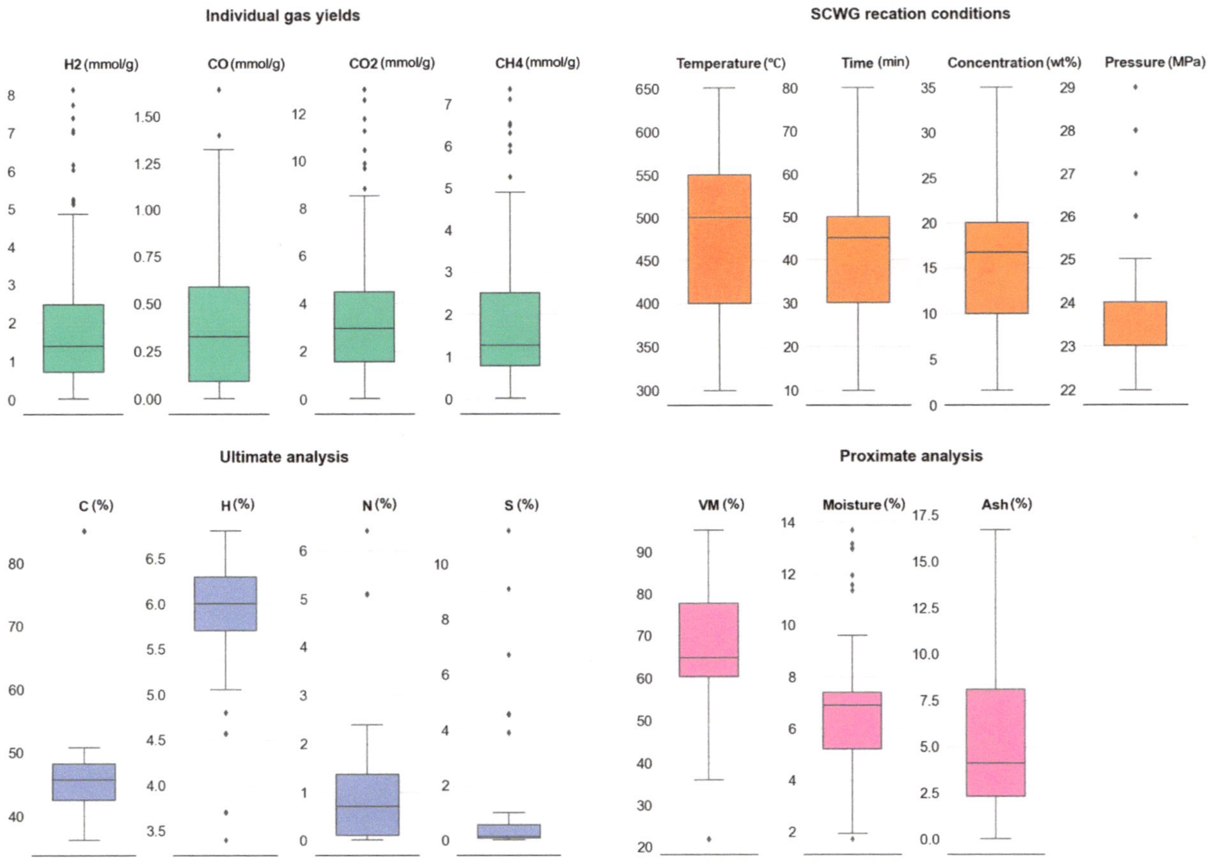

Figure 1. Box plot representation of input and output features.

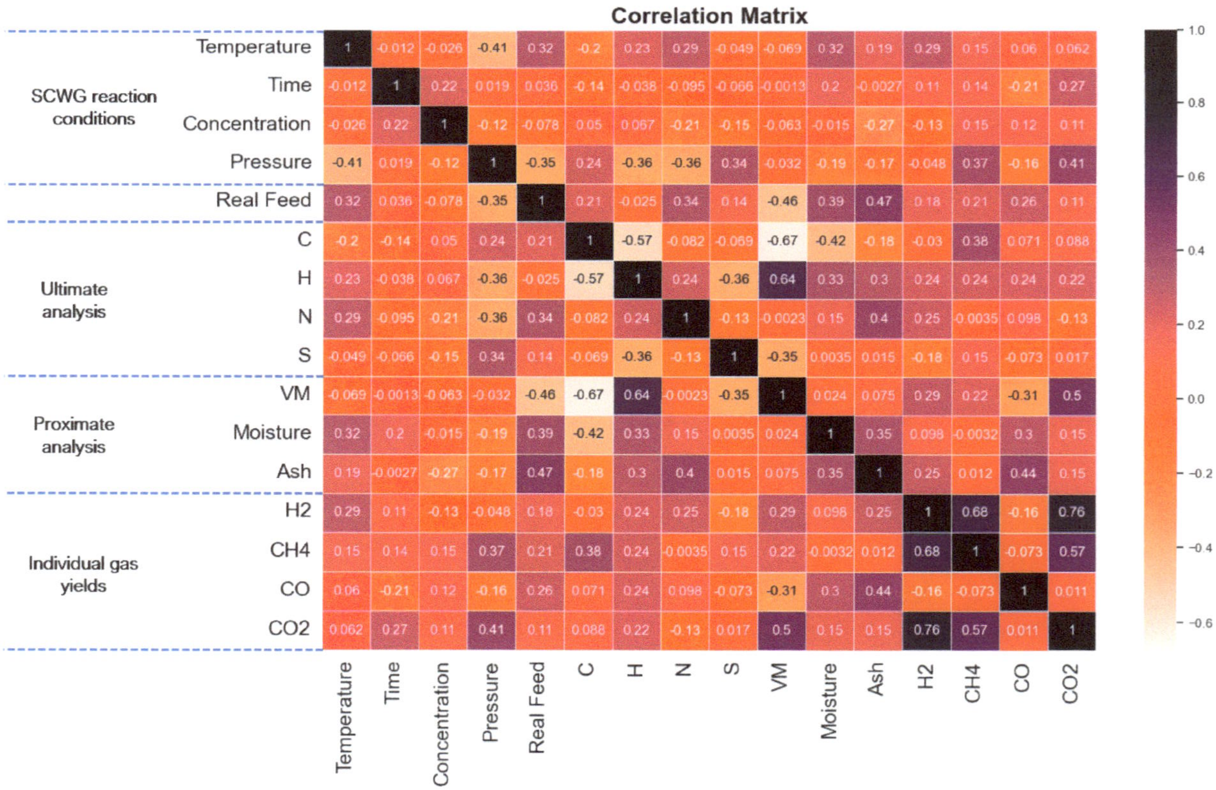

Figure 2. Pearson correlation coefficient (PCC) plot of input and output features.

The LR machine learning model is simplest, and is based on simple regression techniques. The LR model explains linear relationships quite well; however, it often fails to capture the complex non-linear relationships of the thermochemical processes such as SCWG. Nevertheless, the LR model is used as the baseline to compare and evaluate the performance of other more advanced machine learning algorithms to model the complex non-linear processes. GPR is a non-parametric probabilistic and Bayesian regressor [21]. It is highly flexible and can model complex processes, and the choice of kernel function (covariance function) during the gradient process can control the smoothness, periodicity, and other properties of the GPR model. ANN models are inspired by the functioning and structure of the human brain. A typical ANN is composed of nodes (neurons) organized in layers and each node in one layer connects to nodes in the next layer through weighted links [22]. ANNs are the most commonly used machine learning models for thermochemical processes [17–19]. On the other hand, SVMs classify input vectors into distinct categories within a high-dimensional space, and are adept at handling non-linear separations through the use of kernels that project data into higher dimensions where they can be linearly separated [23].

DT, RF, CatBoost, and XGB machine learning models fall under the category of tree-based models. These tree-based models have high accuracy and can handle both categorical and numerical variables having different ranges of values without needing to preprocess the data. The decision tree (DT) model is the simplest and consists of a single tree model based on feature splits on a Boolean condition (True or False) that results in the highest information gain. However, a simple decision tree model is susceptible to overfitting and noise. The RF model improves upon this by creating an ensemble of multiple decision trees in a bagging fashion, each trained on a random subset of the data and features, the results of which are averaged to reduce variance [24], whereas CatBoost and XGB are gradient boosting models that build trees sequentially, where each new tree corrects the errors of the previous trees. The CatBoost model provides an optimized approach and it is a permutation-driven ordered boosting algorithm [25]. XGB is an advanced form of gradient boosting, which offers a performance-oriented architecture that can efficiently handle large-scale data [26]. However, it also requires intricate hyperparameter tuning to achieve optimal performance.

Machine learning models have various parameters for the architect of the models, such as the number of neurons and layers in ANN models, the number of nodes and maximum depth in the case of the decision tree algorithm, or the type of kernel used in the SVM algorithm. The performance of machine learning models is highly dependent on these parameters. For example, a smaller number of neurons in an ANN model can lead to the underfitting of the data, whereas a higher number of neurons can also lead to overfitting and diminishing return, leading to intensive computing. Therefore, optimization of these parameters, commonly referred to as hyperparameters of machine learning models, is needed to ensure the optimal performance of machine learning models while ensuring high efficiency.

In this study, two optimizer algorithms, particle swarm optimization (PSO) and the genetic algorithm (GA), are utilized for hyperparameter tuning of machine learning models. The genetic algorithm (GA) is based on the principle of genetics and natural selection for the survival of the fittest [27]. GA algorithms first start with the initial population of the solutions as analogues to genes or chromosomes, coded as the string of the binary (1 or 0) values. The GA algorithm utilizes genetic methodologies such as selection, mutation, and crossover (recombination) to iteratively update the new generation of the population to adapt to the changing environment, with aim of evolving the population towards the optimal solutions [28]. Crossover and mutations introduce genetic diversity, enabling the algorithm to explore new regions of the solution space. Over various populations, GA converges to provide the population with high-quality genes or solutions.

Particle swarm optimization (PSO), on the other hand, is a heuristic algorithm, based on the collective behavior of animals for feeding [29]. In PSO, each solution is analogous to the food and prey, and is referred to as a particle. A group or swarm of these particles is initiated with a random position in the solution space [24]. These particles then iteratively update their positions to hunt for the best position based on their current position and surrounding points. The aim is to find the global optimum in the solution space. This gives PSO the ability to handle non-linear and multidimensional optimization problems, which is a key asset for optimization of the hyperparameters of the complex machine learning models having multiple parameters to be optimized [30]. Thus, PSO effectively leverages both position and velocity attributes in its search for optimal solutions.

For development of machine learning models, normalization of the clean and pre-processed dataset was achieved using the StandardScaler operation of the sklearn library, which normalizes the dataset around the mean of 0 and standard deviation of 1. This helps to improve the prediction by scaling the input parameters and normalizing their weightings to eliminate the effects of the range of input variables on prediction models. These normalized data were then fed to machine learning models for development. Machine learning models were developed exclusively for each gas yield. The normalized dataset was split into a training dataset and a test dataset. Splitting was performed using a ratio of 80:20 for the training and testing datasets, using train_test_split from the sklearn library of Python. Hyperparameters of each machine learning model for each gas yield were optimized using PSO and GA algorithms, and the effectiveness of the optimizer was compared with unoptimized machine learning models.

For evaluation of the predictive performance of each developed model, the coefficient of determination (R^2) and the mean squared error (MSE) were used as evaluation matrices. R^2 is the measurement of the explainability of the variation in target features (individual gas yields) by the input features [31]. A model with a high value of R^2 represents high explainability of the target features, indicating a good fit by the developed machine learning model. MSE is the measurement of the square of the difference between the predicted value of the target feature by the machine learning model and the actual value of the target feature in the experimental sample [32]. This represents the absolute measurement of the fit of the dataset by machine learning models. The calculation equations of R^2 and MSE are provided in Equations (1) and (2).

$$\text{Coefficient of determination } (R^2) = 1 - \frac{\sum_{n=1}^{N}(y_n - \hat{y_n})^2}{\sum_{n=1}^{N}(y_n - \overline{y})^2} \quad (1)$$

$$\text{Mean squared error (MSE)} = \frac{1}{N}\,(y_i - \hat{y_i})^2 \quad (2)$$

Here, N represents the total number of data points for output y. y_n represents the experimental value of output y for the nth datapoint and $\hat{y_n}$ represents the predicted value of output by the machine learning model for the nth datapoint, whereas $\overline{y}$ is the average or mean of all predicted values of output y for all datapoints (N datapoints).

After the development and selection of the best performing model for each individual gas yield, interpretation of machine learning is essential to understand how each input feature contributes to and influences the prediction of the model. Shapley additive explanation (SHAP) was used to explain the machine learning models. SHAP is based on game theory, which allocates a contribution value to each input feature [33]. Traditionally, Shapley values were used in corporate games among employees for the distribution of the prizes based on their contribution. In SHAP, each feature is analogous to a player, which, with other players (features), contributes to the prediction of the target feature. The SHAP value is calculated by averaging the marginal contribution of a particular feature across all possible combinations of input features. It helps to develop the relationship and interactive influence of the input feature for each gas yield to understand degradation of complex lignocellulosic biomass in SCWG. It also enables the interpretability of the machine learning model to overcome their black-box nature.

3. Results and Discussion

3.1. Exploratory Data Analysis (EDA)

EDA of dataset was conducted for preprocessing and assessment of the dataset. Analysis of the distribution and relationship between input features and target variables was conducted to identify the outliers, handling of missing values, and cleaning of the dataset. Missing values for some biomass properties features were filled using the other literature, which had value of missing datapoint for the same feedstock from the same research group. The final preprocessed and cleaned dataset consisted of 166 datapoints for hydrogen yield prediction and 118 datapoints each for CO, CO_2, and CH_4 gas yield predictions.

The distribution of dataset was visualized using the box plot presented in Figure 1. From Figure 1, it can be observed that the ranges of H_2, CO, CO_2, and CH_4 gas yields were (0.02–8.13 mmol/g), (0.00–1.64 mmol/g), (0.03–13.01 mmol/g), and (0.02–7.35 mmol/g), respectively, while their means were 1.84, 0.38, 3.53, and 1.86 mmol/g, respectively. Similarly, ranges of 'Temperature', 'Time', 'Concentration', 'Pressure', 'C', 'H', 'N', 'S', 'VM', 'Moisture', and 'Ash' input features were (300–651 °C), (10–80 min), (1.64–35.00 wt%), (22–29 MPa), (36.10–85.00%), (3.39–6.80%), (0.00–6.40%), (0.00–11.20%), (21.90–95.00%), (1.71–13.69%), and (0.00–16.70%), respectively. The dataset consisted of a wide range of SCWG process conditions utilized for SCWG of lignocellulosic biomass for improving the scope of the machine learning models. Furthermore, the dataset consisted of both model compounds and real feed for lignocellulosic biomass, incorporating 28 different types of lignocellulosic feedstocks comprising of cellulose, xylose, lignin, kraft lignin, soybean straw, flax straw, canola straw, rice straw, cotton stalk, wheat straw, canola hull, canola meal, pinewood, orange peel, aloe vera rind, banana peel, coconut shell, lemon peel, pineapple peel, sugarcane bagasse, timothy grass, horse manure, pinecone, canola hull fuel pellet, canola meal fuel pellet, oat hull fuel pellet, barley straw fuel pellet, and partially burnt wood.

Correlations between pairs of two input features or pairs of two output variable, and also between pairs of input features and output variables, are visualized using the PCC correlation matrix (Figure 2). From Figure 2, both temperature and volatile matter (VM) have the highest PCC coefficient of 0.29 for hydrogen yield. This shows that the increment in temperature significantly increases hydrogen yield. It can be also observed that hydrogen content (H) of biomass is highly correlated with VM, with PCC of 0.64. Furthermore, hydrogen content has a high correlation with hydrogen yield, with PCC of 0.24. This shows that the high volatile matter containing biomass usually has high hydrogen content, which enhances the hydrogen yield. Among output parameters, hydrogen is strongly positively correlated with CO_2 yield. This is due to the fact that, in SCWG, hydrogen is produced mainly via reforming and the water–gas shift reaction since the reforming reaction mainly produces hydrogen with CO, which further undergoes a water–gas shift reaction to produce more hydrogen or is consumed via a methanation reaction to produce methane. Since CO_2 is also produced via a water–gas shift reaction along with hydrogen; thus, yields of CO_2 and H_2 are correlated.

3.2. Evaluation of Machine Learning Models

LR, GPR, ANN, SVM, DT, RF, XGB, and CatBoost machine learning models were trained on the clean and preprocessed dataset for prediction of gas yields of SCWG of lignocellulosic biomass. Hyperparameters of these machine learning models for each gas yields were optimized using GA and PSO optimizer algorithms. The parameters of GA and PSO optimizer algorithms are presented in Tables 1 and 2. Hyperparameters and their ranges for optimization of each machine learning model for hydrogen yield are presented in Table 3. The results of the optimized hyperparameters of each machine learning model for hydrogen yield by GA and PSO optimizer algorithms are also presented in Table 3. It can be observed that despite being heuristic optimization algorithms, both GA and PSO algorithms are solved for different optimized hyperparameters. This is due to differences in the search mechanisms of both algorithms to find optimal hyperparameters.

Hyperparameters of each machine learning model for the prediction of other gas yields were also optimized using GA and PSO optimizers.

Table 1. Parameters of the genetic algorithm (GA) for hyperparameter tuning of machine learning models.

Parameters	Numbers
Generation (generations)	50
Crossover probability (crossover_prob)	0.8
Population size (population_size)	100
Mutation probability (mutation_prob)	0.2

Table 2. Parameters of the particle swarm optimization (PSO) algorithm for hyperparameter tuning of machine learning models.

Parameters	Numbers
Number of particles (n_particles)	30
Number of iterations (max_iter)	10
Inertial weight (alpha)	0.5
Personal attachment (beta)	1.5
Global attraction (gamma)	1.5

Table 3. Optimized hyperparameters of machine learning models by GA and PSO optimizers and the search range for the prediction of hydrogen yield.

Hyperparameter	Range	Unoptimized	GA	PSO
		Random forest (RF)		
n_estimators	10–500	100	187	10
max_depth	1–50	NaN *	13	41
min_samples_split	2–10	2	3	2
min_samples_leaf	1–10	1	1	1
		extreme gradient boosting (XGB)		
learning_rate	0.01–0.5	NaN	0.402014	0.268485
n_estimators	50–500	NaN	408	500
max_depth	3–10	NaN	9	5
min_child_weight	1–7	NaN	5	1
gamma	0–0.5	NaN	0.141899	0.267668
subsample	0.5–1	NaN	0.989653	0.715084
colsample_bytree	0.5–1	NaN	0.975411	0.500000

Table 3. *Cont.*

Hyperparameter	Range	Unoptimized	GA	PSO
		Decision tree (DT)		
max_depth	1–50	NaN	49	50
min_samples_split	2–50	NaN	8	4
min_samples_leaf	1–50	NaN	7	1
		Support vector machine (SVM)		
C	0.1–1000	1	462.771600	74.618075
epsilon	0.01–1	0.1	0.424860	0.141974
gamma	0.1–1		0.035418	0.045675
		Categorical boosting regressor (CatBoost)		
learning_rate	0.01–0.5	0.03	0.252239	0.439795
depth	4–10	6	7	5
l2_leaf_reg	1–10	3	7.786251	9.244647
		Artificial neural network (ANN)		
learning_rate_init	0.0001–0.1	0.001	0.010889	0.035365
hidden_layer_sizes	5–100	100	81.428966	36.580111
activation_function	identity, logistic, tanh, relu	relu	tanh	logistic
		Gaussian process regression (GPR)		
sigma	0.0001–55	1	0.315774	0.050060
kernel_function	RBF *, Matern	RBF	RBF	RBF

* NaN: Not a Number; RBF: Radial Basis Function.

Unoptimized, GA-, and PSO-optimized machine learning models were compared and evaluated using values of R^2 and MSE of the respective machine learning models. The results of R^2 and MSE values during training and testing of the machine learning model are presented in Figures 3–6. From Figure 3, it can be observed that for prediction of hydrogen yield, the LR model demonstrated poor performance, with the lowest test R^2 of just 0.14 and a very high test MSE of 1.79. This is due to the fact that the LR model utilizes a linear regression mechanism during the learning of the machine learning model. It explains the linear relationships between input features and output very well. However, it is not capable of processing complex and non-linear datasets. The poor performance of the LR model signifies the non-linear relationship between input features and hydrogen yield. Similarly, GPR and SVM models also demonstrated moderate performance with a test R^2 of 0.50 and 0.56, respectively. This is due to the fact that the GPR model is also based on regression analysis, which also suffered due to non-linear relationship of SCWG features. The low performance of the SVM model was also due to non-linearity of the SCWG process, which affects the hyperplane separation and thus affects the performance of the SVM model.

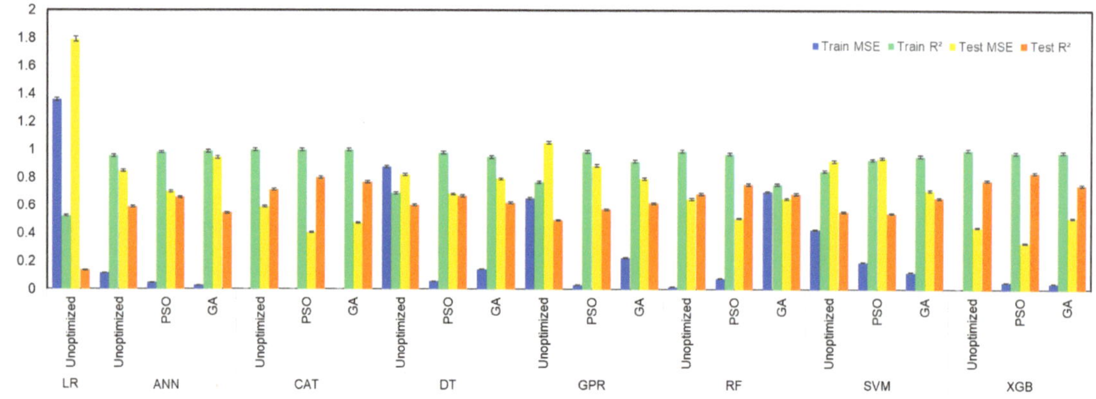

Figure 3. Performance of machine learning models for prediction of gas yields of H_2 gas yield.

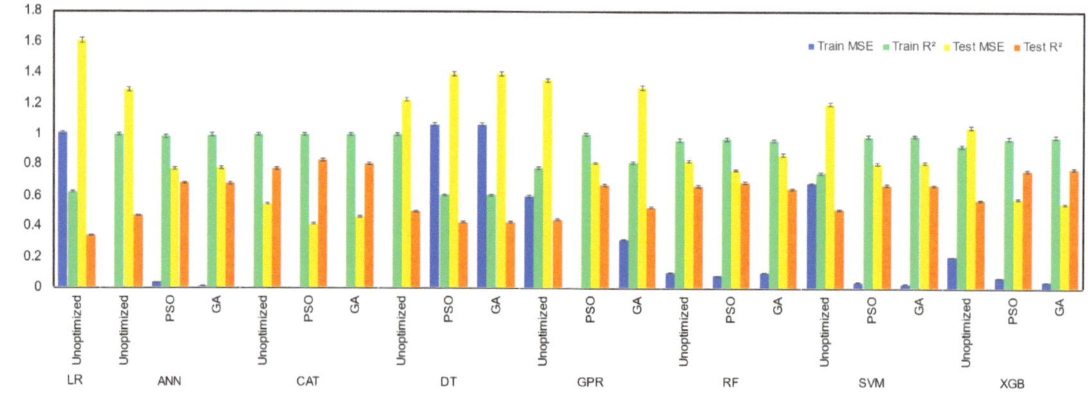

Figure 4. Performance of machine learning models for prediction of gas yields of CH_4.

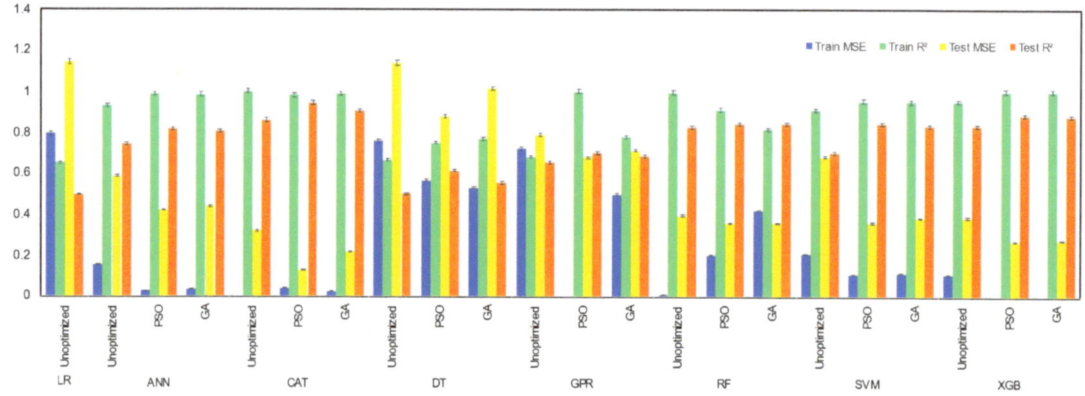

Figure 5. Performance of machine learning models for prediction of gas yields of CO gas yield.

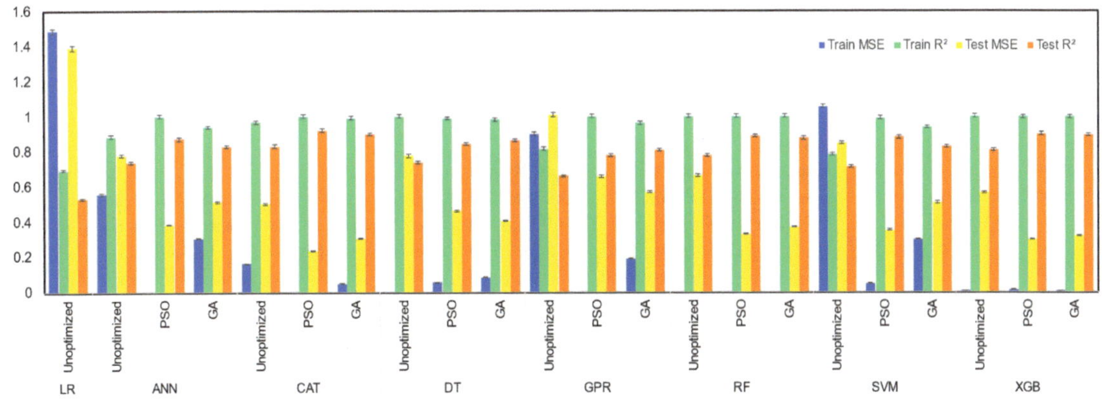

Figure 6. Performance of machine learning models for prediction of gas yields of CO_2 gas yield.

The ANN model usually shows relatively good performance for thermochemical processes. However, the ANN model demonstrated moderate performance, with its test R^2 of 0.59. This is due to the fact that simple ANN models are susceptible to overfitting of small datasets, especially those having non-linear complex relationships [34]. This is also confirmed by its relatively high training R^2 of 0.96 during the training of the model. This highlights the overfitting and biasedness of the ANN model for prediction of the hydrogen yield. A literature study also confirmed the susceptibility of the ANN model for overfitting for prediction of the hydrogen yield. Zhao et al. [35] reported that for prediction of the hydrogen yield from SCWG, ANN and GPR models suffered from overfitting due to the use of a single training process, and these models do not utilize statical averaging or bootstrap sampling compared to ensemble tree models. The SVM model is also susceptible to overfitting due to similar reasons.

Among all unoptimized machine learning models, tree-based models demonstrated high predictive power for hydrogen yield. The unoptimized XGB model showed a high test R^2 of 0.78 and low test MSE of 0.45. The XGB model also demonstrated high prediction capabilities during training of the model, with a high training R^2 of 0.999. This shows the balanced performance of the unoptimized XGB model in both training and testing. The unoptimized CatBoost model also performed well, with its test R^2 of 0.72 followed by a test R^2 of 0.69 of the unoptimized RF model. Among tree-based models, the simple DT model demonstrated the lowest test R^2 of 0.61 compared to ensemble-based tree models. A simple decision tree model is susceptible to overfitting of the dataset, which is minimized in ensemble tree models. Ensemble tree models usually utilize a group of simple decision tree models and either average the prediction of each tree model in the case of the RF model or correct the error of the preceding tree sequentially in the case of the XGB and CatBoost models [36]. This eliminates the biasedness of a single decision tree model and minimizes the overfitting of the dataset by the machine learning model.

The use of GA and PSO optimizers improved the prediction power of nearly all machine learning models. In general, hyperparameter-tuned machine learning models optimized by the PSO algorithm outperformed the GA algorithm-optimized machine learning models. This is due to the difference in the search mechanism of both algorithms for optimal solutions, which resulted in the different optimized hyperparameters selected by both algorithms. Since the performance of a machine learning model is dictated by its hyperparameters, GA- and PSO-optimized models differ in their prediction capabilities. Among all unoptimized, GA-, and PSO-optimized machine learning models, the PSO-optimized XGB model demonstrated the highest test R^2 of 0.84 and the lowest test MSE of 0.34. Interestingly, while the use of the PSO optimizer improved the test R^2 of the XGB model, it also reduced the training R^2 to 0.98. In contrast, the training R^2 scores were 0.999 for the unoptimized model and 0.98 for the GA-optimized model. This highlights the effectiveness of the PSO optimizer algorithm in improving the robustness of the model by minimizing the biasedness and overfitting of the model, which resulted in improved performance on the test dataset. PSO-optimized CatBoost also performed well, with its high test R^2 of 0.80 and low test MSE of 0.41. The order of test R^2 values among PSO-optimized machine learning models was XGB-PSO > CAT-PSO > RF-PSO > DT-PSO > ANN-PSO > GPR-PSO > SVM-PSO. On the other hand, the order of R^2 among GA-optimized machine learning models was CAT-GA > XGB-GA > RF-GA > SVM-GA > DT-GA > GPR-GA > ANN-GA.

For prediction of CH_4 yield, among unoptimized machine learning models, the CatBoost model showed superior performance, with its high test R^2 of 0.78 for prediction of methane yield from SCWG of lignocellulosic biomass (Figure 4). The unoptimized RF model also performed well, with its test R^2 of 0.66. Similar to the prediction of hydrogen yield, the LR machine learning model performed worst among all machine learning models for prediction of methane yield, with its lowest test R^2 of 0.34 and highest test MSE of 1.60. The performance of machine learning models was improved by the use of PSO and GA optimizer algorithms. However, in general, the improvement in the prediction power of optimized machine learning models from unoptimized machine learning models was higher with the PSO optimizer compared to the GA optimizer. Among all machine learning models, CatBoost models were the top three best performing models, with the PSO-optimized CatBoost machine learning model resulting in the highest test R^2 of 0.83, followed by 0.81 of the GA-optimized CatBoost model and 0.78 of the unoptimized CatBoost model. GA- and PSO-optimized XGB models also showed comparable performance, with test R^2 of 0.77 and 0.76. However, LR, GPR, and DT machine learning models demonstrated the worst performance among all machine learning models.

Similar to the prediction of hydrogen yield and methane yield, CatBoost and XGB were the best performing machine learning models for prediction of CO yield of SCWG of lignocellulosic biomass (Figure 5). Among unoptimized machine learning models, CatBoost models had the highest test R^2 of 0.86, followed by 0.83 of XGB and 0.83 of the RF model. The extent of improvement in the performance of machine learning models by the PSO optimizer algorithm was also higher compared to the GA optimizer algorithm from unoptimized machine learning models for the prediction of CO yield. The PSO-optimized CatBoost model was the best performing machine learning model for prediction of CO yield, with its highest test R^2 of 0.94, followed by the GA-optimized CatBoost and PSO-optimized XGB machine learning model. Among all machine learning models, LR and GPR were the worst performing machine learning models.

The CatBoost model was also able to predict the CO_2 gas yield of the SCWG process (Figure 6). Among unoptimized machine learning models, the CatBoost model demonstrated high prediction performance, with its test R^2 of 0.83, followed by test R^2 of 0.81 of XGB and 0.78 of RF. The PSO optimizer further improved the performance of machine learning models, and the CatBoost-PSO model had the highest test R^2 of 0.92 followed by 0.90 of XGB-PSO among all machine learning models. Similar to the prediction of other gas yields, the PSO optimizer performed better compared to the GA optimizer in improving the performance of unoptimized machine learning models. SVM, GPR, and LR had the worst performance in the prediction of CO_2 gas yield.

Thus, boosting ensemble-based machine learning models such as CatBoost and XGB were clearly the best performing machine learning models for the prediction of gas yields of the SCWG process. This is due to the use of the tabular and structured dataset, for which ensemble tree-based model tends to perform the best [37]. Boosting ensemble tree models have also demonstrated their superior prediction power in other thermochemical processes such as pyrolysis [38], hydrothermal liquefaction [39], and hydrothermal carbonization [40]. Moreover, these boosting models utilize the decision of a group of multiple simple tree models, which learn from the preceding tree specifically for the misclassified instances. This limits overfitting, especially for smaller dataset. Studies also showed that the ensemble boosting algorithms outperform even the deep learning models for a variety of tabular datasets [41].

In conclusion, XGB-PSO and CatBoost-PSO models demonstrated the highest prediction power for the yields of H_2, and CH_4, CO, and CO_2, respectively, for SCWG of lignocellulosic biomass. Overall, the effectiveness of the PSO optimizer for hyperparameter tuning of machine learning models was highest compared to the GA optimizer. Due to the superior performance of these machine learning models, XGB-PSO and CatBoost-PSO were selected for further analysis of the prediction of H_2, and CH_4, CO, and CO_2 gas yields, respectively.

3.3. Feature Analysis and Summary Plots

The impact of input features and their relative importance for prediction of gas yields of SCWG process in machine learning models were studied using SHAP analysis. SHAP analysis helps to overcome the black-box nature of machine learning models. SHAP values quantify the contribution of each feature towards the prediction of a machine learning model. The 'base case' refers to the model's prediction using no feature information. Thus, a positive SHAP value indicates that a feature has increased the prediction from the base case, while a negative value indicates a decrease in the prediction power of the machine learning model. A SHAP value of zero suggests the feature has no impact from the base case prediction. This metric offers an intuitive means to interpret complex model predictions.

Feature importance plots, summary plots, and heat maps of SHAP values of input features for prediction of H_2, CH_4, CO, and CO_2 yields are presented in Figures 7–10. From Figure 7, it can be observed that the temperature was the most dominant feature, with feature importance of 21.93%, followed by (15.38%) hydrogen content (H), (11.68%) ash content (ash), (10.84%) time, (9.41%) concentration, and (7.30%) carbon content (C) for prediction of hydrogen yield by the XGB-PSO model. The high feature importance of temperature for hydrogen yield is due to the endothermic nature of reforming, hydrolysis, and water–gas shift reactions, which are favored at high reaction temperatures, enhancing the hydrogen yield [42]. The SHAP summary plot also shows that an increase in the feature value of temperature increased the SHAP values of hydrogen yield and shifted the SHAP value points to right side (more positive). A more detailed analysis of each instance is provided in the heat map of the SHAP value plot for hydrogen yield.

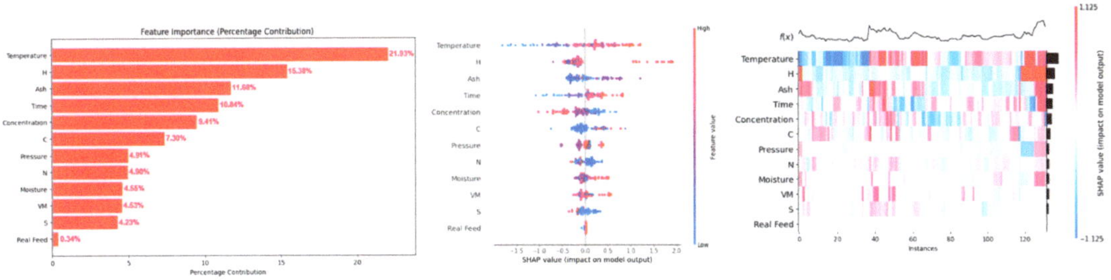

Figure 7. SHAP feature importance, summary plot, and heat map of input features on prediction of H_2 yield in the PSO-optimized XGB model.

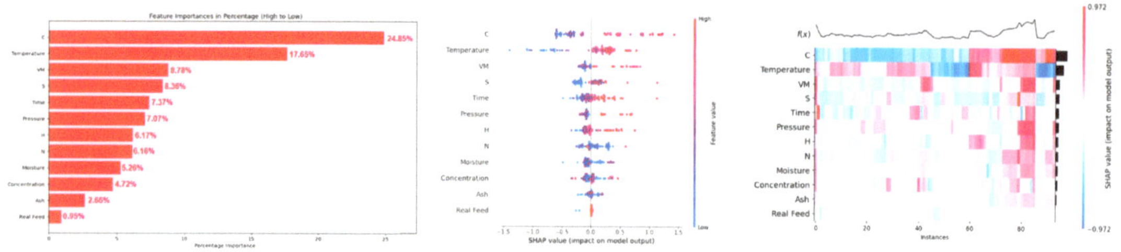

Figure 8. SHAP feature importance, summary plot, and heat map of input features on prediction of CH_4 yield in the PSO-optimized CatBoost model.

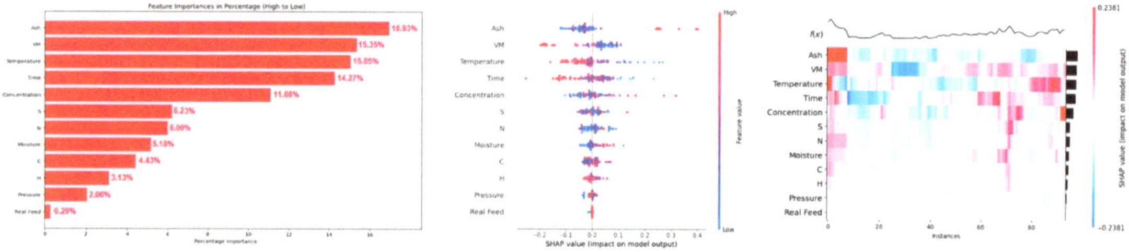

Figure 9. SHAP feature importance, summary plot and heat map of input features on prediction of CO yield in the PSO-optimized CatBoost model.

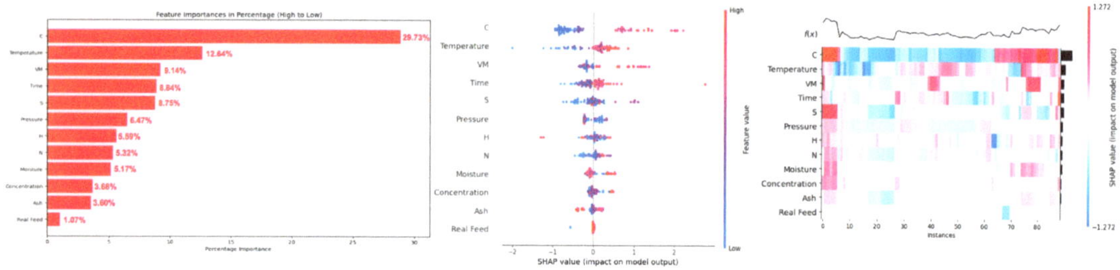

Figure 10. SHAP feature importance, summary plot and heat map of input features on prediction of CO_2 yield in the PSO-optimized CatBoost model.

Hydrogen content was the second most dominant feature for prediction of hydrogen yield, which increased hydrogen yield with an increase in the value of hydrogen content. This shows that hydrogen content of biomass contributes to the hydrogen yield, and biomass with higher hydrogen content is recommended to achieve high hydrogen yield. An increase in ash content also improved hydrogen yield. This is due to the presence of alkali and alkaline earth metals (AAEMs) in the ash content of lignocellulosic biomass [43]. These AAEMs have catalytic effects in promoting reforming, hydrolysis, and water–gas shift reactions, which improves the hydrogen yield [44]. However, higher values of ash content also decreased the hydrogen yield, as indicated in the summary plot. Thus, only the optimum amount of ash content in biomass is beneficial for hydrogen yield. Different nature of components of ash content should also be considered as some of these components in ash content may be less or more active in their catalytic activity for SCWG. For example, silica (SiO_2) present in ash content has very little catalytic activity compared to highly active potassium (K) for the SCWG reaction.

Increase in reaction time (time) also increased the hydrogen yield since longer reaction time allows sufficient time for hydrolysis and reforming reactions to take place that produce hydrogen and CO, enabling the water–gas shift reaction of CO in excess water for production of more hydrogen [45]. Therefore, higher reaction time is beneficial for high hydrogen yield. However, an increase in feedstock concentration decreased the hydrogen yield since hydrogen is also produced from supercritical water (SCW), which acts as a reactant in reforming, hydrolysis, and water–gas shift reactions. Additionally, feedstock concentration measures the amount of biomass in a mixture of water and biomass, where a high concentration indicates a greater amount of biomass in a relatively low amount of water. Therefore, at high feed concentrations or at less water content in feedstock mixture, reforming, hydrolysis, and water–gas shift reactions diminish as per Le Chatelier's principle, which decreases the hydrogen yield at high feedstock concentrations [46]. An increase in carbon content (C) improved the hydrogen yield up to a certain extent; however, a further increase in carbon content did not improve hydrogen yield. Thus, biomass with high carbon content is only beneficial up to a certain extent.

These results are in agreement with the reported literature of SCWG of lignocellulosic biomass, where the temperature is the most important parameter for the SCWG process, followed by reaction time and feedstock concentration, while reaction pressure is the least influential reaction parameter for SCWG of lignocellulosic biomass [47]. Among biomass properties, hydrogen content (H), ash content (ash), and carbon content (C) were the most dominant features, which indicates biomass with high hydrogen content and a moderate amount of ash and carbon content is recommended for high hydrogen yield. Overall, biomass properties had a high feature importance of 52.91% compared to feature importance of 47.09% of SCWG reaction process parameters for the prediction of hydrogen yield. This highlights the importance of screening most suitable lignocellulosic biomass feedstocks to achieve high hydrogen yield. Thus, although optimization of SCWG conditions is necessary for improving hydrogen yield, more attention should be paid to the selection of suitable lignocellulose biomass for holistic optimization of the SCWG process for maximization of hydrogen yield.

Feature analysis based on CatBoost-PSO for methane yield showed that the carbon content (C) of the biomass was the most dominant feature, with its highest feature importance of 24.85%, followed by (17.65%) temperature and (8.78%) volatile matter (VM) of biomass (Figure 8). An increase in the carbon content of the biomass increased the methane yield. This is due to the fact that carbon molecules in SCWG product gas are facilitated by the carbon content of the biomass. An increase in reaction temperature also increased the methane yield, which is due to the hydrogenation and methanation of produced CO and CO_2 at high reaction temperatures, which increased the methane yield at high reaction temperatures [48]. Similarly, an increase in volatile matter (VM) of biomass enhanced methane yield due to the ease of gasification of volatile matters in SCWG. These volatile matters represent the alcohols, ketones, aldehydes, and organic acids. These are the intermediates of the gasification in SCW, which are easily converted into gaseous products such as methane, H_2, and CO_2 [13]. Therefore, an increase in the volatile matter of biomass increased methane yield. Similar to hydrogen yield, biomass properties had cumulative feature importance of 63.19% compared to 36.81% feature importance of SCWG reaction conditions.

For CO gas yield, feature analysis revealed that the ash content (ash), volatile matter (VM), temperature, time, and concentration were the most influential features, with feature importance of 16.93, 15.35, 15.05, 14.27, and 11.08%, respectively (Figure 9). Low to moderate ash content had a negative impact, which is due to the fact that even though AAEMs in ash content enhances reforming reactions, due to enhancement of water–gas shift reactions, most of the produced CO is consumed for hydrogen production. Only at a really high ash content, where the water–gas shift reaction attains equilibrium and an increased content of AAEMs, the water–gas shift reaction is no longer enhanced, leading to the increase in CO yield. This was also observed in hydrogen yield where really high ash content actually decreased the hydrogen yield. Similarly, an increase in volatile matter decreased the CO yield as high volatile matter promotes the further conversion of CO gas into methane and hydrogen via water–gas shift, methanation, and hydrogenation reactions. Similarly, an increase in temperature and time also decreases CO yield, as at high reaction temperature, water–gas shift reactions dominate and consume the CO. CO yield is high at short reaction times as a short reaction time does not allow sufficient time for further conversion of produced CO by consecutive water–gas shift and methanation reactions, which are enabled at longer reaction times. This led to the decrement in CO yield at longer reaction times. However, an increase in feedstock concentration increased the CO yield. This is due to diminished activity of water–gas shift, methanation, and hydrogenation reactions at high feedstock concentrations, which result in unutilized CO gas and thus increases the yield of CO gas at higher feedstock concentrations [49].

Similar to methane yield, carbon content (C) of the biomass was the most dominant feature for prediction of CO_2 gas yield, having feature importance of 29.73%, which is followed by (12.64%) temperature, (9.14%) volatile matter (VM), and (8.84%) time (Figure 10). This is because most of the carbon content comes from the biomass itself, which resulted in its highest feature importance for prediction of CO_2 yield. Thus, an increase in carbon content of biomass increased the CO_2 gas yield. An increase in temperature increases the conversion of CO to CO_2 and hydrogen by enhancing the water–gas shift reaction. Similarly, an increase in volatile matter of biomass favors the production of gaseous products due to the ease of gasification of volatile matter resulting in an increase in yield of CO_2 [50].

An increase in time also allows sufficient time for further conversion of produced CO gas by enhanced water–gas shift reactions at longer reaction time, which increases the CO_2 gas yield.

It can be observed that even though temperature has a high influence among SCWG process features on gas yields of the SCWG process, biomass properties as a whole have feature importance of 52.91, 63.19, 57.54, and 68.37% compared to 47.09, 36.81, 42.46, and 31.63% feature importance of SCWG process parameters for prediction of H_2, CH_4, CO, and CO_2 gas yields, respectively. Thus, biomass characteristic plays a key role in the SCWG degradation mechanism of lignocellulosic biomass, which influences the gas distribution of the SCWG process. The characteristics of biomass should be considered while optimizing SCWG process parameters. These biomass properties and the SCWG process also have interactive effects during the gasification of lignocellulosic biomass in SCW. Hence, study of the interactive effects of these input features on gas yields of SCWG is important to understand the degradation mechanism of lignocellulosic biomass in SCWG.

3.4. *Two-Way SHAP Analysis*

SHAP dependency plots for investigating the influence of interactive effects of the most dominant input features for the prediction of gas yields are presented in Figures 11–14. In a SHAP two-way dependency plot, the x-axis shows the value of feature 1 and the primary y-axis represents the effect as the function of the SHAP values of the target variable (gas yields). The effect of feature 2 is represented using the secondary y-axis and values are represented using a gradient. This helps to visualize the interactive effects of the two input variables on the SHAP values of the gas yields.

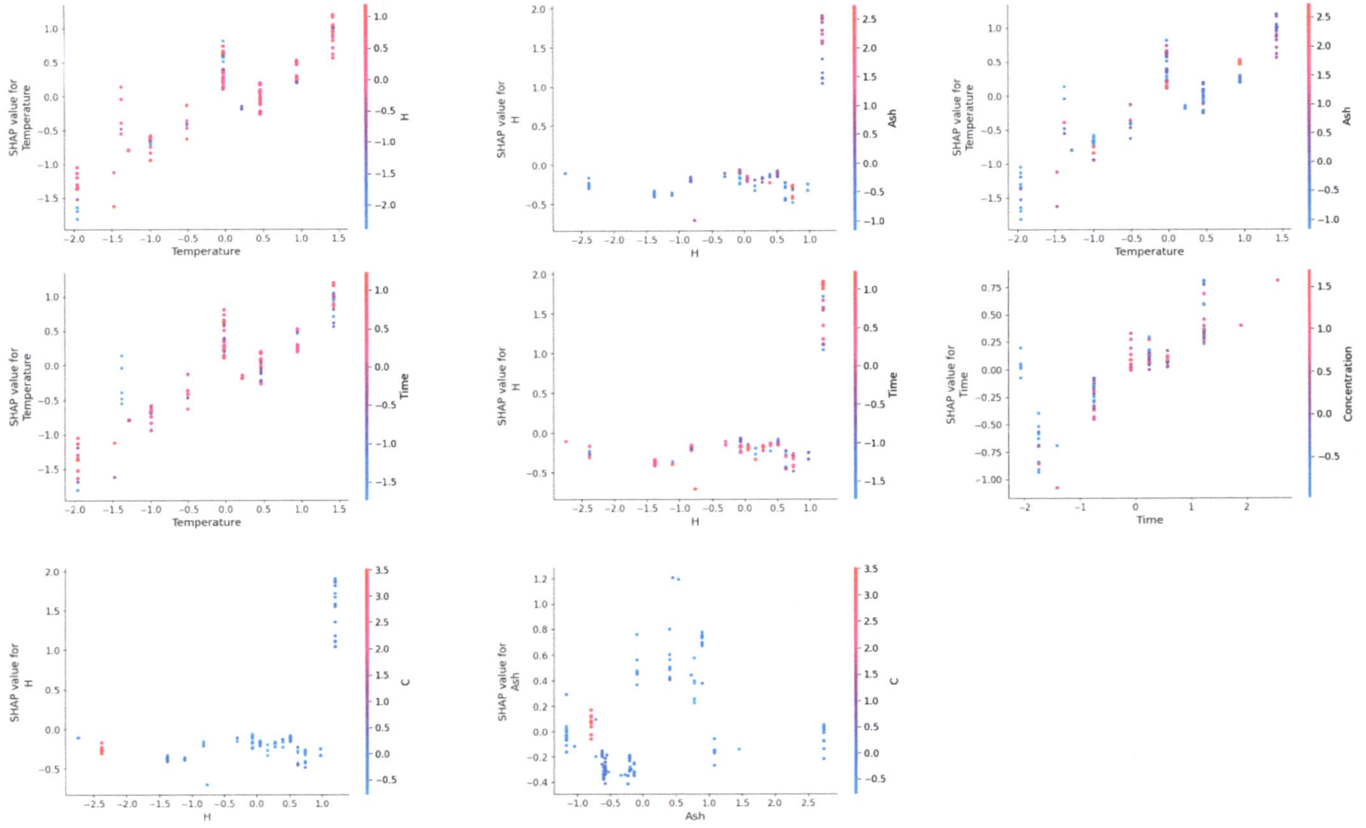

Figure 11. SHAP two-way plot of interactive behavior of the most important input features on prediction of H_2 yield in the PSO-optimized XGB model.

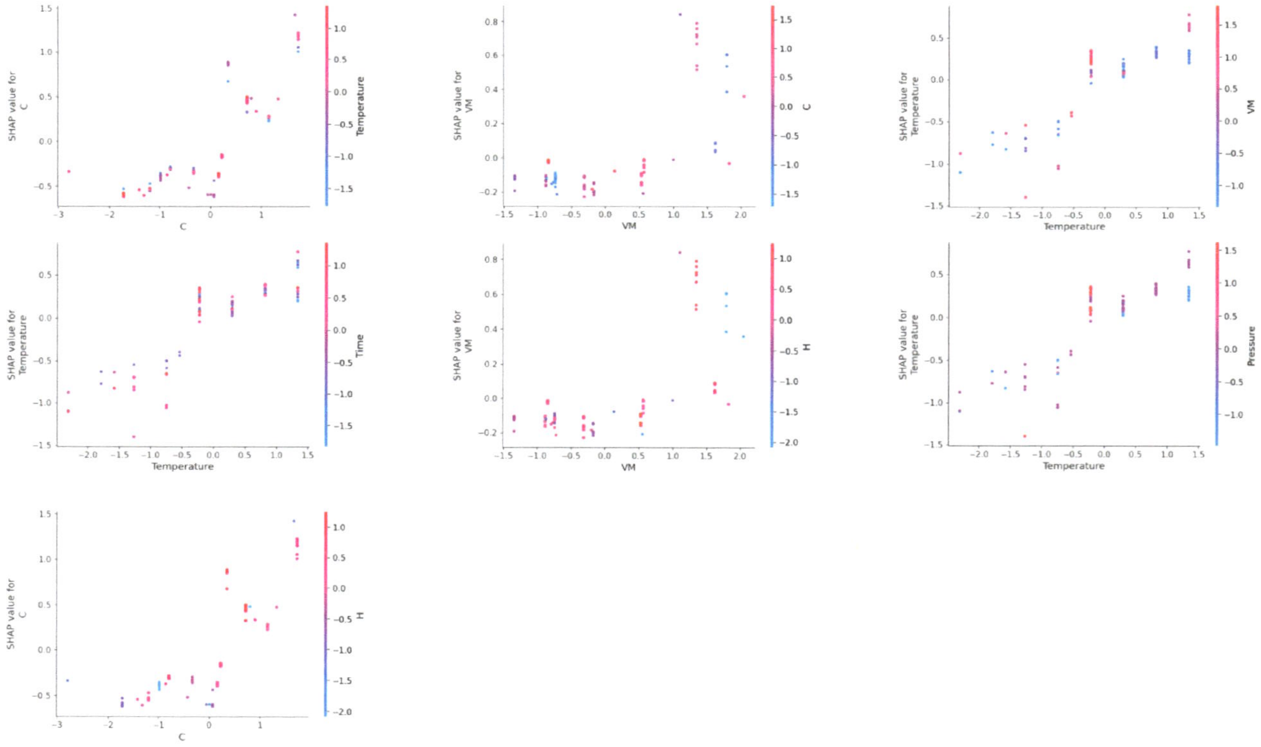

Figure 12. SHAP two-way plot of interactive behavior of the most important input features on prediction of CH_4 yield in the PSO-optimized CatBoost model.

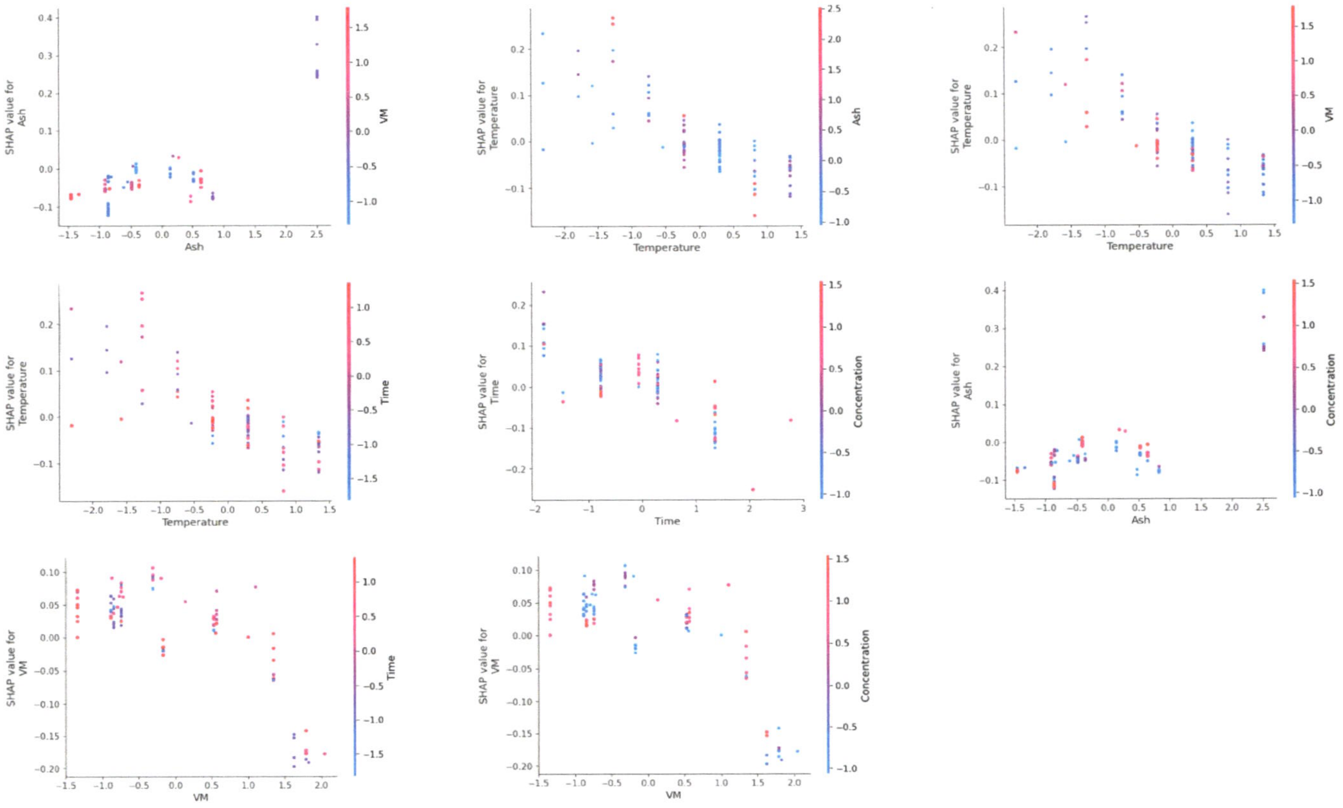

Figure 13. SHAP two-way plot of interactive behavior of the most important input features on prediction of CO yield in the PSO-optimized CatBoost model.

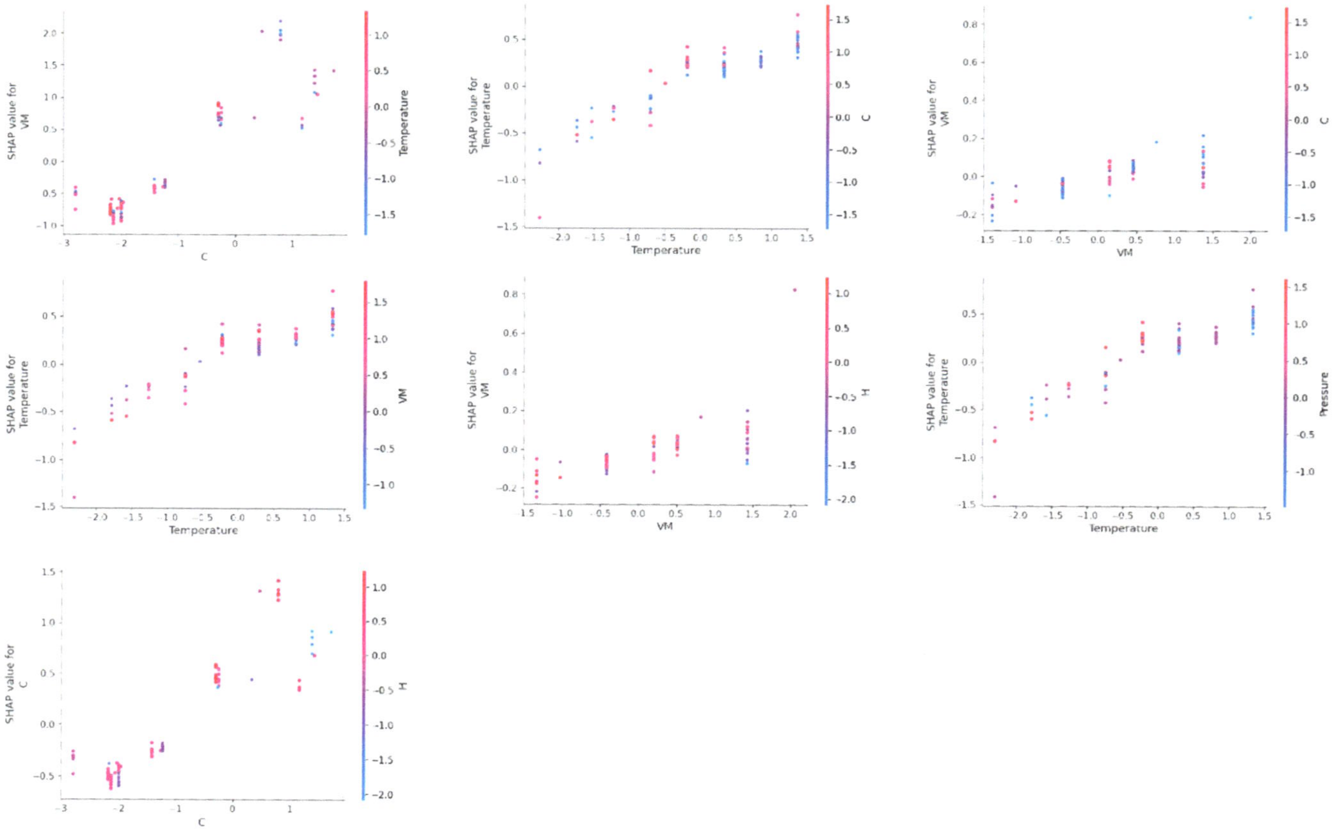

Figure 14. SHAP two-way plot of interactive behavior of the most important input features on prediction of CO_2 yield in the PSO-optimized CatBoost model.

From Figure 11, it can be observed that the input features had interactive effects on the prediction of hydrogen yield. An increase in temperature for high hydrogen content containing biomass resulted in the highest SHAP values for hydrogen yield. High SHAP values for hydrogen yield can also be achieved for moderate hydrogen containing biomass at high reaction temperatures. However, high hydrogen content at low reaction temperatures does not necessarily translate into high hydrogen yield. Similarly, modest ash content helped to achieve high hydrogen yields at high reaction temperatures. However, low reaction temperature even at optimum Ash content does not result in high hydrogen yield. For hydrogen content (H) and ash content of biomass, these features did not show much interaction at a low hydrogen content and low ash content of biomass. Only at an optimum ash content of biomass did an increase in hydrogen content of biomass result in the highest hydrogen yield.

Similarly, for time and hydrogen content, the highest hydrogen yield was obtained at highest hydrogen content and longer reaction time. However, at shorter reaction times and low high hydrogen content, these features did now show much interaction for hydrogen yield. This indicates that for efficient conversion of the hydrogen content of biomass into hydrogen gas during gasification, higher reaction temperature, longer reaction time, and optimum amount of ash content are required. Reaction temperature and reaction time showed interactive behavior, where highest hydrogen yields were obtained at high reaction temperature and longer reaction time. However, a comparable hydrogen yield can also be obtained even at short reaction times at high reaction temperatures. Similarly, reaction time and concentration also showed interactive behavior, and the highest hydrogen yield was obtained at low feedstock concentration at longer reaction times. A high hydrogen yield was obtained even at moderate to high concentrations at longer reaction times.

For prediction of methane yield, carbon content (C) and temperature showed high interactive behavior (Figure 12). The highest SHAP values for methane yield were obtained at high carbon content and high reaction temperatures. Strong interactive behavior was observed at moderate to high values of carbon content and at high reaction temperatures. Volatile matter (VM) and carbon content also showed high interactive behavior, and biomass having high volatile matter usually had moderate to high carbon content, which resulted in the highest methane yield. Temperature and VM of biomass also had strong interactive behavior, and high methane yield was obtained at a high reaction temperature and high amount of volatile matter. However, high SHAP values of methane yield were also observed at moderate VM at high temperatures or, also at moderate temperatures for biomass having high VM content. Temperature and time had interactive behavior for SHAP values of methane yield at high temperature and longer reaction time. Similarly, hydrogen and carbon content of biomass had an interactive effect on methane yield at high carbon and high hydrogen content, which resulted in the highest SHAP values of methane yield.

Interestingly, ash content and volatile matter (VM) of biomass had interactive behavior on SHAP values of CO yield; only at high ash content and moderate volatile matter were the highest SHAP values of CO gas yield observed (Figure 13). However, temperature had strong interactive behavior with ash content, volatile matter, and time. High SHAP values of CO yield were obtained at low temperatures and low ash content of biomass. Similarly, low temperature and moderate values of volatile matter resulted in the highest SHAP values of CO yield. However, high values of CO yields were obtained at low reaction times at low temperatures. Time also had strong interactive behavior with concentration and volatile matter of biomass for CO yield, where high SHAP values of CO yield were observed at short reaction times and high concentrations. However, low values of volatile matter at short reaction times resulted in high SHAP values for CO yield. Volatile matter and concentration themselves also had a strong interactive influence on CO yield, where high SHAP values of CO yield were obtained at low volatile matter and high feedstock concentration.

Two-way SHAP analysis for prediction of CO_2 yield showed that the carbon content (C) of biomass and temperature had a strong interactive influence on SHAP values of CO_2 yield. The highest values of CO_2 yield were observed at high carbon content and high temperatures (Figure 14). Similarly, carbon content of biomass also had interactive behavior with volatile matter on CO_2 yield, where an increase in volatile matter and carbon content increased the SHAP values of CO_2 yield. Temperature and volatile matter also had interactive effects at high values of temperature and volatile matter, where an increase in volatile matter at high reaction temperature increased the SHAP values of CO_2 yield. However, high comparable values of SHAP values were also observed at moderate volatile matter at high temperatures.

Temperature and time also had a strong interactive influence on CO_2 yield. Longer reaction time and high temperature had the highest SHAP values of CO_2 yield. Hydrogen content also demonstrated strong interactive behavior with volatile matter and carbon content of biomass on SHAP values of CO_2 yield. High volatile matter and high hydrogen content showed the highest values of CO_2 yield. This is due to the relationship between volatile matter and hydrogen content of biomass, as volatile matter of biomass represents high quantities of organic acids, hydrocarbons, alcohols, aldehydes, and ketones. These compounds usually have higher amounts of hydrogen atoms; thus, an increase in volatile matter also represents an increase in hydrogen content, which had a positive interactive influence on the SHAP values of CO_2 yield. High values of hydrogen content and carbon content also resulted in high values of CO_2 yield.

Thus, the degradation of lignocellulosic biomass follows a complex reaction mechanism, and input variables such as SCWG reaction conditions and biomass properties have an interactive influence during SCWG of lignocellulosic biomass. These interactive influences of input features have an effect on the product distribution and individual gas yields of the SCWG process. Two-way SHAP analysis highlighted the strong interactive influence of the most dominant features on yields of H_2, CH_4, CO, and CO_2. This shows that the optimization of SCWG of lignocellulosic biomass is a complex process and requires careful simultaneous tuning of various parameters to maximize the hydrogen yield of the SCWG process.

Thus, this study presented a novel and comprehensive application of machine learning models for SCWG of lignocellulosic biomass to elucidate the interactive effects of input features and their complex relationship with gas yields. Utilization of only lab-scale batch reactor data helped to better capture the relationships between input variables and gas yields with minimum influence of other unaccounted-for variables. However, it also resulted in limited scope of the prediction models only to a lab-scale batch reactor. Nevertheless, the main objective of this study was to understand the complex degradation behavior of lignocellulosic biomass during SCWG and interactive effects of input variables on gas yields. This study presented a groundwork for comprehensive optimization of SCWG reaction conditions and selection of suitable biomass, especially at industrial scale, to maximize hydrogen gas yields with a high degree of certainty. This will foster efforts being made for commercialization of SCWG at an industrial scale.

4. Conclusions

In conclusion, this study showed the successful implementation of machine learning models for prediction of gas yields of SCWG of lignocellulosic biomass. Among the eight screened machine learning model for eight prediction of gas yield of SCWG, boosting ensemble tree models such as XGB and CatBoost models demonstrated superior prediction power. For prediction of H_2 yield, PSO-optimized XGB showed eight highest test R^2 of 0.84, whereas, for prediction of CH_4, CO, and CO_2 gas yields, PSO-optimized CatBoost showed highest test R^2 of 0.83, 0.94, and 0.92, respectively. This is due to the use of series of multiple simple decision tree models by boosting ensemble tree models, which improves upon the preceding tree and prevents the overfitting of the dataset. Even though both GA and PSO are heuristic algorithms, the use of the PSO optimizer was more effective compared to the GA optimizer in hyperparameter tuning of machine learning models for improving their prediction power. This is due to the difference in the search mechanisms of these algorithms.

Feature analysis based on PSO-optimized XGB showed that temperature was the most dominant feature for prediction of H_2 yield, with its highest feature importance of 21.93%. For prediction of CH_4, CO, and CO_2 gas yields by the PSO-optimized CatBoost model, carbon content (C), ash content (ash), and carbon content (C) were the most dominant features, with feature importance of 24.85, 16.93, and 29.73%, respectively. Among SCWG reaction conditions and biomass characteristics, biomass characteristics were most dominant as a whole, with cumulative feature importance of 52.91, 63.19, 57.54, and 68.37% for prediction of H_2, CH_4, CO, and CO_2 gas yields from SCWG of lignocellulosic biomass, respectively. SHAP two-way analysis revealed strong interactive behavior of input features for the prediction of H_2, CH_4, CO, and CO_2 gas yields and their effect on the SCWG reaction mechanism. This also confirmed the complex and non-linear nature of the SCWG process. Furthermore, the high importance of biomass characteristics highlights the importance of the selection of suitable feedstocks and understanding the interactive behavior of components of hydrogenous biomass for efficient conversion of lignocellulosic biomass into high hydrogen yield. This work has laid the framework for a comprehensive optimization of the SCWG process, which can be a key asset for the commercialization of the SCWG process at an industrial scale.

Author Contributions: Conceptualization, K.K. and A.K.D.; methodology, K.K.; software, K.K. and A.K.D.; validation, K.K.; formal analysis, K.K.; investigation, K.K.; resources, A.K.D.; data curation, K.K.; writing—original draft preparation, K.K.; writing—review and editing, K.K. and A.K.D.; visualization, K.K.; supervision, A.K.D.; project administration, A.K.D.; funding acquisition, A.K.D. All authors have read and agreed to the published version of the manuscript.

Funding: The authors acknowledge the Natural Sciences and Engineering Research Council of Canada (NSERC), Canada Research Chair (CRC), and MITACS programs for supporting this research.

Institutional Review Board Statement: Not applicable.

Informed Consent Statement: Not applicable.

Data Availability Statement: Data available upon request.

Conflicts of Interest: The authors declare no conflict of interest.

References

1. World Energy Consumption Statistics | Enerdata n.d. Available online: https://yearbook.enerdata.net/total-energy/world-consumption-statistics.html (accessed on 13 December 2023).
2. Global Primary Energy Consumption by Fuel 2022 | Statista n.d. Available online: https://www.statista.com/statistics/265619/primary-energy-consumption-worldwide-by-fuel/ (accessed on 13 December 2023).
3. Opeyemi, B.M. Path to sustainable energy consumption: The possibility of substituting renewable energy for non-renewable energy. *Energy* **2021**, *228*, 120519. [CrossRef]
4. Fatma, S.; Hameed, A.; Noman, M.; Ahmed, T.; Shahid, M.; Tariq, M.; Sohail, I.; Tabassum, R. Lignocellulosic Biomass: A Sustainable Bioenergy Source for the Future. *Protein Pept. Lett.* **2018**, *25*, 148–163. [CrossRef] [PubMed]
5. Alizadeh, R.; Lund, P.D.; Soltanisehat, L. Outlook on biofuels in future studies: A systematic literature review. *Renew. Sustain. Energy Rev.* **2020**, *134*, 110326. [CrossRef]
6. COP28: Global Renewables and Energy Efficiency Pledge n.d. Available online: https://www.cop28.com/en/global-renewables-and-energy-efficiency-pledge (accessed on 13 December 2023).
7. Summary of Global Climate Action at COP 28 n.d. Available online: https://unfccc.int/sites/default/files/resource/Summary_GCA_COP28.pdf (accessed on 13 December 2023).
8. Beig, B.; Riaz, M.; Naqvi, S.R.; Hassan, M.; Zheng, Z.; Karimi, K.; Pugazhendhi, A.; Atabani, A.E.; Chi, N.T.L. Current challenges and innovative developments in pretreatment of lignocellulosic residues for biofuel production: A review. *Fuel* **2021**, *287*, 119670. [CrossRef]
9. Lee, C.S.; Conradie, A.V.; Lester, E. Review of supercritical water gasification with lignocellulosic real biomass as the feedstocks: Process parameters, biomass composition, catalyst development, reactor design and its challenges. *Chem. Eng. J.* **2021**, *415*, 128837. [CrossRef]
10. Heeley, K.; Orozco, R.L.; Macaskie, L.E.; Love, J.; Al-Duri, B. Supercritical water gasification of microalgal biomass for hydrogen production-A review. *Int. J. Hydrogen Energy* **2023**, *49*, 310–336. [CrossRef]
11. Tarhan, C.; Çil, M.A. A study on hydrogen, the clean energy of the future: Hydrogen storage methods. *J. Energy Storage* **2021**, *40*, 102676. [CrossRef]

12. Zou, C.; Li, J.; Zhang, X.; Jin, X.; Xiong, B.; Yu, H.; Liu, X.; Wang, S.; Li, Y.; Zhang, L.; et al. Industrial status, technological progress, challenges, and prospects of hydrogen energy. *Nat. Gas Ind. B* **2022**, *9*, 427–447. [CrossRef]
13. Khandelwal, K.; Nanda, S.; Boahene, P.; Dalai, A.K. Conversion of biomass into hydrogen by supercritical water gasification: A review. *Environ. Chem. Lett.* **2023**, *21*, 2619–2638. [CrossRef]
14. Ali, A.; Razak, S.A.; Othman, S.H.; Eisa, T.A.E.; Al-Dhaqm, A.; Nasser, M.; Elhassan, T.; Elshafie, H.; Saif, A. Financial Fraud Detection Based on Machine Learning: A Systematic Literature Review. *Appl. Sci.* **2022**, *12*, 9637. [CrossRef]
15. Revathy, G.; Alghamdi, S.A.; Alahmari, S.M.; Yonbawi, S.R.; Kumar, A.; Haq, M.A. Sentiment analysis using machine learning: Progress in the machine intelligence for data science. *Sustain. Energy Technol. Assess.* **2022**, *53*, 102557. [CrossRef]
16. Iwendi, C.; Ibeke, E.; Eggoni, H.; Velagala, S.; Srivastava, G. Pointer-Based Item-to-Item Collaborative Filtering Recommendation System Using a Machine Learning Model. *Int. J. Inf. Technol. Decis. Mak.* **2022**, *21*, 463–484. [CrossRef]
17. Sharma, V.; Tsai, M.-L.; Chen, C.-W.; Sun, P.-P.; Nargotra, P.; Dong, C.-D. Advances in machine learning technology for sustainable biofuel production systems in lignocellulosic biorefineries. *Sci. Total. Environ.* **2023**, *886*, 163972. [CrossRef] [PubMed]
18. Umenweke, G.C.; Afolabi, I.C.; Epelle, E.I.; Okolie, J.A. Machine learning methods for modeling conventional and hydrothermal gasification of waste biomass: A review. *Bioresour. Technol. Rep.* **2022**, *17*, 100976. [CrossRef]
19. Khan, M.; Naqvi, S.R.; Ullah, Z.; Taqvi, S.A.A.; Khan, M.N.A.; Farooq, W.; Mehran, M.T.; Juchelková, D.; Štěpanec, L. Applications of machine learning in thermochemical conversion of biomass-A review. *Fuel* **2023**, *332*, 126055. [CrossRef]
20. Demertzis, K.; Kostinakis, K.; Morfidis, K.; Iliadis, L. An interpretable machine learning method for the prediction of R/C buildings' seismic response. *J. Build. Eng.* **2023**, *63*, 105493. [CrossRef]
21. Heng, J.; Hong, Y.; Hu, J.; Wang, S. Probabilistic and deterministic wind speed forecasting based on non-parametric approaches and wind characteristics information. *Appl. Energy* **2021**, *306*, 118029. [CrossRef]
22. Chand, E.L.; Cheema, S.S.; Kaur, M. Understanding Neural Networks. In *Factories of the Future: Technological Advancements in the Manufacturing Industry*; Wiley Publishing Company: New York, NY, USA, 2023; pp. 83–102. [CrossRef]
23. Hussain, S.F. A novel robust kernel for classifying high-dimensional data using Support Vector Machines. *Expert Syst. Appl.* **2019**, *131*, 116–131. [CrossRef]
24. Zhang, Y.; Wang, S.; Ji, G. A Comprehensive Survey on Particle Swarm Optimization Algorithm and Its Applications. *Math. Probl. Eng.* **2015**, *2015*, 931256. [CrossRef]
25. Gonzalez, S.; Garcia, S.; Del Ser, J.; Rokach, L.; Herrera, F. A practical tutorial on bagging and boosting based ensembles for machine learning: Algorithms, software tools, performance study, practical perspectives and opportunities. *Inf. Fusion* **2020**, *64*, 205–237. [CrossRef]
26. Ait Hammou, B.; Ait Lahcen, A.; Mouline, S. A distributed group recommendation system based on extreme gradient boosting and big data technologies. *Appl. Intell.* **2019**, *49*, 4128–4149. [CrossRef]
27. Sohail, A. Genetic Algorithms in the Fields of Artificial Intelligence and Data Sciences. *Ann. Data Sci.* **2023**, *10*, 1007–1018. [CrossRef]
28. Alhijawi, B.; Awajan, A. Genetic algorithms: Theory, genetic operators, solutions, and applications. *Evol. Intell.* **2023**, *1*, 1–12. [CrossRef]
29. Valdez, F. Swarm intelligence: A review of optimization algorithms based on animal behavior. *Stud. Comput. Intell.* **2021**, *915*, 273–298. [CrossRef]
30. Ahmad, Z.; Li, J.; Mahmood, T. Adaptive Hyperparameter Fine-Tuning for Boosting the Robustness and Quality of the Particle Swarm Optimization Algorithm for Non-Linear RBF Neural Network Modelling and Its Applications. *Mathematics* **2023**, *11*, 242. [CrossRef]
31. Chicco, D.; Warrens, M.J.; Jurman, G. The coefficient of determination R-squared is more informative than SMAPE, MAE, MAPE, MSE and RMSE in regression analysis evaluation. *PeerJ. Comput. Sci.* **2021**, *7*, e623. [CrossRef]
32. Zhu, M.; Yang, Y.; Feng, X.; Du, Z.; Yang, J. Robust modeling method for thermal error of CNC machine tools based on random forest algorithm. *J. Intell. Manuf.* **2023**, *34*, 2013–2026. [CrossRef]
33. Merrick, L.; Taly, A. The Explanation Game: Explaining Machine Learning Models Using Shapley Values. In *Lecture Notes in Computer Science (Including Subseries Lecture Notes in Artificial Intelligence and Lecture Notes in Bioinformatics)—LNCS*; Springer International Publishing: Berlin/Heidelberg, Germany, 2020; Volume 12279, pp. 17–38. [CrossRef]
34. Pradhan, N.; Rani, G.; Dhaka, V.S.; Poonia, R.C. Diabetes prediction using artificial neural network. *Deep. Learn. Tech. Biomed. Health Inform.* **2020**, 327–339. [CrossRef]
35. Zhao, S.; Li, J.; Chen, C.; Yan, B.; Tao, J.; Chen, G. Interpretable machine learning for predicting and evaluating hydrogen production via supercritical water gasification of biomass. *J. Clean. Prod.* **2021**, *316*, 128244. [CrossRef]
36. Demir, S.; Sahin, E.K. An investigation of feature selection methods for soil liquefaction prediction based on tree-based ensemble algorithms using AdaBoost, gradient boosting, and XGBoost. *Neural Comput. Appl.* **2023**, *35*, 3173–3190. [CrossRef]
37. Ampomah, E.K.; Qin, Z.; Nyame, G. Evaluation of Tree-Based Ensemble Machine Learning Models in Predicting Stock Price Direction of Movement. *Information* **2020**, *11*, 332. [CrossRef]
38. Pathy, A.; Meher, S.; Balasubramanian, P. Predicting algal biochar yield using eXtreme Gradient Boosting (XGB) algorithm of machine learning methods. *Algal Res.* **2020**, *50*, 102006. [CrossRef]
39. Katongtung, T.; Onsree, T.; Tippayawong, N. Machine learning prediction of biocrude yields and higher heating values from hydrothermal liquefaction of wet biomass and wastes. *Bioresour. Technol.* **2022**, *344*, 126278. [CrossRef] [PubMed]

40. Liu, Q.; Zhang, G.; Yu, J.; Kong, G.; Cao, T.; Ji, G.; Zhang, X.; Han, L. Machine learning-aided hydrothermal carbonization of biomass for coal-like hydrochar production: Parameters optimization and experimental verification. *Bioresour. Technol.* **2024**, *393*, 130073. [CrossRef] [PubMed]
41. Shwartz-Ziv, R.; Armon, A. Tabular data: Deep learning is not all you need. *Inf. Fusion* **2022**, *81*, 84–90. [CrossRef]
42. Khandelwal, K.; Dalai, A.K. Integration of hydrothermal gasification with biorefinery processes for efficient production of biofuels and biochemicals. *Int. J. Hydrogen Energy* **2023**, *49*, 577–592. [CrossRef]
43. Seraj, S.; Azargohar, R.; Borugadda, V.B.; Dalai, A.K. Energy recovery from agro-forest wastes through hydrothermal carbonization coupled with hydrothermal co-gasification: Effects of succinic acid on hydrochars and H2 production. *Chemosphere* **2023**, *337*, 139390. [CrossRef] [PubMed]
44. Khandelwal, K.; Boahene, P.; Nanda, S.; Dalai, A.K. Hydrogen Production from Supercritical Water Gasification of Model Compounds of Crude Glycerol from Biodiesel Industries. *Energies* **2023**, *16*, 3746. [CrossRef]
45. Chen, Y.; Guo, L.; Cao, W.; Jin, H.; Guo, S.; Zhang, X. Hydrogen production by sewage sludge gasification in supercritical water with a fluidized bed reactor. *Int. J. Hydrogen Energy* **2013**, *38*, 12991–12999. [CrossRef]
46. Farobie, O.; Matsumura, Y.; Syaftika, N.; Amrullah, A.; Hartulistiyoso, E.; Bayu, A.; Moheimani, N.R.; Karnjanakom, S.; Saefurahman, G. Recent advancement on hydrogen production from macroalgae via supercritical water gasification. *Bioresour. Technol. Rep.* **2021**, *16*, 100844. [CrossRef]
47. Correa, C.R.; Kruse, A. Supercritical water gasification of biomass for hydrogen production—Review. *J. Supercrit. Fluids* **2018**, *133*, 573–590. [CrossRef]
48. Su, W.; Cai, C.; Liu, P.; Lin, W.; Liang, B.; Zhang, H.; Ma, Z.; Ma, H.; Xing, Y.; Liu, W. Supercritical water gasification of food waste: Effect of parameters on hydrogen production. *Int. J. Hydrogen Energy* **2020**, *45*, 14744–14755. [CrossRef]
49. Norouzi, O.; Safari, F.; Jafarian, S.; Tavasoli, A.; Karimi, A. Hydrothermal gasification performance of Enteromorpha intestinalis as an algal biomass for hydrogen-rich gas production using Ru promoted Fe–Ni/γ-Al_2O_3 nanocatalysts. *Energy Convers. Manag.* **2017**, *141*, 63–71. [CrossRef]
50. Wang, Q.; Zhang, X.; Cui, D.; Bai, J.; Wang, Z.; Xu, F.; Wang, Z. Advances in supercritical water gasification of ligno-cellulosic biomass for hydrogen production. *J. Anal. Appl. Pyrolysis* **2023**, *170*, 105934. [CrossRef]

molecules

Review

Renewable Energy Potential: Second-Generation Biomass as Feedstock for Bioethanol Production

Chidiebere Millicent Igwebuike , Sary Awad * and Yves Andrès

IMT Atlantique, GEPEA, UMR CNRS 6144, 4 Rue Alfred Kastler, F-44000 Nantes, France; chidieberem.igwebuike@gmail.com (C.M.I.); yves.andres@imt-atlantique.fr (Y.A.)
* Correspondence: sary.awad@gmail.com

Abstract: Biofuels are clean and renewable energy resources gaining increased attention as a potential replacement for non-renewable petroleum-based fuels. They are derived from biomass that could either be animal-based or belong to any of the three generations of plant biomass (agricultural crops, lignocellulosic materials, or algae). Over 130 studies including experimental research, case studies, literature reviews, and website publications related to bioethanol production were evaluated; different methods and techniques have been tested by scientists and researchers in this field, and the most optimal conditions have been adopted for the generation of biofuels from biomass. This has ultimately led to a subsequent scale-up of procedures and the establishment of pilot, demo, and large-scale plants/biorefineries in some regions of the world. Nevertheless, there are still challenges associated with the production of bioethanol from lignocellulosic biomass, such as recalcitrance of the cell wall, multiple pretreatment steps, prolonged hydrolysis time, degradation product formation, cost, etc., which have impeded the implementation of its large-scale production, which needs to be addressed. This review gives an overview of biomass and bioenergy, the structure and composition of lignocellulosic biomass, biofuel classification, bioethanol as an energy source, bioethanol production processes, different pretreatment and hydrolysis techniques, inhibitory product formation, fermentation strategies/process, the microorganisms used for fermentation, distillation, legislation in support of advanced biofuel, and industrial projects on advanced bioethanol. The ultimate objective is still to find the best conditions and technology possible to sustainably and inexpensively produce a high bioethanol yield.

Keywords: feedstock; biomass conversion; fermentation; biofuels; bioethanol; projects

Citation: Igwebuike, C.M.; Awad, S.; Andrès, Y. Renewable Energy Potential: Second-Generation Biomass as Feedstock for Bioethanol Production. *Molecules* **2024**, *29*, 1619. https://doi.org/10.3390/molecules29071619

Academic Editor: Mohamad Nasir Mohamad Ibrahim

Received: 7 March 2024
Revised: 25 March 2024
Accepted: 30 March 2024
Published: 4 April 2024

1. Introduction

Currently, emphasis has shifted to the use of renewable sources of energy both in the developed and developing countries of the world. This is due to the steady growth in population and industrialization [1], the decline in known reserves, the rising uncertainty of petroleum supplies as a result of increasing demand, and concerns over climate change (greenhouse gas emissions and global warming), which are linked to the use of fossil fuels [2]. In a bid to satisfy the world demand for energy, lower fossil fuel consumption, reduce the emissions of CO_2, and maintain or develop agricultural activities (by utilizing bio-resources for energy, food, and material), governments across the world are encouraging the exploitation of renewable resources and energies such as biomass, wind, solar and hydroelectricity [3]. Technological advancement in biofuel production supports greener modes of transportation, contributing to a sustainable energy future. Among the available alternative energy sources, biofuels stand out in their general compatibility with existing liquid transport fuels. They are considered a viable alternative to fossil-based fuels. Bioethanol, a type of liquid biofuel, plays a vital role in the global energy transition by reducing greenhouse gas emissions and enhancing energy security. Its use can promote rural development, drive economic growth, and satisfy international targets for renewable energy. To produce bioethanol, lignocellulosic waste plant biomass could be exploited. It is a second-generation feedstock and it is regarded as the most abundantly available, renewable, and inexpensive energy source for the production of bioethanol [4], owing to its high holocellulose composition [5]. This biomass stands out because it does not serve as food for human consumption. Hence, there is no competition in the food market compared to the case of first-generation biomass (e.g., sugarcane and maize). Lignocellulose accounts for more than 90% of worldwide plant biomass production. It amounts to approximately 200 billion tons per year [2].

Lack of knowledge about the importance of agricultural wastes, inadequate finance for the purchase of needed facilities/equipment, and lack of technical know-how or skills for biofuel production are the major causes of incessant disposal of agricultural waste (e.g., cassava peels, sugar beet pulp, *Ulva lactuca*, sugarcane bagasse, corn straw, etc.) in the environment, especially in the rural areas of developing countries where agricultural activities are prevalent, and these wastes are usually rich in cellulose and hemicellulose components and can be broken down into simpler components for the production of useful fuels. The various activities practiced by humans in the disposal of these wastes have caused the deterioration of the environment and the loss of its aesthetic value. They have also caused several health challenges. The conversion of these wastes to useful fuels like bioethanol can help to reduce the occurrence of these environmental pollutants in our environment and also meet the energy demand of the populace. Notwithstanding the advantages inherent in the utilization of lignocellulosic biomass for bioethanol production, there exists a drawback due to the presence of lignin in the cell wall, which confers upon the material its recalcitrance nature and renders it not easily degradable. As a result of this, such material needs to be subjected to different treatment processes to release fermentable sugars. This review, therefore, aims to provide insights and, at the same time, explore the current state of the research on bioethanol production and identify the key challenges and opportunities for advancing the technology.

The study strategy involved conducting a comprehensive literature search using the Google Search Engine and relevant academic databases, such as ScienceDirect, Google Scholar, Researchgate, and PubMed. Keywords and phrases related to second-generation biomass, bioethanol production, and relevant technologies were used to search for potential articles, and the search results were screened based on titles and abstracts to identify relevant articles. Full-text articles were then reviewed to determine their eligibility for inclusion in this review. The quality of the included articles was evaluated based on the credibility of the reports/findings, and key insights were identified and analyzed.

2. Biomass and Bioenergy

For ages, humans have relied on the use of traditional bioenergy. Over 85% of biomass energy is being utilized as solid fuels for heating, cooking, and lighting at present, but these methods have low efficiency. Fuelwood and charcoal are categorized as traditional bioenergy, which only provides heat, and are said to dominate the consumption of bioenergy, especially in the developing world where about 95% of national energy consumption depends on biomass. This biomass will keep on being an essential source of bioenergy in many parts of the world. Up to now, wood fuels stand as the main source of bioenergy across different regions, as they offer energy security services for large divisions of society. Modern bioenergy depends on efficient conversion technologies for domestic use and utilization at both industrial scales and small businesses. The significance of bioenergy is demonstrated worldwide. In North America, the use of ethanol derived from corn reduces reliance on fossil fuels [6]; in Scandinavia, district heating using biomass reduces carbon emissions [7]; in Africa, the use of biofuels for off-grid power improves access to energy [8]; and in Southeast Asia, palm oil biodiesel industry tackles climate change while driving economic growth [9]. Biomass could be processed into more convenient energy carriers, such as liquid fuels (biodiesel, bioethanol, bio-oil), solid fuels (wood chips, firewood, charcoal, briquettes, pellets), gaseous fuels (hydrogen, synthesis gas, biogas), or heat directly from the production process [10]. The combustion of biomass is regarded as a carbon-neutral process since the carbon dioxide emitted has been absorbed by the plants from the atmosphere beforehand. So far, organic wastes and residues are the major biomass sources; nevertheless, energy crops such as poplar, willow, and eucalyptus are gaining significance and market share. Biomass resources comprise wood wastes from industry and forestry, agricultural residues, residues from paper and food industries, animal manure, dedicated energy crops, sewage sludge, municipal green wastes, starch crops (wheat, corn), sugar crops (beet, sugarcane, sorghum), oil crops (oilseed rape, soy, sunflower, palm oil, jatropha), and grasses (Miscanthus) [11].

At present, biomass constitutes only a small fraction of the total carbon use despite often being applied as raw material and a source of energy. Its use is limited to the large-scale production of bioethanol and low-volume products. In the coming years, there is expected to be a continual shift from the present fossil-based to a future bio-based carbon economy, thus causing a gradual effect on all process industries. There will be a constant transition to more complex bio-renewable feedstock, such as algae, agricultural residues, green plants, industrial wastes, or wood, and eventually, bio-based products will replace the petrochemical product tree. This transition is not to be regarded as a threat but a chance to redesign the industrial value chain from renewable material sources to new products and for this to be achieved, the rich molecular structure of renewable biomaterials is to be greatly exploited [12].

3. Structure and Composition of Lignocellulosic Biomass

Lignocellulosic biomass is majorly composed of three polymers: cellulose, hemicellulose, and lignin, including a few amounts of protein, pectin, ash, and extractives. The proportion of these components can differ from one biomass species to another. For instance, hardwoods have more amounts of cellulose, while leaves and wheat straws have more hemicellulose. The proportion of these components is also not often the same within a single plant species due to the stage of growth, age, and certain other conditions. These polymers are linked with each other in a hetero-matrix to different degrees and varying relative compositions depending on the type, species, and source of the biomass material [13]. The relative amounts of cellulose, hemicellulose, and lignin are, among others, important determining factors in identifying the suitability of plant species for use as energy crops [14]. Figure 1 gives the structural presentation of lignocellulosic biomass.

Cellulose is the most abundant fraction in lignocellulosic materials and represents about 40–50% of the biomass by weight. It is a polymer of glucose, comprising linear chains of (1,4)-D-glucopyranose units, whereby the units are linked C_1–C_4 oxygen bridges with the removal of water in the β-configuration (β-glycosidic bonds) with an average molecular weight of around 100,000 [14]. Cellulose is a white solid material that may exist either in crystalline or amorphous states. The crystalline state of cellulose confers its ability to be resistant to chemical attack and degradation. The high strength of cellulose fibers is a consequence of the hydrogen bonding that exists between cellulose molecules [15].

After cellulose, hemicellulose ranks as the second largest carbohydrate in the world. Daily production per person was estimated to be about 20 kg, and about 45,000 million tons are produced on a yearly basis [16]. Hemicellulose can be categorized into four general classes of structurally different cell wall polysaccharide types: xylans, mannans, β-glucans with mixed linkages, and xyloglucans. They exist in structural variations differing in side-chain types, distribution, localization, and/or types and distribution of glycoside linkages in the macro-molecular backbone [17]. The hemicellulose fraction represents 20–40% of the biomass by weight. It is a mixture of polysaccharides that comprises mostly sugars, like glucose, xylose, arabinose, mannose, methylglucuronic, and galacturonic acids. Hemicellulose has an average molecular weight of <30,000 [14]. In appearance, hemicelluloses are also white solid materials and are amorphous in nature [15].

Lignin refers to a group of amorphous that have high molecular weight (over 10,000), are cross-linked, and are chemically related complex polymer compounds that form an essential part of the secondary plant cell wall. It is rich in aromatic subunits and relatively hydrophobic. It hinders the free access of cellulolytic enzymes due to its cross-linkages with other components of the cell wall. It has a very slow rate of decomposition but contributes a significant aspect to the materials that form humus [18]. Lignin comprises three basic monomers: *p*-coumaryl alcohol, coniferyl alcohol, and sinapyl alcohol. All three lignin monomers are found in straws and grasses. Coniferyl alcohol is found in softwoods (gymnosperms e.g., cycads and conifers), while both coniferyl alcohol (50–75%) and sinapyl alcohol (25–50%) are present in hardwoods (dicotyledonous angiosperms) [15].

Lignocellulosic materials have great potential to generate second-generation biofuels, including bio-sourced materials and chemicals, without negatively impacting the world's food security. However, the main drawback of these materials' valorization is their recalcitrance to enzymatic hydrolysis due to the heterogeneous multi-scale structure of plant cell walls. The factors affecting the recalcitrance of these materials are strongly interconnected and not easily dissociated. These factors can be classified into chemical factors (composition and content in lignin, hemicelluloses, acetyl groups) and structural factors (cellulose crystallinity, cellulose specific surface area, degree of polymerization, pore size, and volume) [19]. The structural composition of different lignocellulosic biomass is given in Table 1.

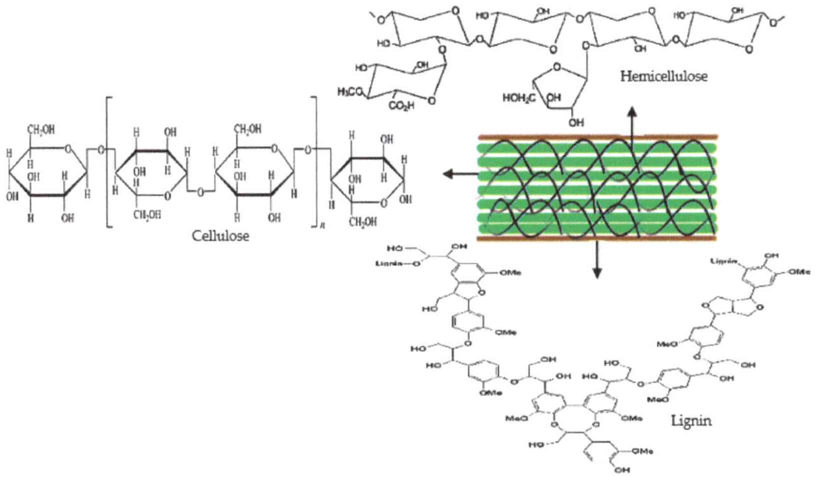

Figure 1. Structure and composition of lignocellulosic biomass [20,21].

Table 1. Compositions of selected lignocellulosic biomass.

Biomass	Cellulose (%)	Hemicellulose (%)	Lignin (%)	References
Oak wood	49.3	25.9	21.7	[22]
Sugar beet pulp	20.71	14.98	3.96	[23]
Sugarcane bagasse	50	25	25	[24]
Rice straw	34.6	27.7	17.6	[25]
Rice husk	33.4	22.1	22.8	[25]
Wheat straw	33.5	24.6	19	[25]
Oil palm empty fruit bunches	39.13	23.04	34.37	[5]
Corncobs	22.1	9.6	6.0	[26]

Table 1. *Cont.*

Biomass	Cellulose (%)	Hemicellulose (%)	Lignin (%)	References
Banana rachis	26.1	11.2	10.8	[27]
Banana pseudostem	20.1	9.6	10.1	[27]
Cassava peels	9.05	7.50	9.16	[28]
Tygra hemp	50.82	27.79	14.68	[29]
Groundnut shell	35.7	18.7	30.2	[30]
Corn stover	36.1	21.4	17.2	[31]
Poplar	42.34	15.23	25.40	[32]
Waste from urban greening	22.96	6.86	22.73	[33]
Spring leaves	21.06	6.00	27.74	[33]
Autumn leaves	14.54	8.45	11.16	[33]
Jerusalem artichoke	25.99	4.50	5.70	[18]
Energy grass	37.85	27.33	9.65	[18]
Sunflower	34.06	5.18	7.72	[18]
Silage	39.27	25.96	9.02	[18]
*Miscanthus sacchariflori*s	42.00	30.15	7.00	[18]
Reed	49.40	31.50	8.74	[18]

4. Groups of Biofuels

Depending on the processing before utilization, biofuels can be classified into primary and secondary biofuels (Figure 2).

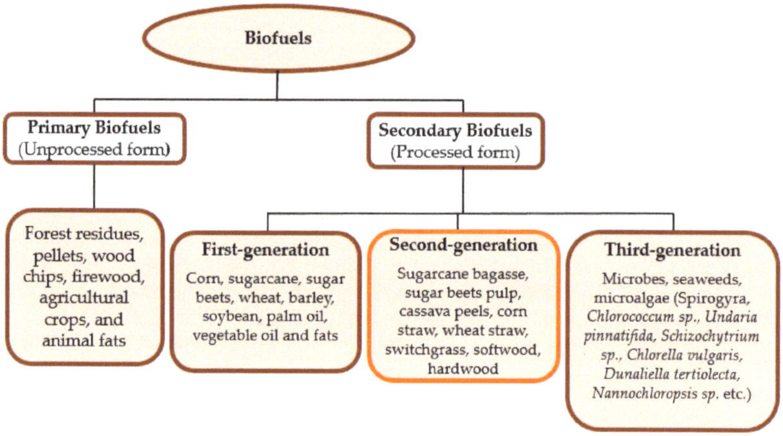

Figure 2. Biofuel classification and examples [34–36].

4.1. Primary Biofuels

Primary biofuels are used in their unprocessed form, i.e., the organic material is used in its natural form, as harvested. Examples of such primary biofuels include forest residues, pellets, wood chips, firewood, agricultural crops, and animal fats. They are primarily used for cooking, heating, agricultural needs, and the production of electricity in small- and large-scale industrial applications. They are common in developing countries. Primary biofuels are also referred to as traditional biomass. The field of application of this biofuel is small, and it does not require processing resource expenses. Energy derived from traditional biomass accounted for about 9% of all energy consumed globally for the year 2013 [37–41].

4.2. Secondary Biofuels

Secondary biofuels are processed materials. They are produced by processing biomass (the primary biofuel). Secondary biofuels can be in the form of solids (charcoal), liquids (bio-oil, ethanol, biodiesel), or gases (hydrogen, biogas, synthesis gas), which can be used in a wider range of applications, such as high-temperature industrial processes and transport, to substitute fossil fuel. Based on the type of raw materials, the historical sequence of the fuel's appearance on the world energy market and the processing technology employed in production, secondary biofuels can further be divided into three generations: first-generation biofuels, second-generation biofuels, and third-generation biofuels [37–40,42]. Table 2 gives some benefits and issues associated with the distinct biofuel generations.

Table 2. Benefits and issues associated with secondary biofuels.

Secondary Biofuels	Benefits	Issues
First generation	• Enhance energy, social, and economic security. • Eco-friendly.	• Impact on food security, land use, biodiversity, and carbon balances. • High production cost. • Partly blended with petroleum-based fuels.
Second generation	• Lower impact on the food sector. • Lower cost of feedstock. • Enhance energy, social, and economic security. • Better land-use efficiency. • Eco-friendly.	• Recalcitrance of cell walls. • May incur a high cost of production. • Infancy stage of cell wall polysaccharide biosynthetic machinery and its regulation.
Third generation	• Mitigates greenhouse gases. • Higher yields. • Shorter harvesting cycle (1–10 days) • High lipid content. • Rapid growth rate. • Reduced land use. • Higher CO_2 tolerance.	• Algae require large amounts of water, phosphorus, and nitrogen to grow. • Algae biomass requires dewatering before the extraction of lipids. • Oil from algae tends to be more volatile (unsaturated), especially at high temperatures, and hence, more likely to degrade. • Higher cost of cultivation. • Higher energy consumption during harvesting.

4.2.1. First-Generation Biofuels

First-generation biofuels are fuels that have been produced from biomass that are generally edible, for example, corn and sugarcane. Sugar beets, barley, potato wastes, and whey are also some of the marginal feedstocks that are used or considered to produce first-generation bioethanol [43]. From the environmental and economic outlook, sugarcane is an ideal feedstock for the production of ethanol, but it is limited to certain regions due to soil and weather conditions [44]. First-generation biofuels can help to enhance domestic energy security and can offer some CO_2 benefits [45]. However, the major concern about these biofuels is their inefficiency and sustainability since the viability of the production of such biofuel is questionable as a result of the conflict with food supply [46], including sourcing of feedstock, land use, deforestation, and the impact it may have on biodiversity. Nowadays, first-generation biofuel production is commercial, with an annual production of about 50 billion liters. Biofuels (first generation) like bioethanol, biodiesel, and biogas are categorized by their ability to be blended with petroleum-based fuels, combusted in existing internal combustion engines, and distributed through existing infrastructure, or by their use in existing alternative vehicle technology, such as natural gas vehicles or flexible-fuel vehicles (FFVs) [45]. First-generation biofuels compete with food and accrue high costs of production. Certain crops and foodstuffs have become expensive due to the fast expansion of global biofuel production from sugar, grain, and oilseed crops; hence, with these drawbacks, there is, therefore, a need to search for non-edible biomass for biofuel production [39].

4.2.2. Second-Generation Biofuels

Second-generation biofuels are produced from lignocellulosic materials, such as corn straw, wheat straw, sugarcane bagasse, sugar beet pulp, cassava peels, and switchgrass, including softwood and hardwood. They rely on the use of biomass that is not suitable for being used as food (non-edible biomass). Second-generation biofuels comprise either plants that are mainly grown for energy production, i.e., bioenergy crops on marginal lands (lands that are unsuitable for food production) or non-edible parts of crops and forest trees, which should be processed efficiently for bioenergy by improving existing technology [47]. Lignocellulose, which makes up the cell walls of plant biomass, is divided into three main components, which include cellulose (30–50%), hemicellulose (10–40%), and lignin (5–20%) [14]. Different authors have certain values for all of these components; however, the extraction process of a particular component, particularly cellulose, is somewhat difficult [40]. The development of second-generation biofuels is generally seen as a sustainable response to the rising controversy surrounding first-generation biofuels [48]. The potential for bioethanol production will be influenced by the chemical composition of the organic compounds involved [37]. Second-generation liquid biofuels are commonly produced by two different methods, which include biological or thermochemical processing from lignocellulosic agricultural biomass. The main benefit of producing second-generation biofuels from inedible feedstock is that it curtails the direct food versus fuel competition connected with first-generation biofuels. Furthermore, in comparison with first-generation biofuels, second-generation biofuels are associated with an increase in land use efficiency and reduced pressure on biodiversity [39]. According to Markus et al. [49], the cell walls of plant biomass represent one of the most abundant renewable resources on earth. At present, only 2% of this biomass is utilized by man in spite of its abundance. This calls for the need to research the feasibility of utilizing plant cell walls in the production of inexpensive biofuels. The major drawback in the use of lignocellulosic materials is the recalcitrance of cell walls to degrade efficiently into simple fermentable sugars. The addition of wall structure-altering agents or the manipulation of the wall polysaccharide biosynthetic machinery should make the tailoring of wall composition and architecture possible in order to improve sugar yields for biofuel production. However, the main challenge is that the study of biosynthetic machinery and its regulation is still in its early stages.

4.2.3. Third-Generation Biofuels

Third-generation biofuels commonly refer to biofuels produced from algal biomass [50]. Microalgae biomass as a candidate for biofuel production is becoming popular owing to its rapid growth rate, high lipid and starch content, ease of cultivation [51], low land usage, and high carbon dioxide absorption [50]. It offers a potential solution to one of the pressing issues faced by modern societies today (the development of renewable energy for transportation) owing to its high surface biomass productivity, ability to grow on marginal lands, and efficient conversion of solar energy to chemical energy [52]. Algae has the ability to produce higher yields with lesser input resources than other biomass; hence, it is separately classified from second-generation biofuels. In terms of the potential of fuel production, with regard to quantity or diversity, no feedstock can compete with algae. The two attributes of algae with respect to the diversity of fuel it can produce include the following; (i) it produces oil that can be easily refined into diesel or certain gasoline components and (ii) it can be genetically manipulated to produce fuels like ethanol, butanol, diesel, and gasoline. Algae is able to produce outstanding yields and has produced about 9000 gallons of biofuel per acre, and this is about 10-fold more than what the best traditional feedstock has been known to produce. There has been a suggestion by those who work closely with algae that yields as high as 20,000 gallons per acre are achievable. Notwithstanding, algae biomass has some drawbacks, and one of these drawbacks is that it requires large volumes of water, phosphorus, and nitrogen to grow, even when grown in wastewater. Also, biofuels produced from algae tend to be less stable than those produced from other sources. The reason for this is that the oil present in algae tends to be highly unsaturated, especially at high temperatures, and, hence, more liable to degradation [50]. Furthermore, microalgae biofuels are not yet commercially sustainable. There are still challenges with regard to the improvement of microalgae strains and cultivation technologies [52].

5. Bioethanol as an Energy Source

In recent years, there has been an increase in the global production and use of biofuels, for example, from 18.2 billion liters to 60.6 billion liters to 162 billion liters in 2000, 2007, and 2019, respectively, with about 85% of this being bioethanol [2,53]. Bioethanol can be produced from a variety of cheap substrates, and it is reported to be one of the important and most widely used liquid biofuels worldwide, especially in transportation [8]. On average, ethanol has about 33% less energy content than gasoline [54]. Depending on the feedstock used, it has been estimated that bioethanol is able to lower the emission of greenhouse gas by approximately 30–85% compared to gasoline [2,55]. An 85% ethanol–15% gasoline blend can cut greenhouse gas emissions by 60–80%. A mixture of 10% ethanol and 90% gasoline can cut emissions by up to 8% [56]. Bioethanol and biodiesel are the two most common biofuels. While bioethanol is produced by the fermentation of biomass rich in carbohydrates, biodiesel can be produced from animal fats, vegetable oils, algae, or recycled cooking greases. Ethanol can be used as a fuel additive and biodiesel can be used either in its pure form or as a diesel additive to fuel a vehicle [57].

Ethanol has been described as being perhaps the oldest product obtained through traditional biotechnology [58]. The use of ethanol as a fuel for motors can be traced back to the days of the Model T. The first set of people to identify that the abundant sugars and starches present in plant biomass could be cheaply and easily converted to renewable biofuel were Henry Ford and Alexander Graham Bell. In the year 2016, the United States was the highest producer of ethanol worldwide, producing almost 60% of the global production [59], and this position is maintained to date.

Bioethanol finds application in the transportation sector, beverage and pharmaceutical industries, and electricity generation. Residues from bioethanol production could be used to produce thermal energy, valuable chemicals, and fertilizers. Bioethanol is a possible alternative to fossil-based transportation fuels because it has broader flammability limits, higher octane number, higher heat of vaporization, and higher flame speeds, and these characteristics allow for a shorter burn time, higher compression ratio, and leaner burn engine, which ultimately results in an advantage in theoretical efficiency over gasoline in an internal combustion engine [60,61]. Mixing ethanol with petrol for transportation boosts the performance of the latter. It also enhances fuel combustion in vehicles, lowering the release of unburned hydrocarbons, carbon monoxide, and carcinogens. Nevertheless, the combustion of ethanol also leads to a heightened reaction with nitrogen in the atmosphere, which can cause a marginal increase in nitrogen oxide gases. Compared to petrol, ethanol contains only trace amounts of sulfur. So, when ethanol is mixed with petrol, it will help lower the fuel's sulfur content and, hence, reduce the emissions of sulfur oxide, which is a major component of acid rain and a carcinogen [38].

A combination of sugar beets and wheat is generally used in the production of EU bioethanol. It has been projected that bioethanol production in the EU probably has a greater potential than biodiesel; this is coming from the estimated abundant supplies and production potential for sugar beets and cereals, but the cost of production of EU biofuels is a consequence of high-priced internal feedstock compared to fossil fuels and remains a major barrier to the market-based expansion of EU biofuel production, especially for bioethanol [62]. At present, France is a front-runner in the EU's attempt to enhance the use of ethanol, accounting for 2% of global production, primarily from wheat and sugar beet [61,63], making France the top producer of fuel ethanol in the European Union. Over

1.2 billion liters of output were expected to be produced in the nation in 2022, an increase of almost 4% from the year before. In contrast, it was anticipated that Germany's ethanol production would total 759 million liters in 2022. In the EU, Germany consumes more ethanol than any other country, with France coming in second [64]. The three types of feedstock used in the production of bioethanol include sucrose-containing feedstock, such as sweet sorghum, sugar beet, and sugarcane, starch-based feedstock such as maize, wheat, and barley, and lignocellulosic biomass, like straw, grasses, and wood [60,65].

6. Processes Involved in Bioethanol Production from Lignocellulosic Biomass

The recalcitrance of plant cell walls due to their complex nature poses some challenges with the use of lignocellulosic biomass for the obtainment of maximum ethanol yield. Therefore, in order to facilitate ethanol productivity and lower production costs, lignocellulosic materials are subjected to different stages of ethanol production processes, which include effective pretreatment processes, hydrolysis, fermentation, and distillation to separate ethanol produced from co-products [66] (Figure 3).

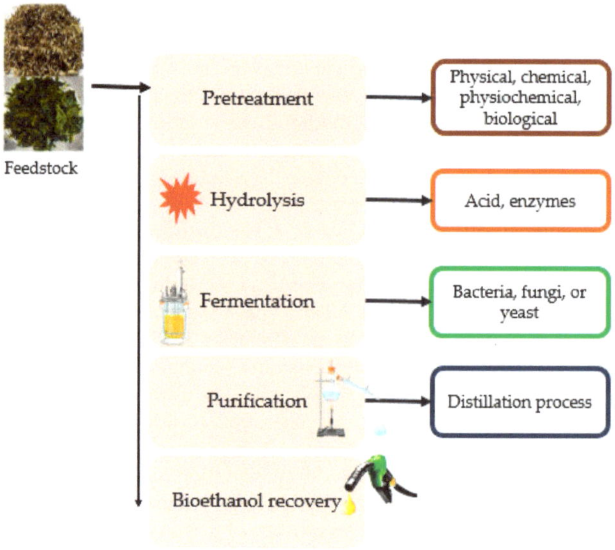

Figure 3. Stepwise process of bioethanol production from lignocellulosic biomass.

6.1. Pretreatment

Pretreatment is an essential stage in bioethanol production from lignocellulosic plant biomass as it aims at altering the complex structure of the material by breaking down the lignin seal to solubilize hemicellulose, reduce the crystallinity, and increase the porosity of the material so as to enhance the accessibility of hydrolyzing agents (enzymes or chemicals), which break down cellulose polymers into simple fermentable sugars [67]. An ideal pretreatment should be inexpensive, effectively de-lignify substrate materials, prevent the loss or deterioration of carbohydrates, produce high sugar yield, and prevent the formation of sugar degradation products [68]. Certain plant biomass, such as cassava, contains high cyanide content [69], which could affect enzymes and microorganisms and ultimately lead to low production of reducing sugars. Hence, Mohammed et al. demonstrated that 24 h of soaking and 120 min of the boiling pretreatment condition is able to reduce the cyanide content in cassava peel waste and improve the total recovery of carbohydrates [70].

In a bid to overcome the challenges inherent to the use of lignocellulosic materials, there has been a shift in pretreatment procedures starting from chemicals and heating methods to biological methods [71], but there is still no satisfactory result from the different pretreatment methods adopted so far in terms of technology for industrial large-scale production, cost-effectiveness, and production of a lower amount of inhibitory products [72]. Several pretreatment processes for lignocellulosic biomass have been proposed and practicalized. They include physical, chemical, physicochemical, and biological pretreatment processes, as shown in Figure 4. The objectives, advantages, and disadvantages of these different pretreatment processes are outlined in Table 3.

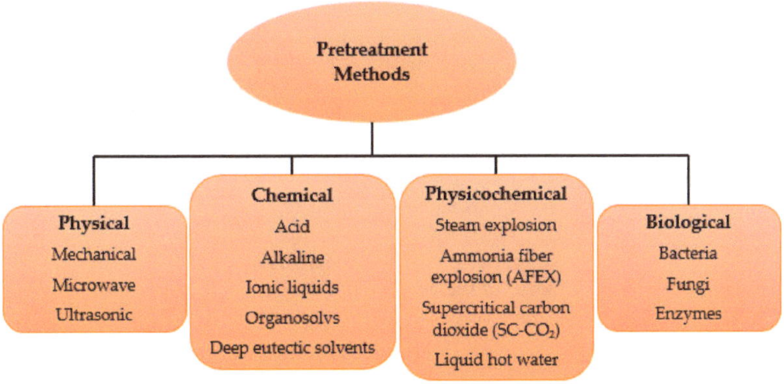

Figure 4. Different pretreatment processes.

Table 3. Advantages and disadvantages of the various pretreatment methods.

Pretreatment Methods	Objectives	Advantages	Disadvantages	References
Physical	To reduce biomass size and decrease crystallinity	Green pretreatment (rarely forms inhibitory product); improves hydrolysis rate	Energy intensive, not economically viable, and unable to remove/alter lignin	[73,74]
Chemical	To break down/solubilize/remove lignin and hemicellulose and increase surface area	Enzymatic hydrolysis might not be necessary (acid hydrolyzes lignocellulosic materials into simple sugars)	Corrosion of equipment, expensive, non-selective, requires high temperatures, chemical recovery issues, requires neutralization, and fermentation inhibitor problems	[60,75]
Physicochemical	To alter lignin, degrade hemicellulose, reduce cellulose crystallinity, and increase the surface area of biomass	Less use of chemicals, requires less energy compared to the mechanical method, high sugar recovery, limited environmental impact, and low cost	Unfinished disruption of lignin–carbohydrate matrix	[17,76]
Biological	To disrupt plant cell walls, selectively remove lignin, and degrade hemicellulose	Mild and eco-friendly, low energy requirement, and no formation of inhibitor byproducts	Relatively slow process and expensive (e.g., GMOs)	[77,78]

6.2. Hydrolysis

After the pretreatment process, the next stage is the hydrolysis procedure. Hydrolysis is a chemical process that involves the use (addition) of water to break down polymers (e.g., cellulose) into monomers (e.g., glucose). It is a chemical reaction that breaks the chemical bonds that exist between two substances and releases energy, in which one molecule of a substance receives an H^+ ion while the other molecule obtains an OH^- group. Hydrolysis is needed to obtain simple fermentable sugars. Acids (HCl, H_2SO_4, etc.) and enzymes are commonly used catalysts in the hydrolysis of biomass.

6.2.1. Acid Hydrolysis

Acid hydrolysis can be achieved by inorganic acids (liquid acid catalyst or solid acid catalyst) or organic acids. Liquid acid hydrolysis is of two types, viz., dilute and concentrated, each of which possesses unique characteristics on biomass. Two reactions are involved with the use of dilute acid hydrolysis. One involves the conversion of cellulose into sugar while the other involves the conversion of sugar into chemicals, and many of these chemicals act as growth inhibitors to fermenting organisms. Concentrated acid hydrolysis, which consists of about 70% acid content, operates at low temperatures (37.8 °C) and pressure. However, dilute acid is the most preferable in terms of economics and effect on biomass [79]. Solid (heterogeneous) acid catalysts have, in recent years, experienced an increase in their use for cellulose hydrolysis into glucose. Examples of these catalysts include H-form zeolites, functionalized silica, immobilized ionic liquids, metal oxides, supported metals, acid resins, heteropoly acids, carbonaceous acids, and magnetic acids [80]. Some of the advantages of heterogeneous catalysts are that they can easily be removed from a reaction mixture through a process like filtration. This is important for industrial manufacturing processes since it makes expensive catalysts simple, efficiently recoverable, and reusable [81]. They are also environmentally friendly and possess good thermal stability [82].

Heterogeneous catalysts also have some disadvantages. When the catalyst's surface has been entirely covered by reactant molecules, the reaction cannot continue until the products have left the surface and some area has, once again, become available for a fresh batch of reactant molecules to adsorb or attach. This explains why the rate-limiting stage in a heterogeneously catalyzed process is frequently the adsorption step [81]. A heterogeneous catalyst is less active and selective compared to a homogeneous catalyst due to the possession of multiple active sites [82]. Also, solid catalysts have lower conversions than homogeneous catalysts and necessitate more extreme reaction conditions to provide the same conversions [83]. Organic acids, e.g., acetic acid, which are generally regarded as weak acids that do not dissociate completely in water, have been examined for the hydrolysis potential of biomass. For example, Kanlaya and Jirasak achieved a 30.36% yield of reducing sugars when 0.25 M acetic acid was used to hydrolyze cassava peels at 135 °C for 90 min [84].

6.2.2. Enzymatic Hydrolysis

Cellulases are a class of enzymes used to catalyze the hydrolysis of cellulose. They are produced by bacteria or fungi. However, there is more interest in the use of cellulase produced by fungi than the ones produced by bacteria. This is because most cellulase-producing bacteria are anaerobes that have a very low growth rate. The release of monomeric sugars from cellulose requires the action of three groups of enzymes, viz., endoglucanases, exoglucanases (cellobiohydrolases), and β-glucosidases [60,85].

1. Endoglucanase: this is one of the enzymes of cellulose deconstruction that acts by splitting the polymer, i.e., the cellulose long chains into shorter molecules (which could be oligosaccharides or smaller polysaccharides units);
2. Exoglucanase: this other group of enzymes frees/releases cellobiose (which is a disaccharide) from either the non-reducing end or the reducing end;
3. β-glucosidase splits cellobiose and other short-chain cello-oligosaccharides into monomer units (glucose).

Other enzymes that have been used for the hydrolysis of plant biomass include xylanases that hydrolyze the major component of hemicellulose (xylan), amylases for the digestion of starch, etc. [86]. Xylan is a heterogenous/complex compound with a backbone consisting of β-1,4-linked xylosyl residues, and so, the xylanolytic enzymes generally consist of a collection of enzymes, such as endoxylanase, β-xylosidase, α-glucuronidase, α-arabinofuranosidase, and acetylxylan esterase [87], which act in collaboration to convert xylan into sugars [88]. Nevertheless, xylanases (endoxylanases) are the most crucial since they are directly involved in cleaving glycosidic bonds and releasing short xylo-oligosaccharides [87].

There are differences in characteristics that exist with the use of acid hydrolysis versus enzymatic hydrolysis, and these are outlined in Table 4.

Table 4. Differences between acid hydrolysis and enzymatic hydrolysis [60,89].

Acid Hydrolysis	Enzymatic Hydrolysis
Corrosive	Non-corrosive
No specificity (selectivity)	More specific
Requires high process temperature (100 °C–160 °C)	Operates in low/milder conditions (44 °C–50 °C, pH 4.8)
Inhibitor formation issues	No inhibitor byproduct issues
Relatively low yield	Relatively high yield
In some instances after hydrolysis, requires neutralization with chemicals, which could be expensive (e.g., NaOH, KOH)	Initial high cost of enzymes. No neutralization needed
Not sensitive to operating conditions	Sensitive to operating conditions
Do not require genetic modification	Could necessitate the genetic modification of enzyme-producing organisms to improve hydrolysis
Non-environmentally friendly	More eco-friendly
Faster process (in minutes)	Takes longer process time (in hours)

State-of-the-art development in bioethanol production mainly focuses on technical advancement to obtain high ethanol yield, reduce the time and cost of processing, and, most importantly, minimize the number of steps involved in lignocellulosic biomass conversion. The technological approaches encompass a variety of techniques, such as process development, genetic modification of feedstocks, cellulase enzyme-based robust hydrolysis technique development, cell immobilization, recombinant microorganism development, fermentation under conditions of high solid load, solid-state fermentation development, and integration of different process steps, amongst others [90,91].

6.2.3. Sugar Degradation Products/Fermentation Inhibitors

The treatment of hemicellulose by dilute acid results in the formation of toxic compounds, like furfural from pentose degradation and soluble aromatic aldehydes from lignin. On a weight basis, aromatic aldehydes are twice as toxic as HMF or furfural [92]. HMF is produced from the dehydration/degradation of hexose sugars during acid pretreatment or hydrolysis. These compounds inhibit the fermentation process, which is needed to produce valuable ethanol from sugar by entering the nucleus of the cell and getting attached to the replicating DNA, thus lowering microbial metabolism, reproduction, and enzymatic activities. Acid concentration, temperature, and time are important factors that determine the formation of inhibitors. High acid concentration and low temperature provide an optimum operating condition for acid hydrolysis of potato peels, as shown in Table 5 [93]. In addition to the above, the degradation of sugars also results in the formation of carboxylic acids, such as acetic, propionic, formic, and lactic acids [94]. The further degradation of furans gives rise to the formation of levulinic acid and formic acid and again, the contamination of substrate by microbes can lead to the formation of different acids, such as lactic acid [95]. The threshold of inhibition depends on the strain of microorganism and inhibitor tested [96,97] (Tables 6 and 7). In the measurement of acid toxicity, pH is an essential parameter to be considered. The concentration of HMF beyond 8 g/L and the concentration of furfural beyond 5 g/L hinder the growth of all microbial strains. Notwithstanding, the concentration of furans at 1 to 2 g/L is lethal to the growth of some strains. Also, the growth rate can be inhibited at 15 g/L acetic acid concentration and 10 g/L formic acid concentration, but these concentrations have not been detected to cause a severe inhibitory effect on productivity [97]. In addition to the individual toxicity of compounds, it is also important to take cognizance of the cocktail effect of inhibitor products since the combined effect could elevate the toxicity of compounds. For instance, the interaction between acetic acid and furfural caused a negative effect, as a reduction in ethanol yield, specific growth rate, and biomass yield of *S. cerevisiae* were observed [98]. Furthermore, a more negative result was observed with the interaction of 2 g per liter of furfural and 2 g per liter of HMF than, with 4 g per liter of HMF and 4 g per liter of furfural acting separately [99]. Several measures can be taken to reduce the effect of fermentation inhibitors, viz., substrate concentration, including salts and produced ethanol, should be below the threshold tolerance of the microorganism involved, minimizing/preventing the use of procedures (e.g., chemicals) that lead to inhibitor formation at the time of pretreatment, in situ detoxification by microorganisms used in fermentation, and the modification of organisms either through microbial adaptation or genetic engineering [100].

Table 5. Effects of acid concentration, temperature, and time on inhibitor formation [93].

Acid Concentration (% *w*/*w*)	Temperature (°C)	Time (min)	Sugar Yield (g/100 g Biomass)	Inhibitor Concentration (g/100 g Biomass)	Ratio (Inhibitor: Sugar) (%)
5.0	135	30	26.32	0.6	2.25
5.0	150	15	25.97	2.2	8.4
10	135	8	55.2	1.1	1.9
10	150	8	46.4	1.91	4.1

Table 6. Growth (% of the control) of glucose and xylose-fermenting microorganisms in the presence of inhibitors (adapted from [101]).

Inhibitors	Concentration (g/L)	*S. cerevisiae*	*Z. mobilis*	*P. stipitis*	*C. shehatae*
Furaldehyde	0.5	53	82	75	81
	1	19	81	53	62
	2	10	44	1	9.7
Acetate	5	79	76	63	96
	10	52	44	64	84
	15	56	26	64	79
Hydroxymethylfuraldehyde	1	35	51	95	92
	3	17	69	31	32
	5	11	33	1.4	8
Vanillin	0.5	49	62	12	67
	1	14	37	0.7	9
	2	9	12	1.4	1.6
Hydroxybenzaldehyde	0.5	75	16	57	60
	0.75	47	8	30	23
	1.5	13	8	0	0.8
Syringaldehyde	0.2	100	82	72	89
	0.75	39	72	38	45
	1.5	19	60	3.6	5

Table 7. Concentration (g/L) of inhibitors at which the growth of microbes is completely hindered; σ_i (%) represents standard errors of the estimates (adapted from [96]).

Inhibitors	*S. cerevisiae*	σ_i (%)	*E. coli*	σ_i (%)	*B. subtilis*	σ_i (%)
Hydroxymethylfurfural	2.2	18.0	2.2	20.1	1.9	15.7
Syringaldehyde	2.5	8.2	2.7	13.7	2.0	6.0
Vanillin	1.08	22.9	2.2	12.0	1.84	18.3
2-Butanone	45.0	11.4	17.8	14.4	31.0	9.1
2-Butanol	36.0	12.6	21.0	6.5	20.0	18.7
Methyl propionate	23.0	11.6	13.68	13.4	21.0	6.0
Ethyl acetate	22.0	19.6	19.0	12.6	30.0	14.6

6.3. Fermentation

Fermentation is a metabolic process that involves the breakdown of a substance into a simpler one by microorganisms such as bacteria, fungi, or yeast. In ethylic fermentation, it is a chemical process by which simple sugars are broken down anaerobically into ethanol.

6.3.1. Industrial Fermentation Technology for Ethanol Production

The industrial fermentation technology applied in the production of bioethanol includes the following.

1. Batch fermentation is also referred to as a 'closed system' and is the most common and simplest method for producing ethanol. In this method, fermentation is carried out in separate batches. The fermenter is first loaded with the substrate, after which the microorganisms are added and left to ferment the substrate. Byproducts accumulate, which continuously changes the culture environment. The products are removed at the end of the fermentation process and the fermenter is cleaned and sterilized in preparation for the next round. The microbes in the fermenter show three distinct growth phases, viz., lag, log (exponential), and stationary phases. The batch fermentation method has some advantages, such as less labor demand, ease of operation, low investment cost, quick and easy control methods, complete sterilization, and less risk of contamination [102].
2. Fed-batch fermentation is an improved version of the batch fermentation process. Here, the feeds containing substrate, culture medium, and other vital nutrients are loaded into the fermenter, after which the cultured microorganisms are introduced and left to ferment the substrate. The feed solution is continuously introduced into the fermenter on an incremental basis throughout the fermentation process without the removal of the products formed. The products are only removed/extracted at the end of each fermentation process. The amount of working volume is a limiting factor in this process [102].
3. Semi-continuous fermentation is sometimes referred to as either repeated fed-batch fermentation or a combination of some features and is notable in the batch and continuous fermentation process. Here, the feed solution is loaded into the fermenter at a constant interval, and the products formed are removed intermittently (not regularly). This process usually requires fixed volume, i.e., the volume of fermented (used) medium removed from the fermenter is usually replaced by an equal volume of fresh feeds at a constant time interval. This practice could help to maintain the growth of microbes for some time, as they get to feed on freshly provided nutrients that replace the already exhausted ones and, also, the intermittent removal of formed products could prevent the fermenting organisms from quickly transiting into the inactive/death phase; hence, an increase in product yield could be achieved. This process allows for an extended fermentation time, and the cycle is not usually terminated until a decline in productivity is detected [103].

4. Continuous fermentation, as the name implies, means that the feed solution is continuously loaded into the fermenting vessel and the products formed are constantly removed/extracted. This allows for a longer fermentation time; the cycle is not interrupted like it is in the batch fermentation process. The growth of microorganisms is, therefore, maintained for a long time in the fermenting vessel due to the fresh nutrient supply and the regular removal of products whose accumulation has been reported to be detrimental to fermenting microorganisms. Hence, this process results in higher productivity [104].

6.3.2. Microorganisms for Sugar Fermentation

Ethanol production requires the fermentation of sugars present in biomass by various microorganisms.

Microorganisms for pentose sugar fermentation: pentose sugars, such as xylose and arabinose, have been regarded as 'un-fermentable sugars'. From a broad perspective, this could be correct since most microorganisms, such as yeast, fungi, and bacteria, are unable to effectively utilize pentose sugars. Nevertheless, there exist certain strains that play a significant role in nature's economy that have been reported to be capable of breaking down these five-carbon sugars [105].

- Bacteria: the majority of filamentous fungi and yeast are unable to ferment pentose sugars anaerobically, but bacteria are able to convert xylose to ethanol under anaerobic fermentation [106]. Xylose-fermenting bacteria comprise both native and genetically modified strains. During xylose fermentation, bacteria do not form xylitol; instead, they use its enzyme, 'xylose isomerase', to convert xylose directly into xylulose, and xylulose is then converted into ethanol through the pentose phosphate pathway (PPP) and the Embden–Meyerhof–Parnas pathway [107]. Examples of pentose-fermenting mesophilic bacteria include *Aerobacter hydrophila*, *E. coli*, *Clostridium acetobutylicum*, *Bacillus polymyxa*, *B. macerans*, and *Klebsiella pneumonia* [108]. Thermophilic anaerobic bacteria have been suggested as promising candidates for the conversion of pentose sugars into ethanol. Some of the species that have been studied include *Thermoanaerobacter ethanolicus*, *T. brockii*, *T. thermohydrosulfuricus*, *Thermoanaerobacterium thermosaccharolyticum*, and *Thermoanaerobacterium saccharolyticum* B6A [109]. The benefit of utilizing bacteria, e.g., *E. coli* ATCC 11303 (pLOI297), for ethanol production is that the process does not need aeration to achieve high productivity, but the downside is the high possibility of contamination since it functions at higher pH. Other disadvantages include its high sensitivity to ethanol inhibition and loss of productivity due to plasmid instability in the course of prolonged operation. Successful large-scale application of bacteria in fermentation is not very certain compared to yeast [110].
- Yeast: yeast is a common and suitable organism for the production of ethanol from sugars. This microorganism has been reported to act favorably in the fermentation of hexose sugars compared to pentose sugars [111]. However, certain strains, such as *C. shehatae*, *Kluveromyces marxianus*, *P. tannophilus*, and *P. stipitis*, have been evaluated for their ethanol production potential. Several other species of yeast that are able to utilize the five-carbon sugar (xylose) include *Clavispora* sp., *Schizosaccharomyces* sp., and *Brettanomyces* sp. Also included are *Debaromyces species*, such as *D. nepalensis* and *D. polymorpha*, and *Candida species*, like *C. blankii*, *C. tenius*, *C. utilis*, *C. solani*, *C. tropicalis*, *C. parapsilosis*, and *C. friedrichii* [108]. Most yeasts are incapable of fermenting xylose directly, so they ferment/utilize xylulose, which is an isomer of xylose. The bacteria enzyme 'xylose isomerase' can catalyze the interconversion of xylose and xylulose (isomerization), which is achieved in a single step, whereas yeast utilizes xylose reductase to reduce xylose to xylitol and then makes use of xylitol dehydrogenase to convert xylitol to xylulose. Species of *Candida*, *Kluyveromyces*, *Brettanomyces*, *Torulaspora*, *Pachysolen*, *Saccharomyces*, *Hansenula*, and *Schizosaccharomyces* have been recognized as the best ethanol-producing yeast from xylulose [112]. Nutrient medium composition, temperature, aeration rate, and pH are some of the factors that affect

xylose-fermenting yeast performance. Some of the benefits associated with the utilization of yeast, e.g., *P. stipitis,* for the conversion of xylose is that it has high selectivity for ethanol production, unlike bacteria and fungi, which form co-products with ethanol. It is also relatively tolerant to ethanol and low pH, properties that reduce the risk of bacterial contamination. However, the drawback of this organism (xylose-fermenting yeast) is that it requires a small amount of oxygen ($\leq$2 mMol/L-h) to realize high conversion efficiency; it is relatively easy to achieve micro-aeration on the laboratory scale, but it is not easy to achieve in the industrial scale. Another downside of xylose-utilizing yeast is that it presents low volumetric productivities when compared to those obtained with bacteria or glucose-fermenting yeast [110]. Compared to *S. cerevisiae,* yeast that utilizes pentose sugars is poorly tolerant to ethanol, inhibitor products, and pH, and these attributes can result in low ethanol yield [113,114].

- Filamentous fungi: xylose conversion by fungi has not been extensively studied compared to xylose fermentation by bacteria and yeast [110]. Filamentous fungi, such as *Neurospora crassa, Mucor* sp., *Fusarium oxysporum, Monilia* sp., and *Paecilomyces* sp., have been known to have pentose sugar fermentation potential. One good thing about the fungal process is that it has the capacity to grow on natural plant material, which is usually absent in yeast-based processes. Nonetheless, the fungal system is associated with properties that make its application in ethanol production unpleasant, such as are low volumetric production, the longer time that it takes to ferment (4 days to 8 days), the small oxygen requirement, the high viscosity of fermentation broth, growth in large clumps instead of dispersed single cells, the co-production of acetic acid alongside ethanol as a major end-product, which ultimately leads to reduced ethanol formation, and low tolerance to substrate and product [108].

An examination of bacterial, yeast, and fungal xylose fermentation, in general, is presented in Table 8.

Table 8. Overall evaluation of xylose fermentation by bacteria, yeast, and fungi (adapted from [112]).

	1st Stage	**2nd Stage**	**3rd Stage**
Organism	Xylose to xylulose-5-P	Xylulose-5-P to pyruvate	Pyruvate to the final product(s)
Bacteria	Isomerization	Pentose phosphate + EMP pathway	Ethanol + mixed acids Ethanol + 2,3-butanediol Ethanol + acetone butanol
Yeasts	Oxidation reduction	Pentose phosphate + EMP pathway	Ethanol
Fungi	Oxidation reduction	Pentose phosphate + EMP pathway	Ethanol Acetic and lactic acids

Microorganisms for hexose sugar fermentation: the species of microorganisms that are able to ferment hexose (e.g., glucose) are more than those that have pentose fermentation potential. Usually, microorganisms, e.g., *E. coli,* utilize glucose first until it gets exhausted during co-fermentation involving the mixture of sugars, before converting pentose sugars, e.g., xylose and arabinose, to ethanol. This sequential use of sugars can result in an incomplete or delayed consumption of secondary sugars, which, in turn, leads to a decrease in yield and productivity [115]. A wide range of microorganisms, such as bacteria, e.g., *E. coli, Zymomonas mobilis, Aerobacter hydrophila, Clostridium acetobutylicum, Klebsiella pneumonia, Bacillus polymyxa, B. macerans, Thermoanaerobacter ethanolicus, Clostridium thermosulfurogenes, C. thermocellum, C. thermosaccharolyticum, C. thermohydrosulfuricum,* fungi, e.g., *Mucor indicus, Neurospora crassa, Fusarium oxysporum, Monilia* sp., *Paecilomyces* sp., and yeast, e.g., *S. cerevisiae, Pichia stipitis, P. angophorae, Candida shehatae, Kluyveromyces fagilis, K. marxianus,*

Pachysolen tannophilus, etc., have been used in the fermentation of hexose sugars [108]. But of all microbes that have been employed in the production of ethanol from plant biomass, *S. cerevisiae* is the most famous and commonly used in industrial-scale applications due to its high tolerance to ethanol, wide range of pH, high productivity, and ability to ferment a wide range of sugars [116]. Pyruvate is the first stage in the alcoholic fermentation pathway, and it is obtained by yeast (*S. cerevisiae*) through the Embden–Meyerhoff–Parnas (EMP) pathway, and bacteria (*Zymomonas*) are formed through the Entner–Doudoroff (ED) pathway. The next stage involves the decarboxylation of pyruvate to acetaldehyde; this reaction is catalyzed by the enzyme called pyruvate decarboxylase. The redox balance of alcoholic fermentation is realized via the reproduction of NAD^+ when acetaldehyde is reduced to ethanol by the enzyme alcohol dehydrogenase. Alcoholic fermentation produces 1 mol of ATP through the ED pathway or 2 mol of ATP through the EMP pathway for each mol of glucose oxidized [117].

6.4. Distillation

Distillation is the process of purifying a liquid by separating the components of the liquid mixtures through heating/boiling and condensation. It is an effective purification technique that employs the differences in volatilities of constituents in a mixture [118]. Ethanol and water are soluble in each other, so distillation is required for the separation and concentration of ethanol from the fermentation broth. It is not possible to obtain 100% purity through simple distillation because of the azeotrope between water and ethanol; there is a strong hydrogen bond that exists between water and ethanol, which causes water to be attached to ethanol as it pulls out when heated and, therefore, about 95% of ethanol can be recovered through this process, which finds relevance in the solvent, chemical, cosmetic, and pharmaceutical industries [73]. To obtain 99.9% ethanol, i.e., anhydrous ethanol, further drying of ethanol or a dehydration step is required [37]. Irrespective of the product being recovered, distillation is an energy-intensive technique; it is expensive and consumes approximately 40% of the total energy used in the chemical and petroleum refining industries [119]. Notwithstanding, since distillation remains the main separation technique in process industries, it, therefore, becomes important to enhance its energy efficiency, particularly when applied in the separation of azeotropic mixtures. To this end, several special separation methods have been employed, such as liquid–liquid extraction, azeotropic distillation, pressure swing distillation, extractive distillation, pervaporation using membrane, salt addition, adsorption, etc. [120].

7. EU Legislation Supporting Advanced Biofuels

A 10% minimum goal for renewable energy used in the transportation sector was outlined in the 2009 EU Energy and Climate Change Package and was to be met by all EU member states in their respective nations by the year 2020. For the years 2010 to 2020, the Renewable Energy Directive (RED) outlined specific objectives and requirements for the transportation sector. The Renewable Energy Directive II (REDII) for 2021–2030 was approved by the European Union in 2018 [121]. It included a minimum target of 32% renewable energy consumption across all sectors and a reduction in greenhouse gases of at least 40%. Transportation should be decarbonized as a top priority going forward because, in comparison to an 18% fall or more in all other sectors, greenhouse gas emissions in the European transportation sector have decreased by only 3.8% since 2008. The EU has encouraged the development of advanced biofuels, which are made from non-food feedstock, as well as conventional biofuels, which are based on food, through the use of directives and national laws. The EU Renewable Energy Directive (RED), which was passed in 2009, stipulated that by 2020, 10% of the energy utilized in the transportation sector must originate from renewable sources [122].

The EU Indirect Land Use Change (ILUC) directive changed the RED in 2015 by placing a 7% cap on the amount that food/feed-based biofuels might contribute to the RES-transport target. A non-binding 0.5% goal for advanced biofuels in 2020 was included by the ILUC directive as an additional measure to encourage the use of biofuels made from non-food feedstock and wastes [123]. By agreeing to modify the Renewable Energy Directive (REDII) in June 2018, the EU Commission, Parliament, and Council established a target of 14% renewable energy sources—transportation energy and a sub-target of 3.5% for advanced biofuels by 2030. The capping of traditional food-based biofuels to a maximum of 7% of each member state's 2020 level signifies a target of at least 7% for non-food-based/advanced fuels [122]. Advanced biofuels can be double counted towards the 3.5% target and the 14% target [123].

8. Industrial Projects/Technology on Advanced Bioethanol

There are several initiatives on the development of renewable fuels (bioethanol) from second-generation (2G) plant biomass, and most of these technologies follow certain process sequences, such as the pretreatment of raw biomass, the production of biocatalysts (enzymes/fermentation microbes, e.g., yeast), simultaneous saccharification and fermentation (SSF) of feedstock, simultaneous saccharification and co-fermentation (SSCF), or separate hydrolysis and fermentation (SHF) (Table 9). Two successful projects, POET-DSM Project LIBERTY in the USA [124] and GranBio in Brazil [125], exemplify the triumph over legal, technological, and financial obstacles in lignocellulosic bioethanol production. Through innovative processes and strategic partnerships, these ventures demonstrate the feasibility of large-scale bioethanol production from agricultural residues.

Table 9. Projects/technology on advanced bioethanol production.

Projects/ Technology	Country/ Location	Feedstock	Technology Operation	Products/Production/ Production Aim	References
Futurol^{TM} technology	France	Silvergrass (Miscanthus), agricultural residues, and wood residues	Steam explosion, on-site production of biocatalysts (enzymes and yeasts resistant to inhibitors, particularly acetic acid), enzymatic hydrolysis, co-fermentation (SSCF) of five-carbon and six-carbon sugars, and recovery of 2G ethanol, lignin, and stillage	55,000 tons (or 70 million liters of ethanol) of bioethanol	[126–128]
Sunliquid® technology	Southwestern Romania Straubing, Germany (demonstration plant)	Wheat and other cereal straw	Chopping of feedstock into smaller sizes, steam explosion pretreatment, enzymatic hydrolysis, simultaneous fermentation with the yeast of both C_5 and C_6 sugars, ethanol, and vinasse recovery	50,000 tons of bioethanol on a yearly basis from 250,000 tons of agricultural residues Demonstration plant: 1000 tons of bioethanol from about 4500 tons of wheat straw, corn stover, etc.	[129,130]
Domsjö Fabriker	Sweden	Spruce and pine biomass (about 1.6 million cubic meters annually)	Debarking and chipping of timber logs, feeding into a digester alongside cooking chemicals. Combustion of the bark to generate energy in the form of steam. Washing, bleaching, and drying cellulose after cooking. Fermentation of dissolved hemicellulose and distillation to produce bioethanol, drying of refined lignin, and recycling of cooking chemicals to produce energy	Cellulose, lignin, and bioethanol, carbon dioxide processed into carbonic acid	[131,132]
ProEthanol2G project	Europe and Brazil	Wheat straw, sugarcane bagasse, and straw	Pretreatment and enzymatic hydrolysis to convert molecules into sugars, followed by fermentation with recombinant yeast strain of the sugar solution and distillation	Europe: bioethanol and electricity from 100% wheat straw Brazil: bioethanol, sugar, and electricity from 100% sugarcane crop, bagasse, and straw	[133–135]

Table 9. *Cont.*

Projects/ Technology	Country/ Location	Feedstock	Technology Operation	Products/Production/ Production Aim	References
BALI™ Biorefinery Demo	Sarpsborg, Norway	Spruce, bagasse, willow, straw, wood, and energy crops	Chemical (sulfite) pretreatment, enzymatic hydrolysis, fermentation (conventional fermentation of C_6 sugars, aerobic fermentation, or chemical conversion of C_5 sugars), and chemical modification of lignin	Processing capacity of 1 to 2 MT per day of biomass Products: ethanol, lignin, and various chemicals	[136–139]
Bamboo biorefinery built on Chempolis's patented formicobio™ technology	Assam, Northeast India	Utilization of 300,000 tons of bamboo annually	Selective dissolution of biomass's major components excluding cellulose by biosolvents under low temperature and pressure, purification of cellulose by washing with water, enzymatic hydrolysis of cellulose, fermentation, and distillation. Combustion of lignin-rich biofuel to produce steam and electricity	Production of 60 million liters of bioethanol, 19,000 tons of furfural, 11,000 tons of acetic acid, and 144 gigawatt hours of green energy, yearly	[140,141]
Crescentino biorefinery complex (PROESA® proprietary technology)	Italy	Rice straw, wheat straw, and energy crops, e.g., *Arundo donax* (giant cane)	Characterization of energy crops, steam pretreatment, enzymatic hydrolysis and co-fermentation (SSCF), and valorization of secondary streams and co-products	Plant capacity—40,000 tons of bioethanol per annum from more than 200,000 tons of feedstock (dry mass) Generate about 13 MW of electricity from lignin	[142–146]
MARINER (Macroalgae Research Inspiring Novel Energy Resources) projects	United States (US)	-	Integrated cultivation and harvesting systems, advanced component technologies, computational modeling tools, aquatic monitoring tools, and advanced breeding and genetic tools	Production of seaweed (macroalgae) for biofuel production Estimated production of 500 million dry metric tons of macroalgae per annum, amounting to ~2.7 quadrillion BTUs (quads) of energy (liquid fuel) and ~10% of US yearly transportation energy demand	[147,148]

Table 9. *Cont.*

Projects/Technology	Country/Location	Feedstock	Technology Operation	Products/Production/Production Aim	References
TATA project	India	Rice straw (design feedstock) and maize stalk (check case)	--	Bioethanol plant production capacity—100,000 liters per day	[149]
LignoFlag	Europe	Wheat straw, corn stover, etc.	Utilizes Sunliquid® technology	Aims to increase plant production capacity to 60,000 tons of ethanol per annum and use co-products for energy generation and soil fertilization	[150]
IOCL's (Indian Oil Corporation Limited) 2G Ethanol Bio-Refinery (Praj's technology)	India	Rice straw	Acid and steam explosion pretreatment, enzymatic hydrolysis, co-fermentation with GMOs (genetically modified organisms) type yeast, distillation, and dehydration	30 million liters of ethanol from 200,000 tons of rice straw per annum	[151,152]
AustroCel's bioethanol plant (Valmet's automation technology)	Hallein, Austria	Waste materials from adjacent viscose pulp mill	Sulfite pulping/digestion of wood chips and fermentation of sulfite spent liquor (SSL) with yeast	30 million liters of bioethanol	[153–156]

From the above, it can be deduced that the reviewed technologies employed steam pretreatment or chemical pretreatment, enzymatic hydrolysis, and wild or recombinant/genetically modified yeasts for the fermentation of sugars.

9. Conclusions

The different generations of biofuels have been reviewed, and it is obvious that the utilization of advanced biofuels obtained from second-generation and third-generation feedstock is more sustainable and cost-effective compared to biofuel generated from first-generation biomass. The processes involved in bioethanol production have been discussed; it was shown that the recalcitrance of lignocellulosic biomass compels it to undergo pretreatment steps to make the material more amenable to catalysis agents. It was also observed that although biological pretreatment and enzymatic hydrolysis of biomass are environmentally friendly, it takes longer process time and steps and could accrue high costs, for example, in the case of genetically modified organisms (GMOs). Also highlighted are the pentose and hexose sugar fermentation; here, it is worth noting that in a mixture of sugars, microorganisms consume hexose sugar (glucose) first. It is only after this sugar has been exhausted that it seeks pentose sugars. Furthermore, this study showed that the legislation in the EU supports the development of second-generation bioethanol production. It also observed the presentations of different projects on advanced bioethanol production. The pursuit remains on the search for the most optimal condition and technology for sustainable high ethanol yield with minimal cost. Therefore, for sustainable bioethanol production, the process optimization of abundantly available lignocellulosic materials, which are often considered waste biomass, e.g., cassava peels, sugar beet pulp, and *Ulva lactuca*, needs to be explored while employing an efficient yet environmentally friendly and cost-effective catalyst, like dilute acid, or a heterogeneous catalyst. This should be followed by hydrolysate fermentation using suitable microorganisms, such as *S. cerevisiae* and *P. stipitis* for hexose and pentose sugar, respectively. In addition to exploring different 2G bioethanol feedstock, comprehensive lifecycle analyses should be conducted to enhance the sustainability and economic viability of the biorefinery process.

Author Contributions: Conceptualization, C.M.I., S.A. and Y.A.; writing—original draft, C.M.I.; visualization, C.M.I.; writing—review and editing, S.A. and Y.A. All authors have read and agreed to the published version of the manuscript.

Funding: This research was funded by the Petroleum Technology Development Fund (PTDF), Nigeria.

Conflicts of Interest: The authors declare no conflicts of interest.

References

1. Cherubini, F.; Str, A.H. Principles of Biorefining. In *Biofuels*; Academic Press: Cambridge, MA, USA, 2011. [CrossRef]
2. Saini, J.K.; Saini, R.; Tewari, L. Lignocellulosic Agriculture Wastes as Biomass Feedstocks for Second-Generation Bioethanol Production: Concepts and Recent Developments. *3Biotech* **2015**, *5*, 337–353. [CrossRef] [PubMed]
3. Lange, J.; Solutions, S.G. Lignocellulose Conversion: An Introduction to Chemistry. *Biofpr* **2007**, *1*, 39–48. [CrossRef]
4. Rajendran, K.; Drielak, E.; Sudarshan Varma, V.; Muthusamy, S.; Kumar, G. Updates on the Pretreatment of Lignocellulosic Feedstocks for Bioenergy Production—A Review. *Biomass Convers. Biorefinery* **2018**, *8*, 471–483. [CrossRef]
5. Isroi; Ishola, M.M.; Millati, R.; Syamsiah, S.; Cahyanto, M.N.; Niklasson, C.; Taherzadeh, M.J. Structural Changes of Oil Palm Empty Fruit Bunch (OPEFB) after Fungal and Phosphoric Acid Pretreatment. *Molecules* **2012**, *17*, 14995–15012. [CrossRef] [PubMed]
6. IPCC; Chum, H.; Faaij, A.; Moreira, J.; Berndes, G.; Dhamija, P.; Dong, H.; Gabrielle, B.; Eng, A.G.; Cerutti, O.M.; et al. SRREN—Chapter 2—Bioenergy. In *Bioenergy*; IPCC: Geneva, Switzerland, 2012; pp. 209–332.
7. Johannesson, A. Swedish District Heating: Reducing the Nation's CO_2 Emissions. Available online: https://www.openaccessgovernment.org/swedish-district-heating-reducing-nations-co2-emissions/33387/ (accessed on 24 March 2024).
8. Igwebuike, C.M.; Awad, S.; Olanrewaju, Y.A.; Andrès, Y. The Prospect of Electricity Generation from Biomass in the Developing Countries. *Int. J. Smart Grid Clean Energy* **2021**, *10*, 150–156. [CrossRef]
9. Mukherjee, I.; Sovacool, B.K. Palm Oil-Based Biofuels and Sustainability in Southeast Asia: A Review of Indonesia, Malaysia, and Thailand. *Renew. Sustain. Energy Rev.* **2014**, *37*, 1–12. [CrossRef]
10. FAO/GBEP. *A Review of the Current State of Bioenergy Development in G8+5 Countries*; FAO: Rome, Italy, 2007.

11. International Energy Agency (IEA); Organisation for Economic Co-operation and Development (OECD). Biomass for Power Generation and CHP. In *IEA Energy Technology Essentials;* OECD/IEA: Paris, France, 2007; pp. 1–4.
12. Marquardt, W.; Harwardt, A.; Hechinger, M.; Kraemer, K.; Viell, J.; Voll, A. The Biorenewables Opportunity—Toward Next Generation Process and Product Systems. *AIChE J.* **2010**, *56*, 2228–2235. [CrossRef]
13. Bajpai, P. Pretreatment of Lignocellulosic Biomass for Biofuel Production. *Green Chem. Sustain.* **2016**, *34*, 86. [CrossRef]
14. McKendry, P. Energy Production from Biomass (Part 1): Overview of Biomass. *Bioresour. Technol.* **2002**, *83*, 37–46. [CrossRef]
15. Bajpai, P. *Wood and Fiber Fundamentals;* Elsevier: Amsterdam, The Netherlands, 2018; ISBN 9780128142400. [CrossRef]
16. Holtzapple, M.T. HEMICELLULOSES. In *Encyclopedia of Food Sciences and Nutrition;* Elsevier: Amsterdam, The Netherlands, 2003; pp. 3060–3071.
17. Kang, Q.; Appels, L.; Tan, T.; Dewil, R. Bioethanol from Lignocellulosic Biomass: Current Findings Determine Research Priorities. *Sci. World J.* **2014**, *2014*, 298153. [CrossRef]
18. Tutt, M.; Olt, J. Suitability of Various Plant Species for Bioethanol Production. *Agron. Res.* **2011**, *9*, 261–267.
19. Zoghlami, A.; Paës, G. Lignocellulosic Biomass: Understanding Recalcitrance and Predicting Hydrolysis. *Front. Chem.* **2019**, *7*, 874. [CrossRef]
20. Alam, M.M.; Maniruzzaman, M.; Morshed, M.M. *Application and Advances in Microprocessing of Natural Fiber (Jute)-Based Composites;* Elsevier: Amsterdam, The Netherlands, 2014; Volume 7, ISBN 9780080965338.
21. Walker, G.M. *Bioethanol: Science and Technology of Fuel Alcohol;* Bookboon: Loughborough, UK, 2012; ISBN 9788776816810.
22. Lee, W.G.; Lee, J.S.; Shin, C.S.; Park, S.C.; Chang, H.N.; Chang, Y.K. Ethanol Production Using Concentrated Oak Wood Hydrolysates and Methods to Detoxify. *Appl. Biochem. Biotechnol.-Part A Enzym. Eng. Biotechnol.* **1999**, *77–79*, 547–559. [CrossRef]
23. El-gendy, N.S.; Madian, H.R.; Nassar, H.N. Response Surface Optimization of the Thermal Acid Pretreatment of Sugar Beet Pulp for Bioethanol Production Using Trichoderma Viride and Saccharomyces Cerevisiae. *Recent Pat. Biotechnol.* **2015**, *9*, 50–62. [CrossRef]
24. Rath, S.; Jena, S.; Murugesan, V.P. Direct Ethanol Production by Pretreatment of Lignocellulosic Biomass Using Neurospora Crassa. *Glob. J. Appl. Agric. Res.* **2011**, *1*, 33–41.
25. Kandanelli, R.; Thulluri, C.; Mangala, R.; Rao, P.V.C.; Gandham, S.; Velankar, H.R. A Novel Ternary Combination of Deep Eutectic Solvent-Alcohol (DES-OL) System for Synergistic and Efficient Delignification of Biomass. *Bioresour. Technol.* **2018**, *265*, 573–576. [CrossRef] [PubMed]
26. Chongkhong, S.; Tongurai, C. Optimization of Glucose Production from Corncob by Microwave-Assisted Alkali Pretreatment and Acid Hydrolysis. *Songklanakarin J. Sci. Technol.* **2018**, *40*, 555–562.
27. Guerrero, A.B.; Ballesteros, I.; Ballesteros, M. The Potential of Agricultural Banana Waste for Bioethanol Production. *Fuel* **2018**, *213*, 176–185. [CrossRef]
28. Aruwajoye, G.S.; Faloye, F.D.; Kana, E.G. Soaking Assisted Thermal Pretreatment of Cassava Peels Wastes for Fermentable Sugar Production: Process Modelling and Optimization. *Energy Convers. Manag.* **2017**, *150*, 558–566. [CrossRef]
29. Wawro, A.; Batog, J.; Gieparda, W. Chemical and Enzymatic Treatment of Hemp Biomass for Bioethanol Production. *Appl. Sci.* **2019**, *9*, 5348. [CrossRef]
30. Ramgopal, Y.N. A Study on Production of Pulp from Ground Nut Shells. *Int. J. Sci. Eng. Res.* **2016**, *7*, 423–428.
31. Öhgren, K.; Bura, R.; Lesnicki, G.; Saddler, J.; Zacchi, G. A Comparison between Simultaneous Saccharification and Fermentation and Separate Hydrolysis and Fermentation Using Steam-Pretreated Corn Stover. *Process Biochem.* **2007**, *42*, 834–839. [CrossRef]
32. Liu, X.; Xu, W.; Mao, L.; Zhang, C.; Yan, P.; Xu, Z.; Zhang, Z.C. Lignocellulosic Ethanol Production by Starch-Base Industrial Yeast under PEG Detoxification. *Sci. Rep.* **2016**, *6*, 20361. [CrossRef] [PubMed]
33. Raud, M.; Kesperi, R.; Oja, T.; Olt, J.; Kikas, T. Utilization of Urban Waste in Bioethanol Production: Potential and Technical Solutions. *Agron. Res.* **2014**, *12*, 397–406.
34. Singh, A.; Srivastava, S.; Rathore, D.; Pant, D. *Environmental Microbiology and Biotechnology: Volume 2: Bioenergy and Environmental Health;* Springer: Singapore, 2020; pp. 1–364. [CrossRef]
35. Igwebuike, C.M.; Oyegoke, T. Decarbonizing Our Environment via the Promotion of Biomass Methanation in Developing Nations: A Waste Management Tool. *Pure Appl. Chem.* **2024**, *2023*, 1–20. [CrossRef]
36. Alam, F.; Date, A.; Rasjidin, R.; Mobin, S.; Moria, H.; Baqui, A. Biofuel from Algae-Is It a Viable Alternative? *Procedia Eng.* **2012**, *49*, 221–227. [CrossRef]
37. Meneses, L.R.; Raud, M.; Kikas, T. *Second-Generation Bioethanol Production: A Review of Strategies for Waste Valorisation Second-Generation Bioethanol Production: A Review of Strategies for Waste Valorisation;* CABI: Wallingford, UK, 2017.
38. Fao.org. *The State of Food and Agriculture, 2008;* FAO: Rome, Italy, 2008; ISBN 9789251059807.
39. Nigam, P.S.; Singh, A. Production of Liquid Biofuels from Renewable Resources. *Prog. Energy Combust. Sci.* **2011**, *37*, 52–68. [CrossRef]
40. Voloshin, R.A.; Rodionova, M.V.; Zharmukhamedov, S.K.; Nejat Veziroglu, T.; Allakhverdiev, S.I. Review: Biofuel Production from Plant and Algal Biomass. *Int. J. Hydrogen Energy* **2016**, *41*, 17257–17273. [CrossRef]
41. REN21. *Renewables 2015 Global Status Report;* REN21 Secretariat: Paris, France, 2015.
42. Igwebuike, C.M. Biodiesel: Analysis of Production, Efficiency, Economics and Sustainability in Nigeria. *Clean Technol. Recycl.* **2023**, *3*, 92–106. [CrossRef]

43. Lee, R.A.; Lavoie, J.-M. From First- to Third-Generation Biofuels: Challenges of Producing a Commodity from a Biomass of Increasing Complexity. *Anim. Front.* **2013**, *3*, 6–11. [CrossRef]
44. Naqvi, M.; Yan, J. First-Generation Biofuels. In *Handbook of Clean Energy Systems*; John Wiley & Sons, Inc.: Hoboken, NJ, USA, 2015; pp. 1–18. [CrossRef]
45. Naik, S.N.; Goud, V.V.; Rout, P.K.; Dalai, A.K. Production of First and Second Generation Biofuels: A Comprehensive Review. *Renew. Sustain. Energy Rev.* **2010**, *14*, 578–597. [CrossRef]
46. Patil, V.; Tran, K.Q.; Giselrød, H.R. Towards Sustainable Production of Biofuels from Microalgae. *Int. J. Mol. Sci.* **2008**, *9*, 1188–1195. [CrossRef] [PubMed]
47. Aro, E.M. From First Generation Biofuels to Advanced Solar Biofuels. *Ambio* **2016**, *45*, 24–31. [CrossRef] [PubMed]
48. Mohr, A.; Raman, S. Lessons from First Generation Biofuels and Implications for the Sustainability Appraisal of Second Generation Biofuels. *Effic. Sustain. Biofuel Prod. Environ. Land-Use Res.* **2015**, *63*, 281–310. [CrossRef] [PubMed]
49. Pauly, M.; Keegstra, K. Cell-Wall Carbohydrates and Their Modification as a Resource for Biofuels. *Plant J.* **2008**, *54*, 559–568. [CrossRef] [PubMed]
50. Bajpai, P. *Third Generation Biofuels*; Springer: Singapore, 2019; p. 87. [CrossRef]
51. Alam, F.; Mobin, S.; Chowdhury, H. Third Generation Biofuel from Algae. *Procedia Eng.* **2015**, *105*, 763–768. [CrossRef]
52. Li-Beisson, Y.; Peltier, G. Third-Generation Biofuels: Current and Future Research on Microalgal Lipid Biotechnology. *OCL-Oilseeds Fats Crop. Lipids* **2013**, *20*, D606. [CrossRef]
53. Bac-To-Fuel Global Biofuel Overview during Pandemic. Available online: http://bactofuel.eu/global-biofuel-overview-during-pandemic/ (accessed on 15 December 2022).
54. How Much Ethanol Is in Gasoline and How Does It Affect Fuel Economy? Available online: https://www.eia.gov/tools/faqs/faq.php?id=27&t=4 (accessed on 25 March 2024).
55. Fulton, L. *Biofuels for Transport*; OECD: Paris, France, 2004; p. 216. [CrossRef]
56. Radovanović, M. *Strategic Priorities of Sustainable Energy Development*; Academic Press: Cambridge, MA, USA, 2023; ISBN 9780128210864.
57. Puppán, D. Environmental Evaluation of Biofuels. *Period. Polytech. Soc. Manag. Sci.* **2002**, *10*, 95–116.
58. Neelakandan, T.; Usharani, G.; Nagar, A. Optimization and Production of Bioethanol from Cashew Apple Juice Using Immobilized Yeast Cells by Saccharomyces Cerevisiae. *Am. J. Sustain. Agric.* **2009**, *4*, 85–88.
59. Dinneen, B. Ethanol Industry Outlook 2017: Building Partnerships and Growing Markets, Washington, DC. *RFA 2017*. 2017. Available online: https://d35t1syewk4d42.cloudfront.net/file/27/Ethanol-Industry-Outlook-2017.pdf (accessed on 6 March 2024).
60. Rozenfelde, L.; Puíe, M.; Krûma, I.; Poppele, I.; Matjuðkova, N.; Vederòikovs, N.; Rapoport, A. Enzymatic Hydrolysis of Lignocellulose for Bioethanol Production. *Proc. Latv. Acad. Sci. Sect. B Nat. Exact. Appl. Sci.* **2017**, *71*, 275–279. [CrossRef]
61. Balat, M. Global Bio-Fuel Processing and Production Trends. *Energy Explor. Exploit.* **2007**, *25*, 195–218. [CrossRef]
62. Schnepf, R. CRS Report for Congress and Agriculture: An Overview. Available online: https://www.everycrsreport.com/files/20060316_RS22404_6cc96b94c1bb5920bcd26b8eab4ba16657622307.pdf (accessed on 6 March 2024).
63. Dufey, A. *Biofuels Production, Trade and Sustainable Development: Emerging Issues*; International Institute for Environment and Development: London, UK, 2006.
64. Sönnichsen, N. Leading Producers of Fuel Ethanol in the EU 2021–2022. Available online: https://www.statista.com/statistics/1295937/leading-fuel-ethanol-producers-in-the-eu/#statisticContainer (accessed on 14 October 2022).
65. Kumar NV, L.; Dhavala, P.; Goswami, A.; Maithel, S. Liquid Biofuels in South Asia: Resources and Technologies. *Asian Biotechnol. Dev. Rev.* **2006**, *8*, 31–49.
66. Himmel, M.E.; Picataggio, S.K. Our Challenge Is to Acquire Deeper Understanding of Biomass Recalcitrance and Conversion. In *Biomass Recalcitrance Deconstructing Plant Cell Wall Bioenergy*; John Wiley & Sons, Inc.: Hoboken, NJ, USA, 2009; pp. 1–6. [CrossRef]
67. Mosier, N.; Wyman, C.; Dale, B.; Elander, R.; Lee, Y.Y.; Holtzapple, M.; Ladisch, M. Features of Promising Technologies for Pretreatment of Lignocellulosic Biomass. *Bioresour. Technol.* **2005**, *96*, 673–686. [CrossRef] [PubMed]
68. Sun, Y.; Cheng, J. Hydrolysis of Lignocellulosic Materials for Ethanol Production: A Review. *Bioresour. Technol.* **2002**, *83*, 1–11. [CrossRef] [PubMed]
69. Tivana, L.D. *Cassava Processing: Safety and Protein Fortification*; Lund University: Lund, Sweden, 2012; ISBN 9789197812252.
70. Abdullahi, M.; Oyeleke, S.B.; Egwim, E. Pretreatment and Hydrolysis of Cassava Peels for Fermentable Sugar Production. *Asian J. Biochem.* **2014**, *9*, 65–70. [CrossRef]
71. Hassan, S.S.; Williams, G.A.; Jaiswal, A.K. Emerging Technologies for the Pretreatment of Lignocellulosic Biomass. *Bioresour. Technol.* **2018**, *262*, 310–318. [CrossRef]
72. Kucharska, K.; Rybarczyk, P.; Hołowacz, I.; Łukajtis, R.; Glinka, M.; Kamiński, M. Pretreatment of Lignocellulosic Materials as Substrates for Fermentation Processes. *Molecules* **2018**, *23*, 2937. [CrossRef]
73. Aslanzadeh, S.; Ishola, M.M.; Richards, T.; Taherzadeh, M.J. *An Overview of Existing Individual Unit Operations*; Elsevier B.V.: Amsterdam, The Netherlands, 2014; ISBN 9780444595041.
74. Kumar, P.; Barrett, D.M.; Delwiche, M.J.; Stroeve, P. Methods for Pretreatment of Lignocellulosic Biomass for Efficient Hydrolysis and Biofuel Production. *Ind. Eng. Chem. Res.* **2009**, *48*, 3713–3729. [CrossRef]

75. Peral, C. *Biomass Pretreatment Strategies (Technologies, Environmental Performance, Economic Considerations, Industrial Implementation)*; Elsevier Inc.: Amsterdam, The Netherlands, 2016; ISBN 9780128036488.
76. Pielhop, T.; Amgarten, J.; Von Rohr, P.R.; Studer, M.H. Steam Explosion Pretreatment of Softwood: The Effect of the Explosive Decompression on Enzymatic Digestibility. *Biotechnol. Biofuels* **2016**, *9*, 152. [CrossRef] [PubMed]
77. Velmurugan, B.; Narra, M.; Rudakiya, D.M.; Madamwar, D. *Sweet Sorghum: A Potential Resource for Bioenergy Production*; Elsevier Inc.: Amsterdam, The Netherlands, 2020; ISBN 9780128189962.
78. Dalena, F.; Senatore, A.; Tursi, A.; Basile, A. *Bioenergy Production from Second- and Third-Generation Feedstocks*; Elsevier Ltd.: Amsterdam, The Netherlands, 2017; ISBN 9780081010266.
79. Shahbazi, A.; Zhang, B. Dilute and Concentrated Acid Hydrolysis of Lignocellulosic Biomass. In *Bioalcohol Production: Biochemical Conversion of Lignocellulosic Biomass*; Woodhead Publishing: Cambridge, UK, 2010; pp. 143–158. [CrossRef]
80. Hu, L.; Lin, L.; Wu, Z.; Zhou, S.; Liu, S. Chemocatalytic Hydrolysis of Cellulose into Glucose over Solid Acid Catalysts. *Appl. Catal. B Environ.* **2015**, *174–175*, 225–243. [CrossRef]
81. Boundless Heterogeneous Catalysis. Available online: http://kolibri.teacherinabox.org.au/modules/en-boundless/www.boundless.com/chemistry/textbooks/boundless-chemistry-textbook/chemical-kinetics-13/catalysis-102/heterogeneous-catalysis-430-4700/index.html (accessed on 26 October 2022).
82. Farnetti, E.; Di Monte, R.; Kašpar, J. Homogeneous and Heterogeneous Catalysis. In *Inorganic and Bio-Inorganic Chemistry—Volume II*; Eolss Publishers Co, Ltd.: Oxford, UK, 2009.
83. Ramos, M.; Dias, A.P.S.; Puna, J.F.; Gomes, J.; Bordado, J.C. Biodiesel Production Processes and Sustainable Raw Materials. *Energies* **2019**, *12*, 4408. [CrossRef]
84. Yoonan, K.; Kongkiattikajorn, J. A Study of Optimal Conditions for Reducing Sugars Production from Cassava Peels by Dilute Acid and Enzymes. *Kasetsart J. Nat. Sci.* **2004**, *35*, 29–35.
85. Bhat, M.K.; Bhat, S. Cellulose Degrading Enzymes and Their Potential Industrial Applications. *Biotechnol. Adv.* **1997**, *15*, 583–620. [CrossRef] [PubMed]
86. Godoy, M.G.; Amorim, G.M.; Barreto, M.S.; Freire, D.M.G. *Agricultural Residues as Animal Feed: Protein Enrichment and Detoxification Using Solid-State Fermentation*; Elsevier B.V.: Amsterdam, The Netherlands, 2018; ISBN 9780444639905.
87. Verma, D.; Satyanarayana, T. Molecular Approaches for Ameliorating Microbial Xylanases. *Bioresour. Technol.* **2012**, *117*, 360–367. [CrossRef] [PubMed]
88. Belancic, A.; Scarpa, J.; Peirano, A.; Díaz, R.; Steiner, J.; Eyzaguirre, J. Penicillium Purpurogenum Produces Several Xylanases: Purification and Properties of Two of the Enzymes. *J. Biotechnol.* **1995**, *41*, 71–79. [CrossRef] [PubMed]
89. Ghasemzadeh, K.; Jalilnejad, E.; Basile, A. *Production of Bioalcohol and Biomethane*; Elsevier Ltd.: Amsterdam, The Netherlands, 2017; ISBN 9780081010266.
90. Lamichhane, G.; Acharya, A.; Poudel, D.K.; Aryal, B.; Gyawali, N.; Niraula, P.; Phuyal, S.R.; Budhathoki, P.; Bk, G.; Parajuli, N. Recent Advances in Bioethanol Production from Lignocellulosic Biomass. *Int. J. Green Energy* **2021**, *18*, 731–744. [CrossRef]
91. Zabed, H.; Sahu, J.N.; Suely, A.; Boyce, A.N.; Faruq, G. Bioethanol Production from Renewable Sources: Current Perspectives and Technological Progress. *Renew. Sustain. Energy Rev.* **2017**, *71*, 475–501. [CrossRef]
92. Zaldivar, J.; Martinez, A.; Ingram, L.O. Effect of Selected Aldehydes on the Growth and Fermentation of Ethanologenic *Escherichia coli*. *Biotechnol. Bioeng.* **1999**, *65*, 24–33. [CrossRef]
93. Lenihan, P.; Orozco, A.; Neill, E.O.; Ahmad, M.N.M.; Rooney, D.W.; Walker, G.M. Dilute Acid Hydrolysis of Lignocellulosic Biomass. *Chem. Eng. J.* **2010**, *156*, 395–403. [CrossRef]
94. Bensah, E.C.; Mensah, M.Y. Alkali and Glycerol Pretreatment of West African Biomass for Production of Sugars and Ethanol. *Bioresour. Technol. Rep.* **2019**, *6*, 123–130. [CrossRef]
95. van der Pol, E.; Bakker, R.; van Zeeland, A.; Sanchez Garcia, D.; Punt, A.; Eggink, G. Analysis of By-Product Formation and Sugar Monomerization in Sugarcane Bagasse Pretreated at Pilot Plant Scale: Differences between Autohydrolysis, Alkaline and Acid Pretreatment. *Bioresour. Technol.* **2015**, *181*, 114–123. [CrossRef] [PubMed]
96. Pereira, J.P.C.; Verheijen, P.J.T.; Straathof, A.J.J. Growth Inhibition of *S. cerevisiae*, *B. subtilis*, and *E. coli* by Lignocellulosic and Fermentation Products. *Appl. Microbiol. Biotechnol.* **2016**, *100*, 9069–9080. [CrossRef]
97. van der Pol, E.C.; Bakker, R.R.; Baets, P.; Eggink, G. By-Products Resulting from Lignocellulose Pretreatment and Their Inhibitory Effect on Fermentations for (Bio)Chemicals and Fuels. *Appl. Microbiol. Biotechnol.* **2014**, *98*, 9579–9593. [CrossRef]
98. Palmqvist, E.; Grage, H.; Meinander, N.Q.; Hahn-Hägerdal, B. Main and Interaction Effects of Acetic Acid, Furfural, and p-Hydroxybenzoic Acid on Growth and Ethanol Productivity of Yeasts. *Biotechnol. Bioeng.* **1999**, *63*, 46–55. [CrossRef]
99. Taherzadeh, M.J.; Gustafsson, L.; Niklasson, C.; Lidén, G. Physiological Effects of 5-Hydroxymethylfurfural on Saccharomyces Cerevisiae. *Appl. Microbiol. Biotechnol.* **2000**, *53*, 701–708. [CrossRef]
100. Taherzadeh, M.J.; Karimi, K. *Fermentation Inhibitors in Ethanol Processes and Different Strategies to Reduce Their Effects*, 1st ed.; Elsevier Inc.: Amsterdam, The Netherlands, 2011; ISBN 9780123850997.
101. Delgenes, J.P.; Moletta, R.; Navarro, J.M. Effects of Lignocellulose Degradation Products on Ethanol Fermentations of Glucose and Xylose by Saccharomyces Cerevisiae, Zymomonas Mobilis, Pichia Stipitis, and Candida Shehatae. *Enzyme Microb. Technol.* **1996**, *19*, 220–225. [CrossRef]
102. Yang, Y.; Sha, M. *A Beginner's Guide to Bioprocess Modes—Batch, Fed-Batch and Continuous Fermentation*; Eppendorf: Shanghai, China, 2019; pp. 1–16. Available online: https://www.eppendorf.com/ (accessed on 25 August 2020).

103. Zohri, A.-N.A.; Ragab, S.W.; Mekawi, M.I.; Mostafa, O.A.A. Comparison between Batch, Fed-Batch, Semi-Continuous and Continuous Techniques for Bio-Ethanol Production from a Mixture of Egyptian Cane and Beet Molasses. *Egypt. Sugar J.* **2017**, *9*, 89–111.
104. Ishizaki, H.; Hasumi, K. Ethanol Production from Biomass. In *Research Approaches to Sustainable Biomass Systems*; Elsevier: Amsterdam, The Netherlands, 2014; pp. 243–258. ISBN 9780124046092.
105. Fred, W.H.B.; Peterson, A.; Davenport, E. *Fermentation Characteristics of Certain Pentose-Destroying Bacteria*; Elsevier: Amsterdam, The Netherlands, 1920.
106. Chandel, A.K.; Chandrasekhar, G.; Radhika, K.; Ravinder, R. Bioconversion of Pentose Sugars into Ethanol: A Review and Future Directions. *Biotechnol. Mol. Biol. Rev.* **2011**, *6*, 8–20.
107. Singla, A.; Paroda, S.; Dhamija, S.S.; Goyal, S.; Shekhawat, K.; Amachi, S.; Inubushi, K. Bioethanol Production from Xylose: Problems and Possibilities. *J. Biofuels* **2012**, *3*, 1. [CrossRef]
108. Mishra, A.; Singh, P. *Microbial Pentose Utilization*; Elsevier Science: Amsterdam, The Netherlands, 1993; p. 39.
109. Sommer, P.; Georgieva, T.; Ahring, B.K. Potential for Using Thermophilic Anaerobic Bacteria for Bioethanol Production from Hemicellulose. *Biochem. Soc. Trans.* **2004**, *32*, 283–289. [CrossRef]
110. Mc Millan, J.D. *Xylose Fermentation to Ethanol: A Review*; NREL/TP-421-4944; National Renewable Energy Laboratory: Golden, CO, USA, 1993; p. 51.
111. Saloheimo, A.; Rauta, J.; Stasyk, O.V.; Sibirny, A.A.; Penttilä, M.; Ruohonen, L. Xylose Transport Studies with Xylose-Utilizing Saccharomyces Cerevisiae Strains Expressing Heterologous and Homologous Permeases. *Appl. Microbiol. Biotechnol.* **2007**, *74*, 1041–1052. [CrossRef]
112. Skoog, K.; Hahn-Hägerdal, B. Xylose Fermentation. *Enzyme Microb. Technol.* **1988**, *10*, 66–80. [CrossRef]
113. Hahn-Hägerdal, B.; Karhumaa, K.; Fonseca, C.; Spencer-Martins, I.; Gorwa-Grauslund, M.F. Towards Industrial Pentose-Fermenting Yeast Strains. *Appl. Microbiol. Biotechnol.* **2007**, *74*, 937–953. [CrossRef]
114. Hahn-Hägerdal, B.; Jeppsson, H.; Skoog, K.; Prior, B.A. Biochemistry and Physiology of Xylose Fermentation by Yeasts. *Enzyme Microb. Technol.* **1994**, *16*, 933–943. [CrossRef]
115. Fernández-Sandoval, M.T.; Galíndez-Mayer, J.; Bolívar, F.; Gosset, G.; Ramírez, O.T.; Martinez, A. Xylose-Glucose Co-Fermentation to Ethanol by *Escherichia coli* Strain MS04 Using Single- and Two-Stage Continuous Cultures under Micro-Aerated Conditions. *Microb. Cell Fact.* **2019**, *18*, 145. [CrossRef] [PubMed]
116. Mohd Azhar, S.H.; Abdulla, R.; Jambo, S.A.; Marbawi, H.; Gansau, J.A.; Mohd Faik, A.A.; Rodrigues, K.F. Yeasts in Sustainable Bioethanol Production: A Review. *Biochem. Biophys. Rep.* **2017**, *10*, 52–61. [CrossRef]
117. Ciani, M.; Comitini, F.; Mannazzu, I. Fermentation. In *Encyclopedia of Ecology*; Elsevier: Amsterdam, The Netherlands, 2008; pp. 1548–1557. [CrossRef]
118. Onuki, S.; Koziel, J.A.; Van Leeuwen, J.; Jenks, W.S.; Greweii, D.; Cai, L. Ethanol Production, Purification, and Analysis Techniques: A Review. *Am. Soc. Agric. Biol. Eng. Annu. Int. Meet. ASABE* **2008**, *12*, 7210–7221. [CrossRef]
119. Pribic, P.; Roza, M.; Zuber, L. How to Improve the Energy Savings in Distillation and Hybrid Distillation-Pervaporation Systems. *Sep. Sci. Technol.* **2006**, *41*, 2581–2602. [CrossRef]
120. Iqbal, A.; Ahmad, S.A. Overview of Enhanced Distillations. *Int. J. Adv. Res. Sci. Eng.* **2015**, *4*, 263–270.
121. European Union. *Biofuel Mandates in the EU by Member State—2022*; USDA GAIN: Washington, DC, USA, 2022.
122. Giuntoli, J. *Advanced Biofuel Policies in Select EU Member States: 2018 Update*; ICCT: Washington, DC, USA, 2018.
123. Giuntoli, J. *Final Recast Renewable Energy Directive for 2021–2030 in the European Union*; ICCT: Washington, DC, USA, 2018.
124. Martin, J. *Project Liberty: Launch of an Integrated Bio-Refinery with Eco-Sustainable and Renewable Technologies. Conversion of Corn Stover Biomass to Bio-Ethanol, Final Report*; USDOE Office of Energy Efficiency and Renewable Energy (EERE), Transportation Office; Bioenergy Technologies Office: Golden, CO, USA, 2021.
125. 2G Ethanol. Available online: https://www.granbio.com.br/en/net-zero-solutions/etanol-2g/about-ethanol/ (accessed on 25 March 2024).
126. de Marignan, A.-L.; PonceletP, A.; Fulgoni, P. 2nd Generation Biofuels: An Industrial First For French FuturolTM Technology. Available online: https://www.ifpenergiesnouvelles.com/article/2nd-generation-biofuels-industrial-first-french-futuroltm-technology (accessed on 25 June 2020).
127. Garriga, C. FuturolTM—Simple, Integrated Cellulosic Ethanol Production Technology. Available online: https://www.axens.net/product/process-licensing/20121/futurol.html (accessed on 5 August 2020).
128. Projet-Futurol The Futurol Project. Available online: https://www.etipbioenergy.eu/images/frederic-martel.pdf (accessed on 25 June 2020).
129. Renewables Clariant Constructs Sunliquid® Bioethanol Plant in Romania. Available online: https://www.fuelsandlubes.com/clariant-constructs-sunliquid-bioethanol-plant-romania/ (accessed on 30 June 2020).
130. Sunliquid® Sunliquid®—An Efficient Production Process for Cellulosic Ethanol. Available online: https://sunliquid-project-fp7.eu/wp-content/uploads/2014/09/factsheet_sunliquid_en.pdf (accessed on 30 June 2020).
131. Domsjö Fabriker. Available online: http://www.domsjo.adityabirla.com/en/sidor/Media.aspx (accessed on 6 March 2024).
132. Winter, L. *We Make More from the Tree*; Domsjö Fabriker Biorefinery: Örnsköldsvik, Sweden, 2018.
133. Gírio, F.; Fonseca, C. *Final Report Summary—PROETHANOL2G (Integration of Biology and Engineering into an Economical and Energy-Efficient 2G Bioethanol Biorefinery)*; European Commission: Brussels, Belgium, 2015.

134. ProEthanol2G On to Second-Generation Bioethanol. Available online: https://ec.europa.eu/programmes/horizon2020/en/news/second-generation-bioethanol (accessed on 8 October 2020).
135. Carvalho, A.C. Integration of Biology and Engineering into an Economical and Energy-Efficient 2G Bioethanol Biorefinery. Available online: https://cordis.europa.eu/project/id/251151 (accessed on 18 October 2022).
136. Bredal, T.H. 12 Million More for Borregaard Research. Available online: https://www.borregaard.com/company/news-archive/12-million-more-for-borregaard-research/ (accessed on 2 February 2023).
137. Robek, A. Borregaard's BALI TM Process Reaches Technical Readiness Level 7 as BIOFOREVER Project. Advances. Available online: http://news.bio-based.eu/borregaards-balitm-process-reaches-technical-readiness-level-7-as-bioforever-project-advances/ (accessed on 12 May 2021).
138. Gregg, J.S.; Bolwig, S.; Hansen, T.; Solér, O.; Ben Amer-Allam, S.; Viladecans, J.P.; Klitkou, A.; Fevolden, A. Value Chain Structures That Define European Cellulosic Ethanol Production. *Sustainability* **2017**, *9*, 118. [CrossRef]
139. Rødsrud, G.; Lersch, M.; Sjöde, A. History and Future of World's Most Advanced Biorefinery in Operation. *Biomass Bioenergy* **2012**, *46*, 46–59. [CrossRef]
140. Chempolis Seals Deal for Indian Biorefinery. Available online: https://chempolis.com/chempolis-seals-deal-for-indian-biorefinery/ (accessed on 16 July 2020).
141. Anttila, J. *Indian-European Joint Venture for the First Commercial Scale Cellulosic Biorefinery in Asia*; Chempolis: Oulu, Finland, 2019.
142. Milanese, S.D. Versalis: Biomass Power Plant Restarted at Crescentino and Bioethanol Production Onstream within the First Half of the Year. Available online: https://www.eni.com/en-IT/media/press-release/2020/02/versalis-biomass-power-plant-restarted-at-crescentino-and-bioethanol-production-onstream-within-the-first-half-of-the-year.html (accessed on 1 July 2020).
143. Ferrari, D. Versalis Restarts Biomass Power Plant at Crescentino Biorefinery and Expects Bioethanol Production Later This Year. Available online: https://bioenergyinternational.com/biochemicals-materials/versalis-restarts-biomass-power-plant-at-crescentino-biorefinery-and-expects-bioethanol-production-later-this-year (accessed on 1 July 2020).
144. Pescarolo, S. *Case Study on the First Advanced Industrial Demonstration Bioethanol Plant in the EU, and How It Was Financed*; ETIP Bioenergy: Aarhus, Denmark, 2013; pp. 1–14.
145. Piero, O. *Crescentino Biorefinery—PROESA^{TM}, Italy. Technology Demo/Industrial Scale Implementation*; IEA Bioenergy: Äänekoski, Finland, 2018.
146. Biochemtex. A New Era Begins-Crescentino: World's First Advanced Biofuels Facility. Available online: https://www.etipbioenergy.eu/images/crescentino-presentation.pdf (accessed on 3 July 2020).
147. Research and Development in the Seaweed Industry. Available online: https://news.algaeworld.org/2020/03/seaweed-the-new-ethanol/ (accessed on 6 July 2020).
148. von Keitz, M. Macroalgae Research Inspiring Novel Energy Resources (MARINER) Program Overview. *Arpa·e* **2017**, *313*, 19.
149. TATA. TATA Projects Wins Order for Bioethanol Plant in India. Available online: https://biofuels-news.com/news/tata-projects-wins-order-for-bioethanol-plant-in-india/ (accessed on 25 June 2020).
150. Project Overview. Available online: https://www.lignoflag-project.eu/lignoflag-project/overview/ (accessed on 30 June 2020).
151. Equity Bulls. *IOCL's 2G Ethanol Bio-Refinery Based on Praj's Technology Inaugurated by Prime Minister Modi*; Praj: Pune, India, 2022.
152. Praj Industries Ltd. In *Brief Process Description of 2G Ethanol Process for Indian Oil Corporation Limited, India*; Praj: Pune, India, 2018.
153. Valmet Valmet to Supply Automation to AustroCel' s New Bioethanol Plant in Hallein, Austria. Available online: https://www.globenewswire.com/news-release/2020/05/05/2027274/0/en/Valmet-to-supply-automation-to-AustroCel-s-new-bioethanol-plant-in-Hallein-Austria.html#:~:text=Valmetwillsupplyautomationto,biofueltoreplacefossilfuel (accessed on 1 July 2020).
154. Dugandzic, I. Austrocel to Kick off Bioethanol Production at the End of Year. Available online: https://www.euwid-paper.com/news/markets/austrocel-to-kick-off-bioethanol-production-at-the-end-of-year/ (accessed on 25 February 2023).
155. Bacovsky, D.; Matschegg, D.; Kourkoumpas, D.; Tzelepi, V.; Sagani, A. *BIOFIT Case Study Report AustroCel Hallein*; Bioenergy and Sustainable Technologies GmbH: Wieselburg-Land, Austria, 2021; pp. 1–47.
156. Mäki, E.; Saastamoinen, H.; Melin, K.; Matschegg, D.; Pihkola, H. Drivers and Barriers in Retrofitting Pulp and Paper Industry with Bioenergy for More Efficient Production of Liquid, Solid and Gaseous Biofuels: A Review. *Biomass Bioenergy* **2021**, *148*, 106036. [CrossRef]

molecules

MDPI

Article

Sustainable Recovery of Platinum Group Metals from Spent Automotive Three-Way Catalysts through a Biogenic Thiosulfate-Copper-Ammonia System

Mariacristina Compagnone [1], José Joaquín González-Cortés [1,*], María Pilar Yeste [2], Domingo Cantero [1] and Martín Ramírez [1]

[1] Department of Chemical Engineering and Food Technologies, Wine and Agrifood Research Institute (IVAGRO), Faculty of Sciences, University of Cadiz, Puerto Real, 11510 Cadiz, Spain; mariacristina.compagnone@uca.es (M.C.); martin.ramirez@uca.es (M.R.)

[2] Department of Material Science, Metallurgical Engineering and Inorganic Chemistry, Institute of Research on Electron Microscopy and Materials (IMEYMAT), Faculty of Sciences, University of Cadiz, Puerto Real, 11510 Cadiz, Spain

* Correspondence: joaquin.gonzalez@uca.es

Abstract: This study explores an eco-friendly method for recovering platinum group metals from a synthetic automotive three-way catalyst (TWC). Bioleaching of palladium (Pd) using the thiosulfate-copper-ammonia leaching processes, with biogenic thiosulfate sourced from a bioreactor used for biogas biodesulfurization, is proposed as a sustainable alternative to conventional methods. Biogenic thiosulfate production was optimized in a gas-lift bioreactor by studying the pH (8–10) and operation modes (batch and continuous) under anoxic and microaerobic conditions for 35 d. The maximum concentration of 4.9 g $S_2O_3^{2-}$ L^{-1} of biogenic thiosulfate was reached under optimal conditions (batch mode, pH = 10, and airflow rate 0.033 vvm). To optimize Pd bioleaching from a ground TWC, screening through a Plackett–Burman design determined that oxygen and temperature significantly affected the leaching yield negatively and positively, respectively. Based on these results, an optimization through an experimental design was performed, indicating the optimal conditions to be $Na_2S_2O_3$ 1.2 M, $CuSO_4$ 0.03 M, $(NH_4)_2SO_4$ 1.5 M, Na_2SO_3 0.2 M, pH 8, and 60 °C. A remarkable 96.2 and 93.2% of the total Pd was successfully extracted from the solid at 5% pulp density using both commercially available and biogenic thiosulfate, highlighting the method's versatility for Pd bioleaching from both thiosulfate sources.

Keywords: three-way catalysts; platinum group metals; gas-lift; biogas desulfurization; bioleaching

Citation: Compagnone, M.; González-Cortés, J.J.; Pilar Yeste, M.; Cantero, D.; Ramírez, M. Sustainable Recovery of Platinum Group Metals from Spent Automotive Three-Way Catalysts through a Biogenic Thiosulfate-Copper-Ammonia System. *Molecules* **2023**, *28*, 8078. https://doi.org/10.3390/molecules28248078

Academic Editor: Mohamad Nasir Mohamad Ibrahim

Received: 31 October 2023
Revised: 8 December 2023
Accepted: 12 December 2023
Published: 14 December 2023

1. Introduction

The platinum group metals (PGMs) comprise a group of precious metals, specifically platinum (Pt), palladium (Pd), rhodium (Rh), ruthenium (Ru), iridium (Ir), and osmium (Os). These metals are highly prized for their diverse properties, which make them indispensable in numerous advanced applications across medical, electronic, and industrial domains [1].

These metals are a rarity in the Earth's upper continental crust, with ore deposits typically containing only 3–4 g of PGMs per ton of ore. The bulk of PGM reserves can be found in South Africa (accounting for approximately 95% of global reserves) and Russia (about 2%), resulting in a significant dependence on these sources within the industry [2].

Pt and Pd, in particular, are vital components of modern industry, thanks to their unique physico-chemical properties [3]. According to Johnson Matthey's assessment of PGM supply and demand in 2022, the global gross demand for Pt and Pd was 211.6 and 315.6 tons, respectively [4]. Consequently, the limited availability of PGMs poses a substantial challenge, especially for the European Union (EU). In response, the European Commission included PGMs in the list of critical raw materials for the EU in 2011 [5].

Over the past few decades, the demand for PGMs has seen consistent growth due to their significant industrial value, leading to increasing concerns about sustainability [6]. Producing 1 kg of PGMs from primary mineral ores requires an estimated 20 times more energy and results in higher greenhouse gas (GHG) emissions than recycling them from secondary sources [7]. Hence, there is an urgent need to transform the abundant and sometimes hazardous waste generated by PGM industries, including those in the automotive catalyst, super alloy, electronic, space industry material, biomedical equipment, and jewelry sectors, into opportunities for metal recycling.

Since 1979, the automotive industry has been the largest consumer of PGMs, with estimates suggesting that the demand for PGMs in the automotive sector in 2020 accounted for over 65% of the global demand [8]. Catalytic converters are known to contain significantly higher concentrations of PGMs than what is found in natural ores [9]. Recycling just 2 mg of spent automotive catalysts, for instance, is estimated to spare the need to mine 150 kg of PGM ores [10]. Consequently, three-way catalysts (TWCs) emerge as a particularly rich secondary source of PGMs.

Recovering PGMs from TWCs is crucial for recycling and reducing the financial and environmental impact of mining [1,11]. PGM recovery can be achieved through pyrometallurgical and hydrometallurgical methods. Pyrometallurgy involves pre-treatment steps combined with high-temperature processes that emit greenhouse gases and pollutants, making waste management challenging [7,12,13]. On the other hand, hydrometallurgy involves the use of uses corrosive chemicals, posing environmental and health risks, and requires resource-intensive chemical production [14,15]. Furthermore, proper waste treatment and disposal are crucial in both methods to minimize their environmental footprint.

Hence, ongoing efforts focus on the development of more eco-friendly and efficient methods for PGM recovery from TWCs, such as the use of green solvents [7,16]. Thiosulfate is a chemical compound that has been used in the recovery of precious metals and PGMs from various sources [16–18]. This anionic complexing agent forms stable complexes with metals, allowing them to be solubilized from ore or other sources, primarily through the coordination of sulfur atoms to the metal ions [19]. This process is an alternative to the more commonly used cyanide- or aqua regia-based methods for the recovery of PGMs and other precious metals and is considered to be more environmentally friendly [7,20]. Thiosulfate leaching is generally less toxic, more selective for PGMs, and can be used under milder conditions, making it a potentially safer and more environmentally friendly option [7,20]. Additionally, it can be applied to low-grade or complex ores that are challenging to process using cyanide. Thiosulfate-based processes for metal recovery can be more challenging to implement and optimize compared to cyanide-based methods [21]. Factors such as pH, temperature, and the concentration of various reagents need to be carefully controlled to ensure efficient metal recovery [17,22]. While thiosulfate-based methods have shown promise, they have not yet completely replaced cyanide-based methods in the PGM industry due to the need for further research, development, and optimization [7,23].

During biogas desulfurization, sulfur-containing compounds like hydrogen sulfide (H_2S) are removed from biogas. This process typically involves the use of sulfur-reducing bacteria or other microorganisms that metabolize sulfur compounds [24,25]. As a result of these microbial activities, thiosulfate ($S_2O_3{}^{2-}$) is produced as a generally considered undesirable byproduct [26,27]. Biologically produced thiosulfate has been recently used in leaching processes for the recovery of gold and silver [28–30]. In the study conducted by McNeice et al. [30], they reported gold extraction efficiencies spanning a range from 20 to 60%. This was achieved through the utilization of biogenic thiosulfate, following the production of approximately 400 mg L^{-1} of biogenic thiosulfate with the aid of sodium sulfide (Na_2S) (chemical source) in conjunction with *M. sulfidovorans* (DSMZ 11578). Pourhossein and Mousavi [29] generated a biogenic thiosulfate solution at a concentration of 350 mg L^{-1}, using elemental sulfur (chemical source), with which they successfully recovered 31% of gold and 40% of silver from discarded printed circuit boards with a 1% pulp density within 48 h. The same authors generated 500 mg L^{-1} of biogenic thiosulfate and improved the recovery yield by leaching 65% of gold in 36 h at a 0.5% pulp density of the same matrix [28]. Despite the good results, this approach, which reduces waste and offers an eco-friendly and potentially cost-effective method, has not been tested for PGM leaching from TWCs until now. In addition, the use of a biological bioreactor for biogas desulfurization to generate thiosulfate has also not been studied, as elemental sulfur or sulfate are the main oxidation products [31].

In this pioneering study, we present a novel and sustainable approach to recover PGMs from a palladium/alumina (Pd/Al_2O_3) catalyst by harnessing renewable biogenic thiosulfate derived from a bioreactor performing the biodesulfurization of biogas. This marks the first instance in which biogenic thiosulfate has been proposed as a sustainable and eco-friendly medium for PGM recovery, representing a significant advancement in the field. Our investigation encompasses the optimization of biogenic thiosulfate production within a gas-lift bioreactor, with an in-depth examination of the effects of different operation modes (batch and continuous), pH (8–10), and electron acceptors (O_2, NO_3^- and NO_2^-).

Additionally, to establish a comprehensive understanding of the PGM recovery process, we draw a comparative analysis between commercially available thiosulfate and biogenic thiosulfate, evaluating their respective effectiveness in recovering PGMs from the Pd/Al_2O_3 catalyst. This comparative assessment provides invaluable insights into the advantages and potential of biogenic thiosulfate in the context of sustainable and environmentally responsible PGM recovery.

2. Results and Discussion

2.1. Plackett–Burman Experimental Design

To explore the relationships governing metal leaching while minimizing experimental variance, we employed a synthetic Pd/Al_2O_3 catalyst in a Plackett–Burman design involving 12 runs and 8 factors. Utilizing glass beakers and a solution comprising $Na_2S_2O_3{\cdot}5H_2O$, $(NH_4)_2SO_4$, $CuSO_4{\cdot}5H_2O$, and Na_2SO_3, tests were meticulously conducted with continuous agitation, controlled heating, and real-time monitoring.

The results obtained through the proposed Plackett–Burman (PB) experimental design for the assessment of Pd leaching percentages in various experimental conditions have yielded noteworthy insights. The initial findings reveal a distinct outcome, particularly evident in experiment number 7, which stands out as a promising outlier (Table 1). In this specific trial, 26.0% of Pd was leached by setting the following conditions: $Na_2S_2O_3$ 1.2 M; $CuSO_4$ 0.03 M; $(NH_4)_2SO_4$ 1.5 M; Na_2SO_3 0.1 M; pH 8; T (60 °C); airflow rate (0 vvm). This result contrasted with the remaining experiments where Pd recovery remained below 2.5%. This significant disparity underscores the positive outcome of experiment number 7.

Table 1. PB design for screening of significant factors affecting Pd leaching.

Run No.	$Na_2S_2O_3$ (M)	$CuSO_4$ (M)	$(NH_4)_2SO_4$ (M)	Na_2SO_3 (M)	pH	T (°C)	Airflow Rate (vvm)	Pd Leaching (%)
1	1.2 (+)	0.06 (+)	1.5 (+)	0.0 (−)	12.0 (+)	60.0 (+)	0 (−)	2.0
2	0.6 (−)	0.06 (+)	1.5 (+)	0.0 (−)	12.0 (+)	25.0 (−)	0 (−)	1.1
3	1.2 (+)	0.03 (−)	0.5 (−)	0.0 (−)	12.0 (+)	60.0 (+)	2 (+)	1.0
4	0.6 (−)	0.03 (−)	0.5 (−)	0.1 (+)	12.0 (+)	60.0 (+)	0 (−)	1.4
5	0.6 (−)	0.06 (+)	0.5 (−)	0.0 (−)	8.0 (−)	60.0 (+)	2 (+)	0.1
6	1.2 (+)	0.03 (−)	1.5 (+)	0.0 (−)	8.0 (−)	25.0 (−)	2 (+)	0.1
7	1.2 (+)	0.03 (−)	1.5 (+)	0.1 (+)	8.0 (−)	60.0 (+)	0 (−)	26.0
8	1.2 (+)	0.06 (+)	0.5 (−)	0.1 (+)	12.0 (+)	25.0 (−)	2 (+)	0.6
9	0.6 (−)	0.03 (−)	1.5 (+)	0.1 (+)	12.0 (+)	25.0 (−)	2 (+)	1.0
10	1.2 (+)	0.06 (+)	0.5 (−)	0.1 (+)	8.0 (−)	25.0 (−)	0 (−)	0.1
11	0.6 (−)	0.06 (+)	1.5 (+)	0.1 (+)	8.0 (−)	60.0 (+)	2 (+)	0.1
12	0.6 (−)	0.03 (−)	0.5 (−)	0.0 (−)	8.0 (−)	25.0 (−)	0 (−)	0.1

While the presence of outliers in PB experimental designs, as observed in exp. 7, can have negative consequences, such as potentially leading to misleading results and reduced precision [32], it is important to note that the overall effect of factors is typically consistent. However, the presence of outliers may introduce some level of uncertainty when determining the significance of each individual factor. The analysis of the standardized effect size of each independent variable on the leaching of Pd is shown in Figure S1. Therefore, from this first set of experiments, it was deduced that elevated concentrations of oxygen, copper, and a high pH level demonstrate an adverse effect on Pd leaching. Conversely, higher values of temperature, ammonia, thiosulfate, and sulfite exhibit a beneficial effect, promoting and enhancing the Pd leaching process.

The outcomes observed in this study are likely attributed to various underlying factors. The detrimental impact of oxygen on thiosulfate leaching of gold and nickel from different sources is well-documented [33–36]. Oxygen can have adverse effects due to its tendency to promote the oxidation of valuable minerals and metals. This undesirable oxidation can hinder the efficiency of leaching, potentially leading to reduced yields and lower leaching rates. For example, Zhang et al. [36] investigated the oxidation rate of colloidal gold in ammonia-thiosulfate systems in the absence of copper, specifically focusing on the relationship between this rate and the concentration of dissolved oxygen. Significantly, the research outcomes illuminated that an excess of oxygen (0.5 mM) exerted an adverse impact on the dissolution of gold, while copper emerged as a more efficacious oxidizing agent than oxygen for the dissolution of gold colloid in thiosulfate solutions. Consequently, minimizing oxygen exposure or carefully controlling its presence is often a crucial factor in optimizing leaching processes for enhanced metal recovery.

Sulfite and ammonia can serve as effective reducing agents, facilitating the reduction in certain metal compounds to their soluble forms [20,22,37–39]. This reduction reaction increases the solubility of metals and minerals, leading to improved leaching efficiency. Additionally, sulfite can help mitigate the undesirable effects of excessive oxidation by consuming oxygen or other oxidizing agents, which can further enhance the effectiveness of leaching processes. Consequently, sulfite's role as a reducing agent and oxygen scavenger often contributes to a positive influence on the leaching of valuable metals [20]. Sulfite also acts as a stabilizing agent for thiosulfate, preventing its decomposition and thereby extending its useful lifespan in the leaching solution [37,38]. This stability ensures that thiosulfate remains available for complexation with PGM, increasing the overall leaching efficiency.

Higher concentrations of thiosulfate increase the availability of thiosulfate ions to form stable complexes with PGMs, making them more readily accessible for leaching [7,17,40]. Xu et al. [17] found that Pd extraction from decopperized anode slime was initially around 20% at a thiosulfate concentration of 0.2 M, progressively rising to nearly 40% at 0.6 M. However, it is noteworthy that the extraction efficiency began to decline as the thiosulfate concentration continued to increase. Therefore, the slight increase in thiosulfate concentration accelerates the leaching kinetics, resulting in faster PGM dissolution. A similar trend can be observed in the leaching process of other metals. Chen et al. [41] detected a noteworthy increasing trend in the extraction of gold, which elevated from around 0.8 to 35% extraction, and in the case of silver, an increase from approximately 55 to 75% extraction was demonstrated. This trend coincided with the elevation of thiosulfate concentrations ranging from 0.01 to 0.4 M. Dong et al. [42] reported that the gold leaching percentage exhibited a notable enhancement, increasing from approximately 33% at 0.05 M of thiosulfate to 83.8% when the thiosulfate concentration was raised to 0.2 M. However, the gold leaching percentage declined to 80.3% when the thiosulfate concentration reached 0.25 M, underscoring the critical importance of determining the optimal concentration range for the process. Therefore, with a surplus of thiosulfate in the leaching solution, a larger quantity of metals can be complexed and subsequently recovered, leading to higher metal recovery yields and lower leaching times [17,43,44].

In the area of PGM leaching processes related to automotive catalysts, it is widely recognized that elevated temperatures play a pivotal role in enhancing reaction kinetics. This phenomenon is evident in various approaches, encompassing both conventional methods, such as pyrometallurgical and hydrometallurgical techniques, as well as novel biological methodologies [1,7,18]. Elevated temperatures can improve the solubility of target metals, resulting in higher metal recovery efficiency. For instance, in the context of smelting processes, where temperatures can range from 1100 to 2000 °C, a remarkable recovery rate exceeding 99% for Pt, 99% for Pd, and 97% for Rh has been achieved [18]. The impact of temperature on PGM leaching is further exemplified by research on platinum leaching from catalysts in aqua regia, wherein higher temperatures were found to enhance Pt leaching [45]. Sodium cyanide, renowned for its capacity to form stable metal complexes under elevated temperature and pressure, has been a pivotal lixiviant in PGM leaching from spent automotive catalysts. An investigation into the effect of temperature and pressure on cyanide leaching of PGMs revealed temperature-dependent improvements in leaching efficiency for Pd, Pt, and Rh. Pd increased from 96% at 100 °C to 98% at 160 °C, Pt rose from 92 to 97%, and Rh increased from 89 to 93% over the same temperature range. [46]. Additionally, the role of temperature, pH, and glycine concentration in cyanide production and PGM recovery using the bacterium *C. violaceum* DSM 30,191 has been explored. This two-stage bioleaching process involved biologically producing cyanide and subsequently utilizing it for PGM recovery. The leaching efficiencies displayed a temperature-dependent trend, with recovery rates progressing from 67% at 100 °C to 80% at 150 °C, followed by a subsequent decrease. This trend aligns with prior research findings on the subject [46,47]. This phenomenon underscores the importance of an elevated temperature in achieving higher leaching efficiency, primarily attributable to the robust metallic bonds within PGM compounds.

In the study conducted by Chen et al. [16], the impact of copper concentration on gold and silver extraction within thiosulfate solutions is explored. Notably, gold extraction exhibits a remarkable upsurge from 24 to 66% for gold and from 2 to 26% for silver with a modest increment in copper concentration (0.01 to 0.1 M). Xu et al. [17] observed that palladium extraction from decopperized anode slime initially registered at approximately 28% when the copper concentration was at 0.1 M, gradually increasing to over 40% at 0.3 M. However, it is worth noting that as the copper concentration further rose, the extraction efficiency started to decrease. This observation underscores that a marginal presence of copper expedites the leaching kinetics, leading to faster dissolution of PGMs. This is probably due to the fact that high concentrations of copper may lead to competitive reactions with PGM ions [48,49]. Therefore, copper can enhance PGM solubility and facilitate their leaching by forming stable complexes with thiosulfate ions. These complexes promote PGM dissolution, leading to improved extraction rates. However, high copper concentrations can lead to the formation of excessive copper-thiosulfate complexes, which can compete with metal-thiosulfate complexes for available thiosulfate ions, reducing the efficiency of metal leaching [16,44,50,51].

2.2. Effect of the Main Operational Variables

To enhance Pd leaching, the levels of the chosen factors were adjusted based on the nature of the influence of each factor determined by the PB design, as illustrated in Figure S1. As a result, the concentrations of $Na_2S_2O_3$, $(NH_4)_2SO_4$, and Na_2SO_3 doubled, while the $CuSO_4$ and pH values were decreased. The alterations made to the remaining factors boost metal leaching in experiments A and D (Figure 1).

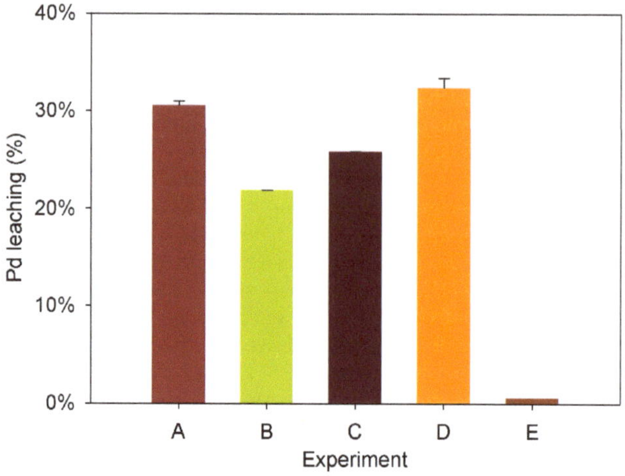

Figure 1. Percentage of palladium leached in the different experiments from the single-variable impact analysis. A–E represent different experimental conditions, as listed in Table 2.

Table 2. Experimental design for analyzing the impact of each variable.

Test	$Na_2S_2O_3$ (M)	$CuSO_4$ (M)	$(NH_4)_2SO_4$ (M)	Na_2SO_3 (M)	pH
A	2.4	0.030	1.5	0.1	8.0
B	1.2	0.015	1.5	0.1	8.0
C	1.2	0.030	3.0	0.1	8.0
D	1.2	0.030	1.5	0.2	8.0
E	1.2	0.030	1.5	0.1	6.0

In these cases, doubling the concentration of $Na_2S_2O_3$ and Na_2SO_3 had a positive impact, elevating Pd leaching from the initial 26.00% (exp.7, Table 1) to 30.64 ± 0.02% and 32.41 ± 0.04%, respectively (exps. A and D, Figure 1). Experiments B and C did not largely affect the Pd leaching, indicating that doubling the NH_4 concentration and halving the Cu concentration do not have a large impact on the process. In contrast, experiment E demonstrated that both lower and higher pH values than 8.0 significantly inhibited the leaching process to a similar extent, corroborating the pH sensitivity of this process to the pH.

These results show that, as discussed in Section 2.1, elevated thiosulfate concentrations enhance the abundance of thiosulfate ions, facilitating the formation of stable complexes with PGMs. This heightened availability renders PGMs more accessible for the leaching process [7,17,40]. Additionally, adding a small amount of sulfite to the leaching solution is a common strategy to partially reduce thiosulfate degradation in thiosulfate-based leaching processes. Thiosulfate is known to be relatively unstable and can decompose over time, which can reduce its effectiveness in recovering valuable metals, such as gold and PGMs [22,52]. Sulfite ions can act as a reducing agent, which can enhance the stability of thiosulfate and its leaching efficiency by reducing the oxidation state of metals [16,19,53].

The critical role played by the pH level in PGM leaching processes has been reported by other authors [17,54–57]. The optimal pH for leaching PGMs can vary significantly depending on the specific leaching process and conditions. Some processes favor highly acidic conditions, while others work best under neutral or slightly alkaline conditions. The choice of pH level is crucial in designing effective PGM leaching processes. Ilyas and Kim [56] document an upward trend in the extraction of Pt and Pd as the pH of the leach liquor decreases from 11.2, resulting in 96.4% Pt and 92.7% Pd extraction, to 10.4, where extraction rates reach 98.2% for Pt and 97.6% for Pd. Bax et al. [55] reported a substantial increase in the recovery of Pt and Pd at pH 1, achieving a 99.7% recovery for Pt and 96.3% for Pd. In contrast, it was observed that at pH 3, the recovery rates significantly dropped to just 11.1% for Pt and 11.8% for Pd. In view of these results, thiosulfate complexation of PGM is pH-dependent, and there is typically an optimal pH range within which the reaction is most efficient and can vary depending on the type of PGM and the associated minerals [18,49,58,59]. Deviating from this range can result in a reduced leaching rate owing to the instability of the thiosulfate complexes formed with PGM, reducing their dissolution and recovery.

Given the obtained results, it was concluded that the best conditions obtained in this study for the leaching of PGMs from TWCs were $Na_2S_2O_3$ 1.2 M, $CuSO_4$ 0.03 M, $(NH_4)_2SO_4$ 1.5 M, Na_2SO_3 0.2 M, pH 8, and 60 °C.

2.3. Biogenic Thiosulfate Production

Biological desulfurization is a well-established technique for effectively eliminating hydrogen sulfide (H_2S) from biogas, a critical prelude to biogas valorization. However, this process typically generates byproducts, such as biogenic sulfur or thiosulfate, which traditionally have limited applications. Consequently, exploring practical uses for these biologically derived waste products is of significant interest from both environmental and economic standpoints. In this study, the production and potential maximization of biogenic thiosulfate derived from a desulfurization bioreactor were studied (Figure 2).

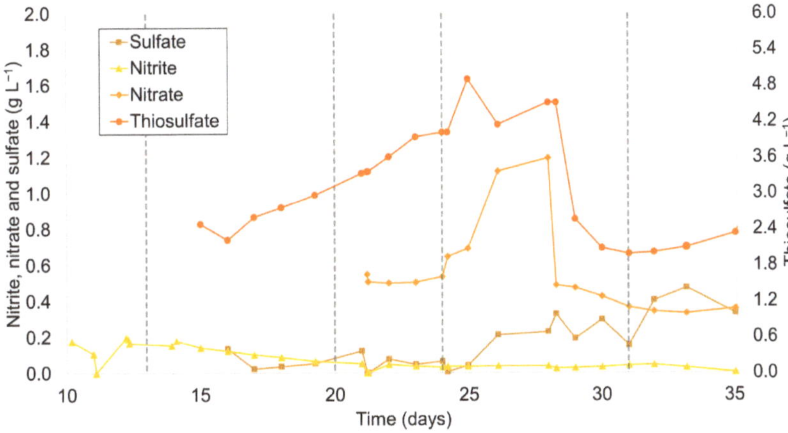

Figure 2. Nitrite, nitrate, sulfate, and thiosulfate concentrations measured throughout the operation of the gas-lift bioreactor.

During Stage I, the objective was to grow the biomass and adapt it to sulfide removal from the biogas at a lower pH (8.5) to avoid excess sulfide accumulation in the liquid. From Stage II, the objective was to maximize thiosulfate production. For this purpose, the pH was raised from 8.5 to 10 and nitrite was used. After 6 days of operation, an average thiosulfate concentration of 3.02 ± 0.29 g $S_2O_3^{2-}$ L^{-1} was achieved. After this stage, on day 21 (Stage III), a significant alteration was made by replacing NO_3^- with NO_2^- as the primary electron acceptor, with a concentration of 1 g N-NO_3^- L^{-1}, to assess its influence on the process. This decision was prompted by the observation that the nitrite concentration, as measured in Stage II, exhibited a slow and protracted decline, suggesting a possible scenario in which microorganisms were not efficiently utilizing nitrite. The primary difference between nitrate and nitrite is their oxidation states: nitrate is more oxidized than nitrite, usually making it a more favorable electron acceptor for microbes, as it yields more energy during its reduction [60,61]. In this particular scenario, the concentration of thiosulfate exhibited a persistent upward trend. It increased from

3.1 g $S_2O_3^{2-}$ L^{-1} of biogenic thiosulfate on day 20 following the introduction of nitrite (marking the conclusion of Stage II) to 4.0 g $S_2O_3^{2-}$ L^{-1} on day 24 after transitioning from nitrite to nitrate (signifying the completion of Stage III). As illustrated in Figure 2, the continuous growth in thiosulfate concentration observed in both Stage II and Stage III underscores the similar ability of nitrite and nitrate to serve as effective electron acceptors in the biogenic thiosulfate production processes.

In order to continue increasing the concentration, the liquid feed of the medium was stopped, and the bioreactor was operated in batch mode (Stage IV). This operation mode allowed the maximum thiosulfate concentration of the entire operation to be reached at 4.9 g $S_2O_3^{2-}$ L^{-1} on day 25. This means that 24.5% of the total dissolved sulfide was present in the form of thiosulfate. These findings align consistently with the research conducted by Van Den Bosch [26], which demonstrated that, under alkaline pH conditions and with an H_2S supply rate of 68 g S-H_2S m^{-3} h^{-1}, approximately 20–22% of the supplied H_2S is transformed into thiosulfate. Notably, our current study yielded a similar conversion rate, despite employing a substantially lower supply rate of 20 g S-Na_2S m^{-3} h^{-1}, which shows that the bioprocess is reproducible and if a higher concentration of thiosulfate is to be obtained, the sulfide load must be increased. This observation underscores the consistent selectivity for biogenic thiosulfate formation, even with varying H_2S supply rates.

While the strategy exhibited effectiveness for the initial days, with steady accumulation of thiosulfate, there was a significant subsequent decline of 60% in the thiosulfate concentration up to 2.0 g $S_2O_3^{2-}$ L^{-1} on day 31. The decline in thiosulfate levels within the bioreactor serves as a clear indicator of its utilization as an energy source [62,63]. Bacteria capable of utilizing sulfide ions (HS^-) for growth can also harness thiosulfate for their metabolic energy needs [62,64]. Transitioning the bioreactor into a batch mode prompted the accumulation of biomass, prompting these microorganisms to explore alternative energy sources in addition to sulfide ions. Consequently, thiosulfate emerged as a viable energy source under these conditions [62]. Additionally, it is important to highlight that a reduced presence of sulfide ions relative to the bacterial population correlates with diminished thiosulfate levels [65]. This correlation arises due to the chemical reactivity of sulfide, resulting in the formation of stable polysulfide compounds, which subsequently lead to thiosulfate production [65–67]

In an attempt to mitigate the observed decline, the introduction of oxygen into the bioreactor was ceased (Stage V). This intervention was implemented to curb the oxidation of accumulating thiosulfate to sulfate, which went from 0.006 g SO_4^{2-} L^{-1} on day 24 to 0.333 g SO_4^{2-} L^{-1} on day 35 as depicted in Figure 2. This correlation also gives support to the hypothesis that thiosulfate has been employed by the bacterial consortium as an energy source once the bioreactor started to operate in batch mode. Unfortunately, the outcomes of this intervention proved to be unsatisfactory, as they failed to elevate the thiosulfate concentration in the bioreactor, which reached 2.1 $\pm$ 0.17 g $S_2O_3^{2-}$ L^{-1}, even following an extended period of 10 days.

The observed lack of success can be ascribed to the requirement that thiosulfate production hinges on the conversion of sulfide into sulfur, leading to the formation of polysulfide compounds and, subsequently, facilitating thiosulfate production. These polysulfides are oxidized to thiosulfate only in the presence of oxygen [68] as shown in Equation (3), in Section 3.4.

The culmination of this study revealed that the highest concentration of biogenic thiosulfate, amounting to 4.9 g L^{-1}, was achieved under ideal conditions. These optimal conditions encompassed the utilization of a batch operational mode, aerobic conditions, a pH level set at 10, and an airflow rate of 0.033 vvm. This outcome showcases the successful fine-tuning of the bioreactor system for enhanced biogenic thiosulfate production.

2.4. Comparative Analysis of PGM Recovery Using Lower Pulp Density with Biogenic and Synthetic Thiosulfate

In the context of PGM recovery from secondary sources, prior studies have indicated that higher recovery yields are achieved when employing lower percentages of solid pulps [12,14,69]. Therefore, a series of experiments was conducted by lowering the solid pulp concentration from 15 to 5%*w/v*, utilizing the previously optimized conditions (exp D, Figure 1). Additionally, these optimal conditions were tested using chemical thiosulfate and biogenic thiosulfate produced using the desulfurization bioreactor operated (Figure 2).

To comprehensively characterize these experiments under optimal conditions, both the liquid and solid samples were analyzed using inductively coupled plasma emission spectroscopy (ICP) and X-ray fluorescence (XRF). The outcomes of these analyses are visualized in Figure 3.

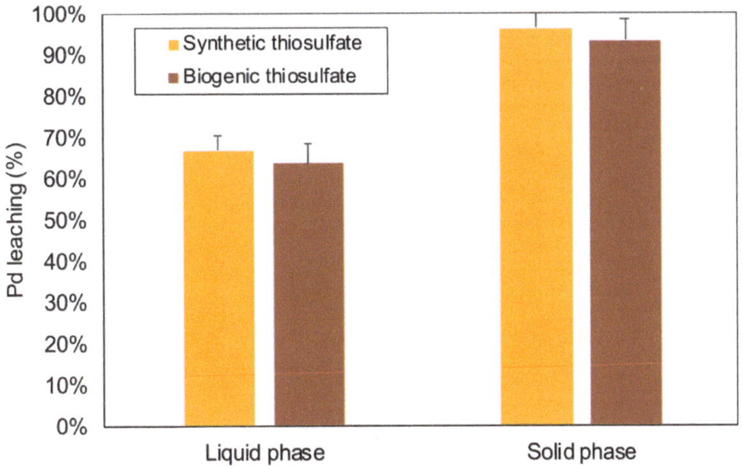

Figure 3. Tests at 5%*w/v* with synthetic and biogenic thiosulfate. The liquid and the solid phases were analyzed with inductively coupled plasma and X-ray fluorescence, respectively.

The leaching of Pd has significantly improved with the use of a solid pulp concentration of 5%*w/v*, compared to the previously tested 15%*w/v*. Specifically, the Pd recovery rate detected in the liquid phase has increased from 32.41 $\pm$ 0.04% to 66.75 $\pm$ 0.04%. This noteworthy enhancement in Pd recovery highlights the substantial impact of the adjusted solid pulp concentration on the leaching process.

The effect of solid pulp concentration on the recovery of PGMs in leaching processes can vary depending on several factors, including the specific leaching method, matrix characteristics, and operational conditions [23,49,70]. Specifically, a decrease in solid pulp concentration can enhance PGM recovery from automotive catalysts in leaching and bioleaching processes [7,18,71,72]. Lowering the solid pulp concentration in the leaching system can lead to a higher liquid-to-solid ratio, which can improve mass transfer and contact between the leaching solution and the PGM-bearing material. This can enhance the dissolution and recovery of PGMs [14]. Also, high solid concentrations can lead to interference or passivation effects, where the surface of PGM particles becomes less reactive due to the accumulation of reaction products. Lower solid concentrations can mitigate these effects and allow for more efficient leaching [14,70,73].

The leaching of Pd using biologically produced thiosulfate achieved a recovery rate in the liquid phase of 63.53 $\pm$ 0.07% (Figure 3). In summary, the comparison of biogenic thiosulfate and synthetic thiosulfate described in the present study revealed nuanced differences in their impacts on the precious metal recovery process. These results underscore the promising potential of biogenic thiosulfate as an effective agent for the recovery of precious metals, emphasizing its role in sustainable and environmentally friendly metal extraction processes.

Analysis of the data presented in Figure 3 also shows that a portion of the leached Pd from the solid phase, which was successfully extracted, did not dissolve into the liquid phase. In fact, upon analyzing the solid pulp after treatment, it was determined that 96.18 ± 0.04% of Pd was extracted from the catalyst when using commercial thiosulfate and 93.22 ± 0.06% when employing biogenic thiosulfate. This occurrence is attributed to a precipitation phenomenon. Significantly, Pd displayed instability, both immediately after the leaching process and subsequent to the filtration stage (as shown in Figure S2). This instability and precipitation is likely a result of the cooling that occurs after the leaching process, causing a portion of the leachate to crystallize as the liquid and solid phases reach room temperature [40,53]. The recovery of precious metals like Pd, Pt, and Rh from the liquid phase after a leaching process is indeed a complex process that often involves adjusting the pH and introducing a suitable precipitating agent to cause the precious metals to form insoluble compounds, which can then be separated from the liquid phase [74,75]. Additionally, as the conditions change, such as controlled cooling or evaporation of the leachate solution, precious metal compounds may precipitate out of the solution in the form of solid crystals [76]. These crystals can be further processed to obtain the pure metal forms.

In the context of our scientific discussion, it is essential to note the works of Pourhossein and Mousavi [28,29], where a biogenic thiosulfate solution generated by *Acidithiobacillus thiooxidans* was employed to recover precious metals from discarded telecommunication printed circuit boards. While Pourhossein and Mousavi [29] achieved a recovery of 31 ± 0.93% of gold and 40 ± 1.43% of silver at a 1% pulp density within a 48 h timeframe, the results were notably improved in their subsequent study [28], where the recovery rate increased to 65 ± 0.78% for gold and achieved this within 36 h at a reduced pulp density of 0.5%*w*/*v*. It is important to compare these outcomes with more conventional methods, such as those employed by Huang et al. [77] that used commercial thiosulfate to recover 94.5% of gold.

3. Materials and Methods

3.1. Synthetic Catalyst (Pd/Al_2O_3)

For the metal bioleaching tests, a synthetic catalyst Pd/Al_2O_3, with the Pd supported on alpha-alumina. The catalyst was prepared by Incipient wetness impregnation using a 0.5M $Pd(NO_3)_2$ solution. Three impregnation cycles were carried out to obtain 0.8% weight of Pd. Finally, the catalyst was calcined in a muffle furnace at 500 °C for one hour with a heating ramp of 5 °C min^{-1} to decompose the palladium precursor to palladium oxide. To obtain the Pd in the metallic state, the catalyst was reduced with a flow of 100 mL min^{-1} of 5%H_2/Ar at 500 °C for 1 h (heating ramp of 5 °C min^{-1}).

3.2. Screening through a Plackett–Burman Design

To explore the relationship of the independent variables (factors) and to minimize the variance of the estimates of these relationships with a limited number of experiments a PB design with 12 runs, 8 factors, and 3 degrees of freedom was conducted.

The leaching tests were carried out in glass beakers in which 15%*w*/*v* of solid samples were placed with a liquid solution containing $Na_2S_2O_3{\cdot}5H_2O$, $(NH_4)_2SO_4$, $CuSO_4{\cdot}5H_2O$, and Na_2SO_3 at the concentrations indicated in Table 1. The tests were conducted with continuous agitation at a fixed speed of 300 rpm, and, when applicable, simultaneous heating was applied to reach 60 °C using a magnetic stirrer. LabVIEW platform (National Instruments™, Austin, TX, USA) with cDAQ Chassis and three modules: a current input module (NI-9209), voltage output module (NI-9264), and digital I/O module (NI-9184) were used for monitoring the tests. pH was adjusted by the addition of 1M NaOH solution, and the value was kept steady by readjustment every 30 min. When the reaction was completed, the solutions were centrifuged at 1610× *g* for 10 min to separate the liquid phases from the solid phases, which were dried in an oven at 80 °C for 24 h.

3.3. *Effect of the Main Operational Variables*

Additional experiments were designed based on the nature of the influence (positive or negative) of the different variables on the percentage of Pd leaching obtained in PB. Thus, the value of the positively affecting variables was increased by 100% and the value of the negatively affecting variables was reduced by 50%, as shown in Table 2.

The temperature was maintained constant at 60 °C and no O_2 was added, considering the detrimental effect of changing these values. The leaching tests were conducted using the identical procedure employed in the PB design experiments.

3.4. *Biogenic Thiosulfate Production*

Thiosulfate production was studied in an inner loop jacketed gas-lift bioreactor (Applikon Biotechnology BV, Delft, The Netherlands) with a working volume of 3L (Figure 4).

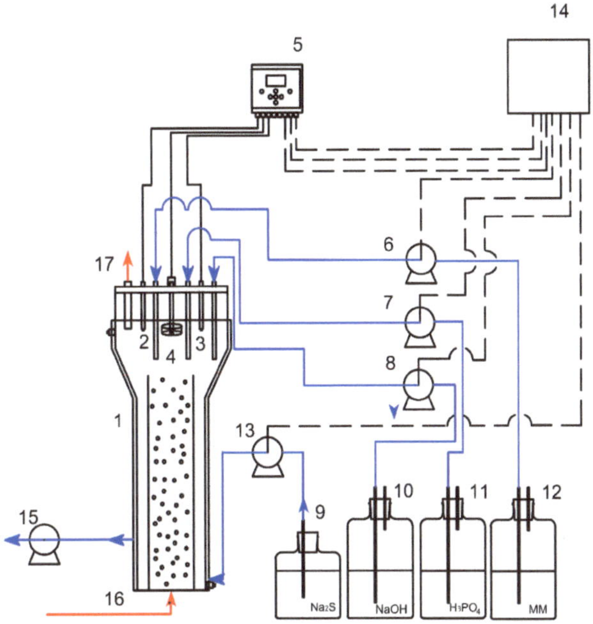

Figure 4. Inner loop jacketed gas-lift bioreactor scheme. 1: Gas-lift bioreactor; 2: pH probe; 3: ORP probe; 4: sensor level; 5: Multimeter; 6: Analog peristaltic pump; 7: H_3PO_4 peristaltic pump; 8: NaOH peristaltic pump; 9: Na_2S container; 10: NaOH container; 11: H_3PO_4 container; 12: Mineral medium container; 13: Na_2S peristaltic pump; 14: PC and control system; 15: Discharge peristaltic pump; 16: Gas flow inlet; 17: Gas flow outlet. Red arrows stand for gaseous streams while blue arrows represent liquid streams.

The medium used had the following composition (g/L): KH_2PO_4 (4); NH_4Cl (2); $MgSO_4{\cdot}7H_2O$ (1.6) and 0.8 mL L^{-1} of solution SL-4 × 5, whose composition was reported by [78]. The volume was kept constant by a sensor level (Vertical Float Switch, Cynergy3, Wimborne, UK) connected to a discharge peristaltic pump. The oxidation−reduction potential (ORP) and pH were measured with a multiparametric analyzer (Crison Multimeter 44, Hach Lange S.L.U, Barcelona, Spain). The pH was controlled by the addition of H_3PO_4 (2 N) and NaOH (2 N). The bioreactor was fed with $Na_2S{\cdot}3H_2O$ to maintain a constant inlet load (IL) of 20 g S-Na_2S m^{-3} h^{-1}. The airflow rate was maintained at 0.033 vvm throughout the entire operation. This specific airflow rate was selected due to the fact that elemental sulfur is an intermediate byproduct of the sulfide-to-sulfate oxidation process catalyzed by sulfur-oxidizing bacteria (SOB) [66]. Elemental sulfur becomes the primary product when oxygen levels are limited (Equation (1)).

$$HS^- + 0.5\ O_2 \leftrightarrow S^0 + OH^- \quad (1)$$

Consequently, lower concentrations of dissolved oxygen were deliberately maintained to promote biogenic sulfur production. This approach allowed the sulfide to react chemically with sulfur, resulting in the formation of stable polysulfide compounds (Equation (2)) [67].

$$HS^- + (x - 1)\ S^0 \leftrightarrow Sx^{2-} + H^+;\ x \leq 6 \quad (2)$$

Under alkaline pH conditions and in the presence of oxygen, these polysulfide compounds are rapidly oxidized to thiosulfate (Equation (3)) [68].

$$Sx^{2-} + O_2 + 5\ H_2O \leftrightarrow 0.5\ S_2O_3^{2-} + (x - 1)\ S^0 + OH^- \quad (3)$$

Sulfide addition was controlled via ORP, where an ORP value (ORP alarm) between −200 and −410 mV served as the set point. When the potential exceeded the ORP alarm threshold, the sulfide feed pump operated for a determined time (4–20 s). The system was monitored and controlled using LabVIEW ™ (National Instruments™, USA).

The bioreactor underwent a 35-day operational process, as detailed in Table 3.

Table 3. Operating parameters of the gas-lift bioreactor.

Stage	Time (days)	Action	pH	O_2 (vvm)	Electron Acceptor Source	Operation Mode	HRT (Days)
I	0–13	Start-up	8.5	0.033	NO_2^-	Continuous	3.5
II	13–20	pH increase	10	0.033	NO_2^-	Continuous	3.5
III	20–24	Anoxic operation (nitrate)	10	0.033	NO_3^-	Continuous	3.5
IV	24–31	Batch mode	10	0.033	NO_3^-	Batch	-
V	31–35	Air supply cut	10	0	NO_3^-	Batch	-

Initially, activated sludge from a conventional wastewater treatment plant located in El Torno, Cádiz, Spain was used to inoculate the bioreactor (20%*v/v*), marking the commencement of the operation. This initial start-up phase, referred to as Stage I, extended for 13 days. During Stage I, the bioreactor operated with NO_2^- serving as the primary electron acceptor. Nitrite was introduced into the bioreactor to achieve a target concentration of 14 g $N\text{-}NO_2^-\ L^{-1}$. At the end of this phase, a pH adjustment from 8.5 to 10 was implemented, favoring the growth of sulfur-oxidizing alkaliphilic microorganisms (Stage II). On the 20th day of operation, as the nitrite neared depletion, a key modification was introduced by substituting NO_2^- with NO_3^- as the primary electron acceptor at a concentration of 1 g $N\text{-}NO_3^-\ L^{-1}$ to investigate its impact on the process (Stage III). From day 24 onwards (Stage IV), the bioreactor was transitioned into batch mode until the end of the operation, with a particular focus on increasing thiosulfate concentration. On day 31 the air supply was intentionally cut, allowing for the assessment of its influence on thiosulfate production during Stage V. Bacterial growth was assessed through cell counting employing an enhanced Neubauer cell counting chamber (Marienfeld Superior™, Lauda-Königshofen, Germany) and an optical microscope (Olympus BH-2, Olympus Europa SE and Co. KG, Hamburg, Germany). Bioreactor samples were collected and subjected to suitable dilution. Triplicate measurements were performed to ensure the reliability and reproducibility of the results.

3.5. *Effect of Lower Pulp Density and Biogenic Thiosulfate*

In this phase, the ability of the thiosulfate-copper-ammonia system to leach Pd from a lower pulp density (5%*w/v*) of synthetic catalyst was studied. For this purpose, biogenic thiosulfate was used to test its leaching ability. The leaching tests were carried out following the same procedure stated in Section 2.2 with the difference of the pulp density (5% *w/v*) and the use of the optimized leaching solution with $Na_2S_2O_3$ 1.2 M, $CuSO_4$ 0.03 M, $(NH_4)_2SO_4$ 1.5 M, Na_2SO_3 0.2 M, pH 8, and 60 °C. To attain the target concentration of 1.2 M for biogenic thiosulfate, the liquid within the bioreactor (3 L) underwent a four-hour evaporation process at 100 °C. Following concentration and cooling, a significant portion of residual salts precipitated out. Subsequently, the liquid underwent filtration using a membrane filter with a pore size of 0.22 µm. This filtration step successfully eliminated these residual salts, resulting in a purified thiosulfate concentration within the culture medium.

3.6. Analytical Techniques

The metal content of the liquid and solid phases of the previous experiments was measured by induction-coupled plasma atomic emission spectroscopy (ICP-AES) of the UCA SC-ICYT (Iris Intrepid, Thermo Scientific, Waltham, MA, USA). Sulfate concentration was measured using the turbidimetric method (4500-SO_4^{2-} E) [79]. Nitrite and nitrate were analyzed by a colorimetric method (4500-NO_2^- B) and an ultraviolet spectrophotometric screening method (4500-NO_3^- B), respectively [79], using a Spectroquant Pharo 300 spectrophotometer (Merck, Darmstadt, Germany). The thiosulfate concentration was measured by iodometric titration [80].

4. Conclusions

In conclusion, this study explored an innovative and environmentally friendly approach for the recovery of PGMs, focusing on the bioleaching of Pd from spent synthetic catalysts. The presented research introduces a novel method involving the utilization of a thiosulfate-copper-ammonia complex, with a unique twist—the incorporation of renewable biogenic thiosulfate obtained from a bioreactor engaged in the biodesulfurization of biogas. To achieve this, the production of biogenic thiosulfate was optimized by studying operational parameters such as pH, electron acceptors, and operation mode. Notably, a transition to the batch mode with aerobic conditions at pH 10 proved most effective for biogenic thiosulfate production. Under the most favorable conditions (batch mode, aerobic conditions, pH = 10, and an airflow rate of 0.033 vvm), a maximum concentration of 4.9 g L^{-1} of biogenic thiosulfate was obtained. However, the biomass growth resulting from the batch operation led to a subsequent decline in thiosulfate levels, indicating the potential utilization of thiosulfate as an energy source. The Pd bioleaching from the ground catalyst was subject to optimization, indicating the ideal conditions for Pd bioleaching to be $Na_2S_2O_3$ 1.2 M, $CuSO_4$ 0.03 M, $(NH_4)_2SO_4$ 1.5 M, Na_2SO_3 0.2 M, pH 8, and a temperature of 60 °C. The feasibility of using biogenic thiosulfate in the Pd bioleaching process was demonstrated achieving extraction rates of 93.2% of the total Pd from the solid. This underscores the viability of the proposed method as a sustainable and effective approach for Pd recovery from spent TWCs, addressing the growing demand for PGMs and contributing to the responsible utilization of critical raw materials in automotive catalysts. Future research should focus on a more detailed exploration of the impact of impurities in real spent automotive three-component catalysts on palladium leaching dynamics. The central aim of upcoming studies is to optimize the recovery of platinum group metals from automotive catalysts, achieving a balance among efficiency, environmental responsibility, and economic feasibility.

Supplementary Materials: The following supporting information can be downloaded at: https://www.mdpi.com/article/10.3390/molecules28248078/s1, Figure S1: Standardized effect size of the studied factors on Pd leaching; Figure S2: Precipitation of Pd after separation of the liquid and the solid phase through filtration with 0.22 µm nitrocellulose membranes (a) in the solid phase and (b) in the liquid phase.

Author Contributions: Conceptualization, J.J.G.-C., M.P.Y., D.C. and M.R.; Formal analysis, M.C.; Funding acquisition, M.R.; Investigation, M.C., J.J.G.-C. and M.P.Y.; Methodology, J.J.G.-C., D.C. and M.R.; Project administration, M.R.; Supervision, J.J.G.-C. and M.R.; Writing—original draft, M.C.; Writing—review and editing, J.J.G.-C., M.P.Y., D.C. and M.R. All authors have read and agreed to the published version of the manuscript.

Funding: This work has been co-financed by the 2014–2020 ERDF Operational Programme and by the Department of Economy, Knowledge, Business and University of the Regional Government of Andalusia. Project reference: FEDER-UCA18-106138. The postdoctoral position at the University of Cadiz for J.J. Gonzalez-Cortes is funded by the Spanish Ministry of Universities through the NextGenerationEU program. The support is provided as a Margarita Salas fellowship (ref. 2021-067/PN/MS-RECUAL/CD).

Institutional Review Board Statement: The study did not require ethical approval.

Informed Consent Statement: Not applicable.

Data Availability Statement: Data will be made available on request.

Conflicts of Interest: The authors declare that they have no known competing financial interests or personal relationships that could have appeared to influence the work reported in this paper.

References

1. Xia, J.; Ghahreman, A. Platinum Group Metals Recycling from Spent Automotive Catalysts: Metallurgical Extraction and Recovery Technologies. *Sep. Purif. Technol.* **2023**, *311*, 123357. [CrossRef]
2. Paiva, A.P. Recycling of Palladium from Spent Catalysts Using Solvent Extraction—Some Critical Points. *Metals* **2017**, *7*, 505. [CrossRef]
3. Dong, H.; Zhao, J.; Chen, J.; Wu, Y.; Li, B. Recovery of Platinum Group Metals from Spent Catalysts: A Review. *Int. J. Miner. Process.* **2015**, *145*, 108–113. [CrossRef]
4. Bloxham, L.; Brown, S.; Cole, L.; Cowley, A.; Fujita, M.; Girardot, N.; Jiang, J.; Raithatha, R.; Ryan, M. PGM Market Report. 2022. Available online: https://matthey.com/pgm-market-report-2022 (accessed on 11 December 2023).
5. European Commission. *Tackling the Challenges in Commodity Markets and on Raw Materials*; European Commission: Brussels, Belgium, 2011.
6. Fajar, A.T.N.; Hanada, T.; Firmansyah, M.L.; Kubota, F.; Goto, M. Selective Separation of Platinum Group Metals via Sequential Transport through Polymer Inclusion Membranes Containing an Ionic Liquid Carrier. *ACS Sustain. Chem. Eng.* **2020**, *8*, 11283–11291. [CrossRef]
7. Grilli, M.L.; Slobozeanu, A.E.; Larosa, C.; Paneva, D.; Yakoumis, I.; Cherkezova-Zheleva, Z. Platinum Group Metals: Green Recovery from Spent Auto-Catalysts and Reuse in New Catalysts—A Review. *Crystals* **2023**, *13*, 550. [CrossRef]
8. Bloxham, L.; Brown, S.; Cole, L.; Cowley, A.; Fujita, M.; Girardot, N.; Jiang, J.; Raithatha, R.; Ryan, M. PGM Market Report: February 2021. 2021. Available online: https://matthey.com/news/2021/pgm-market-report-may-2021-out-now (accessed on 11 December 2023).
9. Thormann, L.; Buchspies, B.; Mbohwa, C.; Kaltschmitt, M. PGE Production in Southern Africa, Part I: Production and Market Trends. *Minerals* **2017**, *7*, 224. [CrossRef]
10. Fornalczyk, A.; Saternus, M. Removal of Platinum Group Metals from the Used Auto Catalytic Converter. *Metalurgija* **2009**, *48*, 133–136.
11. Xu, B.; Chen, Y.; Zhou, Y.; Zhang, B.; Liu, G.; Li, Q.; Yang, Y.; Jiang, T. A Review of Recovery of Palladium from the Spent Automobile Catalysts. *Metals* **2022**, *12*, 533. [CrossRef]
12. Compagnone, M.; González-Cortés, J.J.; del Pilar Yeste, M.; Cantero, D.; Ramírez, M. Bioleaching of the α-Alumina Layer of Spent Three-Way Catalysts as a Pretreatment for the Recovery of Platinum Group Metals. *J. Environ. Manag.* **2023**, *345*, 118825. [CrossRef]
13. Bahaloo-Horeh, N.; Mousavi, S.M. Efficient Extraction of Critical Elements from End-of-Life Automotive Catalytic Converters via Alkaline Pretreatment Followed by Leaching with a Complexing Agent. *J. Clean. Prod.* **2022**, *344*, 131064. [CrossRef]
14. Paiva, A.P.; Piedras, F.V.; Rodrigues, P.G.; Nogueira, C.A. Hydrometallurgical Recovery of Platinum-Group Metals from Spent Auto-Catalysts—Focus on Leaching and Solvent Extraction. *Sep. Purif. Technol.* **2022**, *286*, 120474. [CrossRef]
15. Birloaga, I.; Vegliò, F. An Innovative Hybrid Hydrometallurgical Approach for Precious Metals Recovery from Secondary Resources. *J. Environ. Manag.* **2022**, *307*, 114567. [CrossRef] [PubMed]
16. Chen, J.; Xie, F.; Wang, W.; Fu, Y.; Wang, J. Leaching of Gold and Silver from a Complex Sulfide Concentrate in Copper-Tartrate-Thiosulfate Solutions. *Metals* **2022**, *12*, 1152. [CrossRef]
17. Xu, B.; Yang, Y.; Li, Q.; Yin, W.; Jiang, T.; Li, G. Thiosulfate Leaching of Au, Ag and Pd from a High Sn, Pb and Sb Bearing Decopperized Anode Slime. *Hydrometallurgy* **2016**, *164*, 278–287. [CrossRef]

18. Karim, S.; Ting, Y.P. Recycling Pathways for Platinum Group Metals from Spent Automotive Catalyst: A Review on Conventional Approaches and Bio-Processes. *Resour. Conserv. Recycl.* **2021**, *170*, 105588. [CrossRef]
19. Grosse, A.C.; Dicinoski, G.W.; Shaw, M.J.; Haddad, P.R. Leaching and Recovery of Gold Using Ammoniacal Thiosulfate Leach Liquors (a Review). *Hydrometallurgy* **2003**, *69*, 1–21. [CrossRef]
20. Aylmore, M.G. Thiosulfate as an Alternative Lixiviant to Cyanide for Gold Ores. In *Gold Ore Processing*; Adams, Mike, D., Eds.; Elsevier B.V.: Amsterdam, The Netherlands, 2016; pp. 485–523. ISBN 9780444636584. [CrossRef]
21. Huy Do, M.; Tien Nguyen, G.; Dong Thach, U.; Lee, Y.; Huu Bui, T. Advances in Hydrometallurgical Approaches for Gold Recovery from E-Waste: A Comprehensive Review and Perspectives. *Miner. Eng.* **2023**, *191*, 107977. [CrossRef]
22. Xu, B.; Kong, W.; Li, Q.; Yang, Y.; Jiang, T.; Liu, X. A Review of Thiosulfate Leaching of Gold: Focus on Thiosulfate Consumption and Gold Recovery from Pregnant Solution. *Metals* **2017**, *7*, 222. [CrossRef]
23. Yakoumis, I.; Panou, M.; Moschovi, A.M.; Panias, D. Recovery of Platinum Group Metals from Spent Automotive Catalysts: A Review. *Clean. Eng. Technol.* **2021**, *3*, 100112. [CrossRef]
24. González-Cortés, J.J.; Quijano, G.; Ramírez, M.; Cantero, D. Methane Concentration and Bacterial Communities' Dynamics during the Anoxic Desulfurization of Landfill Biogas under Diverse Nitrate Sources and Hydraulic Residence Times. *J. Environ. Chem. Eng.* **2023**, *11*, 109285. [CrossRef]
25. Almenglo, F.; González-Cortés, J.J.; Ramírez, M.; Cantero, D. Recent Advances in Biological Technologies for Anoxic Biogas Desulfurization. *Chemosphere* **2023**, *321*, 138084. [CrossRef] [PubMed]
26. Van Den Bosch, P.L.F.; Sorokin, D.Y.; Buisman, C.J.N.; Janssen, A.J.H. The Effect of PH on Thiosulfate Formation in a Biotechnological Process for the Removal of Hydrogen Sulfide from Gas Streams. *Environ. Sci. Technol.* **2008**, *42*, 2637–2642. [CrossRef] [PubMed]
27. Sorokin, D.Y.; Kuenen, J.G.; Muyzer, G. The Microbial Sulfur Cycle at Extremely Haloalkaline Conditions of Soda Lakes. *Front. Microbiol.* **2011**, *2*, 44. [CrossRef] [PubMed]
28. Pourhossein, F.; Mousavi, S.M. Improvement of Gold Bioleaching Extraction from Waste Telecommunication Printed Circuit Boards Using Biogenic Thiosulfate by *Acidithiobacillus thiooxidans*. *J. Hazard. Mater.* **2023**, *450*, 131073. [CrossRef] [PubMed]
29. Pourhossein, F.; Mousavi, S.M. A Novel Rapid and Selective Microbially Thiosulfate Bioleaching of Precious Metals from Discarded Telecommunication Printed Circuited Boards (TPCBs). *Resour. Conserv. Recycl.* **2022**, *187*, 106599. [CrossRef]
30. McNeice, J.; Mahandra, H.; Ghahreman, A. Biogenic Production of Thiosulfate from Organic and Inorganic Sulfur Substrates for Application to Gold Leaching. *Sustainability* **2022**, *14*, 16666. [CrossRef]
31. González-Cortés, J.J.; Torres-Herrera, S.; Almenglo, F.; Ramírez, M.; Cantero, D. Anoxic Biogas Biodesulfurization Promoting Elemental Sulfur Production in a Continuous Stirred Tank Bioreactor. *J. Hazard. Mater.* **2021**, *401*, 123785. [CrossRef]
32. Lawson, J. Regression Analysis of Experiments with Complex Confounding Patterns Guided by the Alias Matrix. *Comput. Stat. Data Anal.* **2002**, *39*, 227–241. [CrossRef]
33. Senanayake, G.; Senaputra, A.; Nicol, M.J. Effect of Thiosulfate, Sulfide, Copper(II), Cobalt(II)/(III) and Iron Oxides on the Ammoniacal Carbonate Leaching of Nickel and Ferronickel in the Caron Process. *Hydrometallurgy* **2010**, *105*, 60–68. [CrossRef]
34. Senanayake, G. Gold Leaching by Thiosulphate Solutions: A Critical Review on Copper(II)-Thiosulphate-Oxygen Interactions. *Miner. Eng.* **2005**, *18*, 995–1009. [CrossRef]
35. Yang, Y.; Gao, W.; Xu, B.; Li, Q.; Jiang, T. Study on Oxygen Pressure Thiosulfate Leaching of Gold without the Catalysis of Copper and Ammonia. *Hydrometallurgy* **2019**, *187*, 71–80. [CrossRef]
36. Zhang, X.M.; Senanayake, G.; Nicol, M.J. A Study of the Gold Colloid Dissolution Kinetics in Oxygenated Ammoniacal Thiosulfate Solutions. *Hydrometallurgy* **2004**, *74*, 243–257. [CrossRef]
37. Xia, C.; Yen, W.T.; Deschenes, G. Improvement of Thiosulfate Stability in Gold Leaching. *Miner. Metall. Process.* **2003**, *20*, 68–72. [CrossRef]
38. Junior, F.; Pedrosa, B.; Mano, E.S.; Neto, D. Using Thiosulphate as Gold Leaching Agent in a Brazilian Carbonaceous Ore: Batch Tests Analysis. In Proceedings of the 12th International Conference on Process Hydrometallurgy, Santiago, Chile, 28–30 October 2020.
39. Navarro, P.; Vargas, C.; Villarroel, A.; Alguacil, F.J. On the Use of Ammoniacal/Ammonium Thiosulphate for Gold Extraction from a Concentrate. *Hydrometallurgy* **2002**, *65*, 37–42. [CrossRef]
40. Azaroual, M.; Romand, B.; Freyssinet, P.; Disnar, J.; Azaroual, M.; Romand, B.; Freyssinet, P.; Solubility, J.D. Solubility of Platinum in Aqueous Solutions at 25 °C and PHs 4 to 10 under Oxidizing Conditions. *Geochim. Cosmochim. Acta* **2001**, *65*, 4453–4466. [CrossRef]
41. Chen, Y.; Shang, H.; Ren, Y.; Yue, Y.; Li, H.; Bian, Z. Systematic Assessment of Precious Metal Recovery to Improve Environmental and Resource Protection. *ACS EST Eng.* **2022**, *2*, 1039–1052. [CrossRef]
42. Dong, Z.; Jiang, T.; Xu, B.; Zhang, B.; Liu, G.; Li, Q.; Yang, Y. A Systematic and Comparative Study of Copper, Nickel and Cobalt-Ammonia Catalyzed Thiosulfate Processes for Eco-Friendly and Efficient Gold Extraction from an Oxide Gold Concentrate. *Sep. Purif. Technol.* **2021**, *272*, 118929. [CrossRef]
43. Gui, Q.; Khan, M.I.; Wang, S.; Zhang, L. The Ultrasound Leaching Kinetics of Gold in the Thiosulfate Leaching Process Catalysed by Cobalt Ammonia. *Hydrometallurgy* **2020**, *196*, 105426. [CrossRef]
44. Wan, R.Y.; LeVier, K.M. Solution Chemistry Factors for Gold Thiosulfate Heap Leaching. *Int. J. Miner. Process.* **2003**, *72*, 311–322. [CrossRef]

45. Pietrelli, L.; Fontana, D. Automotive Spent Catalysts Treatment and Platinum Recovery. *Int. J. Environ. Waste Manag.* **2013**, *11*, 222–232. [CrossRef]
46. Chen, J.; Huang, K. A New Technique for Extraction of Platinum Group Metals by Pressure Cyanidation. *Hydrometallurgy* **2006**, *82*, 164–171. [CrossRef]
47. Shams, K.; Beiggy, M.R.; Shirazi, A.G. Platinum Recovery from a Spent Industrial Dehydrogenation Catalyst Using Cyanide Leaching Followed by Ion Exchange. *Appl. Catal. A Gen.* **2004**, *258*, 227–234. [CrossRef]
48. Sun, S.; Jin, C.; He, W.; Li, G.; Zhu, H.; Huang, J. A Review on Management of Waste Three-Way Catalysts and Strategies for Recovery of Platinum Group Metals from Them. *J. Environ. Manag.* **2022**, *305*, 114383. [CrossRef]
49. Mpinga, C.N.; Eksteen, J.J.; Aldrich, C.; Dyer, L. Direct Leach Approaches to Platinum Group Metal (PGM) Ores and Concentrates: A Review. *Miner. Eng.* **2015**, *78*, 93–113. [CrossRef]
50. Li, N.; Liu, T.; Xiao, S.; Yin, W.; Zhang, L.; Chen, J.; Wang, Y.; Zhou, X.; Zhang, Y. Thiosulfate Enhanced Cu(II)-Catalyzed Fenton-like Reaction at Neutral Condition: Critical Role of Sulfidation in Copper Cycle and Cu(III) Production. *J. Hazard. Mater.* **2023**, *445*, 130536. [CrossRef] [PubMed]
51. Han, C.; Wang, G.; Zou, M.; Shi, C. Separation of Ag and Cu from Their Aqueous Thiosulfate Complexes by UV-C Irradiation. *Metals* **2019**, *9*, 1178. [CrossRef]
52. Liu, X.; Xu, B.; Min, X.; Li, Q.; Yang, Y.; Jiang, T.; He, Y.; Zhang, X. Effect of Pyrite on Thiosulfate Leaching of Gold and the Role of Ammonium Alcohol Polyvinyl Phosphate (AAPP). *Metals* **2017**, *7*, 278. [CrossRef]
53. Alvarado-Macías, G.; Fuentes-Aceituno, J.C.; Nava-Alonso, F. Silver Leaching with the Thiosulfate-Nitrite-Sulfite-Copper Alternative System. *Hydrometallurgy* **2015**, *152*, 120–128. [CrossRef]
54. Torres, R.; Lapidus, G.T. Platinum, Palladium and Gold Leaching from Magnetite Ore, with Concentrated Chloride Solutions and Ozone. *Hydrometallurgy* **2016**, *166*, 185–194. [CrossRef]
55. Bax, A.; Dunn, G.M.; Lewins, J.D. Recovery of Platinum Group Metals. U.S. Patent No. 7544231, 9 June 2009.
56. Ilyas, S.; Kim, H. Recovery of Platinum-Group Metals from an Unconventional Source of Catalytic Converter Using Pressure Cyanide Leaching and Ionic Liquid Extraction. *JOM* **2022**, *74*, 1020–1026. [CrossRef]
57. Mwase, J.M.; Petersen, J.; Eksteen, J.J. A Conceptual Flowsheet for Heap Leaching of Platinum Group Metals (PGMs) from a Low-Grade Ore Concentrate. *Hydrometallurgy* **2012**, *111–112*, 129–135. [CrossRef]
58. Fotoohi, B.; Mercier, L. Recovery of Precious Metals from Ammoniacal Thiosulfate Solutions by Hybrid Mesoporous Silica: 2—A Prospect of PGM Adsorption. *Sep. Purif. Technol.* **2015**, *149*, 82–91. [CrossRef]
59. Fotoohi, B.; Mercier, L. Recovery of Precious Metals from Ammoniacal Thiosulfate Solutions by Hybrid Mesoporous Silica: 3—Effect of Contaminants. *Sep. Purif. Technol.* **2015**, *139*, 14–24. [CrossRef]
60. Bao, H.X.; Li, Z.R.; Song, Z.B.; Wang, A.J.; Zhang, X.N.; Qian, Z.M.; Sun, Y.L.; Cheng, H.Y. Mitigating Nitrite Accumulation during S^0-Based Autotrophic Denitrification: Balancing Nitrate-Nitrite Reduction Rate with Thiosulfate as External Electron Donor. *Environ. Res.* **2022**, *204*, 112016. [CrossRef] [PubMed]
61. Qian, J.; Zhou, J.; Zhang, Z.; Liu, R.; Wang, Q. Biological Nitrogen Removal through Nitritation Coupled with Thiosulfate-Driven Denitritation. *Sci. Rep.* **2016**, *6*, 27502. [CrossRef] [PubMed]
62. Sorokin, D.Y.; Van Den Bosch, P.L.F.; Abbas, B.; Janssen, A.J.H.; Muyzer, G. Microbiological Analysis of the Population of Extremely Haloalkaliphilic Sulfur-Oxidizing Bacteria Dominating in Lab-Scale Sulfide-Removing Bioreactors. *Appl. Microbiol. Biotechnol.* **2008**, *80*, 965–975. [CrossRef] [PubMed]
63. Sousa, J.A.B.; Bijmans, M.F.M.; Stams, A.J.M.; Plugge, C.M. Thiosulfate Conversion to Sulfide by a Haloalkaliphilic Microbial Community in a Bioreactor Fed with H_2 Gas. *Environ. Sci. Technol.* **2017**, *51*, 914–923. [CrossRef]
64. Qian, J.; Zhang, M.; Jing, R.; Bai, L.; Zhou, B.; Zhao, M.; Pei, X.; Wei, L.; Chen, G.H. Thiosulfate as the Electron Acceptor in Sulfur Bioconversion-Associated Process (SBAP) for Sewage Treatment. *Water Res.* **2019**, *163*, 114850. [CrossRef]
65. Van Den Bosch, P.L.F. Biological Sulfide Oxidation by Natron-Alkaliphilic Bacteria: Application in Gas Desulfurization. Ph.D. Thesis, Wageningen University, Wageningen, The Netherlands, 2008. ISBN 978-3-540-44855-6.
66. Kleinjan, W.E.; de Keizer, A.; Janssen, A.J.H. Biologically Produced Sulfur. In *Elemental Sulfur and Sulfur-Rich Compounds I*; Steudel, R., Ed.; Springer: Berlin/Heidelberg, Germany, 2003; pp. 167–188. ISBN 978-3-540-44855-6.
67. Hedderich, R.; Klimmek, O.; Kröger, A.; Dirmeier, R.; Keller, M.; Stetter, K.O. Anaerobic Respiration with Elemental Sulfur and with Disulfides. *FEMS Microbiol. Rev.* **1998**, *22*, 353–381. [CrossRef]
68. Chen, K.Y.; Morris, J.C. Kinetics of Oxidation of Aqueous Sulfide by O_2. *Environ. Sci. Technol.* **1972**, *6*, 529–537. [CrossRef]
69. Xolo, L.; Moleko-Boyce, P.; Makelane, H.; Faleni, N.; Tshentu, Z.R. Status of Recovery of Strategic Metals from Spent Secondary Products. *Minerals* **2021**, *11*, 673. [CrossRef]
70. Nguyen, V.T.; Riaño, S.; Aktan, E.; Deferm, C.; Fransaer, J.; Binnemans, K. Solvometallurgical Recovery of Platinum Group Metals from Spent Automotive Catalysts. *ACS Sustain. Chem. Eng.* **2021**, *9*, 337–350. [CrossRef]
71. Trinh, H.B.; Lee, J.C.; Srivastava, R.R.; Kim, S.; Ilyas, S. Eco-Threat Minimization in HCl Leaching of Pgms from Spent Automobile Catalysts by Formic Acid Prereduction. *ACS Sustain. Chem. Eng.* **2017**, *5*, 7302–7309. [CrossRef]
72. Saguru, C.; Ndlovu, S.; Moropeng, D. A Review of Recent Studies into Hydrometallurgical Methods for Recovering PGMs from Used Catalytic Converters. *Hydrometallurgy* **2018**, *182*, 44–56. [CrossRef]

73. Lanaridi, O.; Platzer, S.; Nischkauer, W.; Betanzos, J.H.; Iturbe, A.U.; Del Rio Gaztelurrutia, C.; Sanchez-Cupido, L.; Siriwardana, A.; Schnürch, M.; Limbeck, A.; et al. Benign Recovery of Platinum Group Metals from Spent Automotive Catalysts Using Choline-Based Deep Eutectic Solvents. *Green Chem. Lett. Rev.* **2022**, *15*, 404–414. [CrossRef]
74. Canda, L.; Heput, T.; Ardelean, E. Methods for Recovering Precious Metals from Industrial Waste. *IOP Conf. Ser. Mater. Sci. Eng.* **2016**, *106*, 012020. [CrossRef]
75. Grad, O.; Ciopec, M.; Negrea, A.; Duțeanu, N.; Vlase, G.; Negrea, P.; Dumitrescu, C.; Vlase, T.; Vodă, R. Precious Metals Recovery from Aqueous Solutions Using a New Adsorbent Material. *Sci. Rep.* **2021**, *11*, 2016. [CrossRef]
76. Adeeyo, A.O.; Bello, O.S.; Agboola, O.S.; Adeeyo, R.O.; Oyetade, J.A.; Alabi, M.A.; Edokpayi, J.N.; Makungo, R. Recovery of Precious Metals from Processed Wastewater: Conventional Techniques Nexus Advanced and Pragmatic Alternatives. *Water Reuse* **2023**, *13*, 134–161. [CrossRef]
77. Huang, Z.; Zhao, M.; Wang, S.; Dai, L.; Zhang, L.; Wang, C. Selective Recovery of Gold Ions in Aqueous Solutions by a Novel Trithiocyanuric-Zr Based MOFs Adsorbent. *J. Mol. Liq.* **2020**, *298*, 112090. [CrossRef]
78. Fernández, M.; Ramírez, M.; Gómez, J.M.; Cantero, D. Biogas Biodesulfurization in an Anoxic Biotrickling Filter Packed with Open-Pore Polyurethane Foam. *J. Hazard. Mater.* **2014**, *264*, 529–535. [CrossRef]
79. Clesceri, L.S.; Greenberg, A.E.; Eaton, A.D. *Standard Methods for the Examination of Water and Waste Water*, 20th ed.; APHA: Washington, DC, USA, 1999.
80. Rowley, K.; Swift, E.H. Goulometric Titration of Thiosulfate with Iodine: Application to the Determination of Oxidizing Agents. *Anal. Chem.* **1954**, *26*, 373–375. [CrossRef]

molecules

Article

Pyrolysis of Solid Recovered Fuel Using Fixed and Fluidized Bed Reactors

Myeongjong Lee, Hyeongtak Ko and Seacheon Oh *

Department of Environmental Engineering, Kongju National University, 1223-24 Cheonan-Daero, Seobuk, Cheonan 31080, Chungcheongnam-do, Republic of Korea; dlaudwhd77@naver.com (M.L.); kht3113@naver.com (H.K.)
* Correspondence: ohsec@kongju.ac.kr; Tel.: +82-41-521-9423

Abstract: Currently, most plastic waste stems from packaging materials, with a large proportion of this waste either discarded by incineration or used to derive fuel. Accordingly, there is growing interest in the use of pyrolysis to chemically recycle non-recyclable (i.e., via mechanical means) plastic waste into petrochemical feedstock. This comparative study compared pyrolysis characteristics of two types of reactors, namely fixed and fluidized bed reactors. Kinetic analysis for pyrolysis of SRF was also performed. Based on the kinetic analysis of the pyrolytic reactions using differential and integral methods applied to the TGA results, it was seen that the activation energy was lower in the initial stage of pyrolysis. This trend can be mainly attributed to the initial decomposition of PP components, which was subsequently followed by the decomposition of PE. From the kinetic analysis, the activation energy corresponding to the rate of pyrolysis reaction conversion was obtained. In conclusion, pyrolysis carried out using the fluidized bed reactor resulted in a more active decomposition of SRF. The relatively superior performance of this reactor can be attributed to the increased mass and heat transfer effects caused by fluidizing gases, which result in greater gas yields. Regarding the characteristics of liquid products generated during pyrolysis, it was seen that the hydrogen content in the liquid products obtained from the fluidized bed reactor decreased, leading to the formation of oils with higher molecular weights and higher C/H ratios, because the pyrolysis of SRF in the fluidized bed reactor progressed more rapidly than that in the fixed bed reactor.

Keywords: pyrolysis; solid recovered fuel; fixed bed reactor; fluidized bed reactor; kinetic analysis

Citation: Lee, M.; Ko, H.; Oh, S. Pyrolysis of Solid Recovered Fuel Using Fixed and Fluidized Bed Reactors. *Molecules* **2023**, *28*, 7815. https://doi.org/10.3390/molecules28237815

Academic Editor: Mohamad Nasir Mohamad Ibrahim

Received: 25 October 2023
Revised: 24 November 2023
Accepted: 25 November 2023
Published: 28 November 2023

1. Introduction

The past few decades have seen a continuous increase in the production and demand for plastics. This trend is attributable not only to the diverse physiochemical properties of plastics, which allow for their wide applicability, but also to their low cost. However, this surge in plastic usage has driven a corresponding increase in plastic waste. This finding is reflected in the fact that plastic constitutes more than 50% of waste produced by the average household [1]. Therefore, in light of the long-term persistence of plastic in the environment, recycling plastic waste has been highlighted as an important factor in the large-scale establishment of a circular economy and carbon neutrality [2]. In terms of the specific underlying plastic waste, approximately 40% of all plastic waste is generated from packaging materials, and approximately 60% of this waste is channeled toward energy recovery or discarded [3]. Unlike biomass, plastic waste discarded in landfills does not biodegrade; instead, this plastic undergoes decomposition that can occur for many centuries. In addition, this decomposition does not occur uneventfully, as it has been linked to the accumulation of wide-reaching pollutants in landfills [4,5]. Plastics mainly consist of LDPE (low-density polyethylene), HDPE (low-density polyethylene), PP (polypropylene), PS (polystyrene), PET (polyethylene terephthalate), and PVC (polyvinyl chloride). Among these constituents, PE, PP, and PS constitute 50–70% of most plastics [6,7]. For recycling of plastic wastes, this waste is supposed to be first sorted on the basis of its physiochemical makeup. However, considering the mixed nature of plastic waste, the processes currently used to sort it have numerous limitations; this is a serious issue, especially considering the fact that the efficiency of mechanical recycling for material reuse is dependent on the proper sorting and evaluation of plastic waste [8]. In a situation in which the sorting or separation of already heterogeneous plastic waste is not optimal, another layer of inefficiency is introduced in the already disorganized, wasteful plastic recycling industry. Additionally, mechanical recycling is typically accompanied by the deterioration of plastic properties, so alternative methods to conventional mechanical recycling must be considered, especially for ensuring the sustainable use of plastics in a circular economy [9]. Plastics consist of a wide range of hydrocarbons; this makeup translates into plastics hosting a lot of recoverable chemical energy [6,10]. Thus, chemical recycling through pyrolysis is garnering attention as an important alternative to recycling plastic waste into petrochemical feedstock. Pyrolysis is a reaction that entails the decomposition of high-molecular-weight compounds with long-chain structures into low molecular weight compounds. This process occurs under heating conditions in an oxygen-free atmosphere, with its yields being oil, non-condensable gas and solid residue [11]. Pyrolysis addresses a major drawback of traditional mechanical recycling—its inability to enable continuous recycling—by allowing for the recovery of otherwise non-recyclable waste plastics [12]. It has also been shown that the calorific value of pyrolysis-derived oil produced from plastic waste is comparable to that of conventional diesel fuel, allowing it to serve as a robust replacement. Moreover, the materials generated post-pyrolysis can be reused in existing petrochemical processes, making them an excellent alternative in the energy market [13–18].

In South Korea, a specific recycling rate is mandated for plastic packaging materials by the extended producer responsibility (EPR) system. Accordingly, this waste is either mechanically recycled or used as solid recovered fuel (SRF). In other words, waste plastics that can be mechanically sorted by type are recycled as materials, and those that cannot be mechanically sorted are converted into SRF and recycled as fuel for energy recovery. In addition, SRF is managed by the standards on the heating value, moisture content, ash content, chlorine and sulfur contents, and heavy metal contents (mercury, cadmium, lead, and arsenic) based on the Korean Waste Management Regulation. In this study, an investigation was conducted on the pyrolysis of SRF for chemical recycling to change SRF into plastics or materials that can be used as the raw materials of other products. To achieve this aim, we used both batch fixed bed and continuous fluidized bed reactors. We then comprehensively compared its yield and pyrolytic properties through the two types of reactors. Additionally, using kinetic analyses, we investigated the activation energy required for the pyrolysis of SRF.

2. Results and Discussion

2.1. Kinetic Analysis

Figure 1 shows the TG (thermogravimetric) and DTG (derivative thermogravimetric) curves, illustrating the heating rates under a nitrogen atmosphere for kinetic analysis of the pyrolysis of the SRF. The TG curve in Figure 1a shows that the decomposition temperature rose as the heating rate increased. This occurred likely due to thermal transfer lag caused by the increased heating rate. Additionally, Figure 1b's DTG curve shows more than two peaks, confirming that the SRF used in this study was a heterogeneous mixture of different plastics. The SRF used in this study was primarily made from plastic packaging waste composed of PP and PE. Generally, PP begins to undergo pyrolysis at 400 °C, whereas PE is pyrolyzed at a higher temperature [19,20]. Therefore, the peak (i.e., at ~400 °C) in the DTG curve is likely due to the pyrolysis of PP, whereas the subsequent peak (at 450–500 °C) is likely attributable to PE. The activation energies for each conversion rate were determined by applying the differential method described in Equation (2), and these are presented in Figure 2. Figure 2b shows that the activation energy increased (from 59.9 to 116.3 kJ/mol) as the conversion rate increased to 0.35, after which it remained fairly constant (between 112.9

and 131.3 kJ/mol). This trend can be attributed to the SRF sample used in this study being a composite of plastics mostly made of PE and PP; the initial conversion rates were primarily due to PP, whereas the subsequent ones were a manifestation of the pyrolysis of PE.

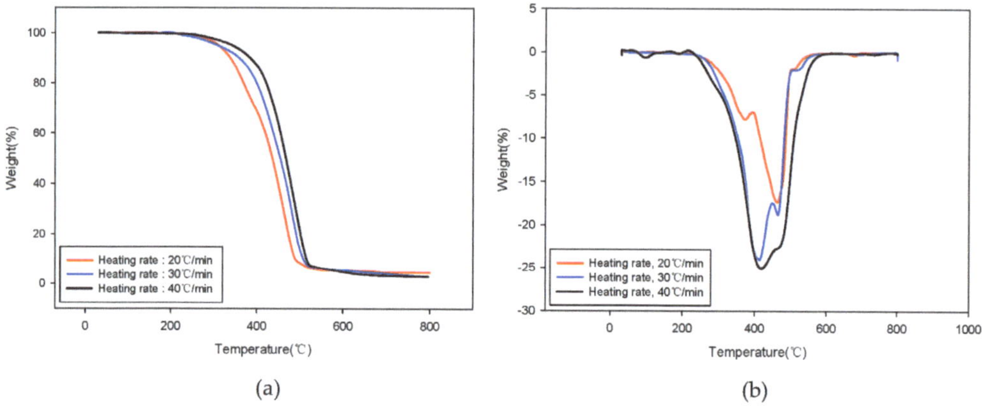

Figure 1. TG (**a**) and DTG (**b**) curves of SRF across various heating rates in pure nitrogen.

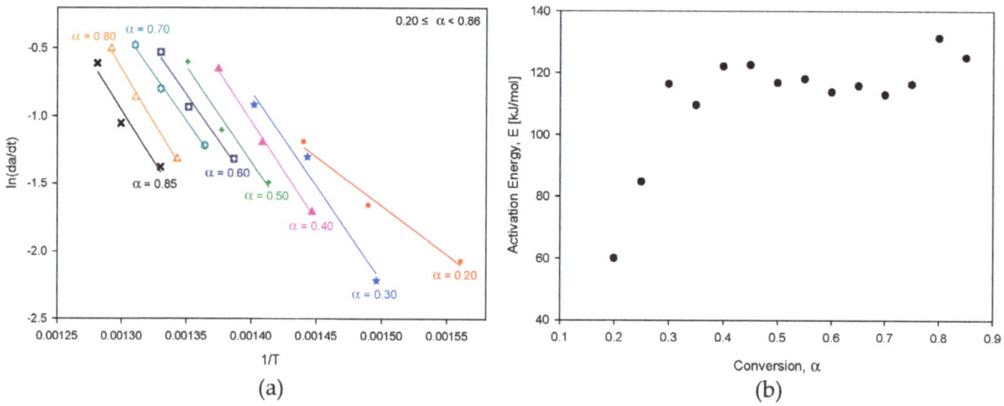

Figure 2. Application of differential method (**a**) and activation energies (**b**) over a range of conversion.

Figure 3 shows the outputs of the integral methods, encapsulated in Equations (10) and (11); these outputs represent activation energies across various heating rates. The range of conversion rates for obtaining activation energy at each heating rate in Figure 3 was set based on the peaks observed in the DTG analysis in Figure 1b, and was applied in two stages. Figure 3 shows that when using the integral method, the activation energies appeared to be lower than those elucidated via the differential method (Figure 2). However, the tendency for the activation energy to be lower in the initial stages of pyrolytic conversion was also observed in results obtained via the differential method.

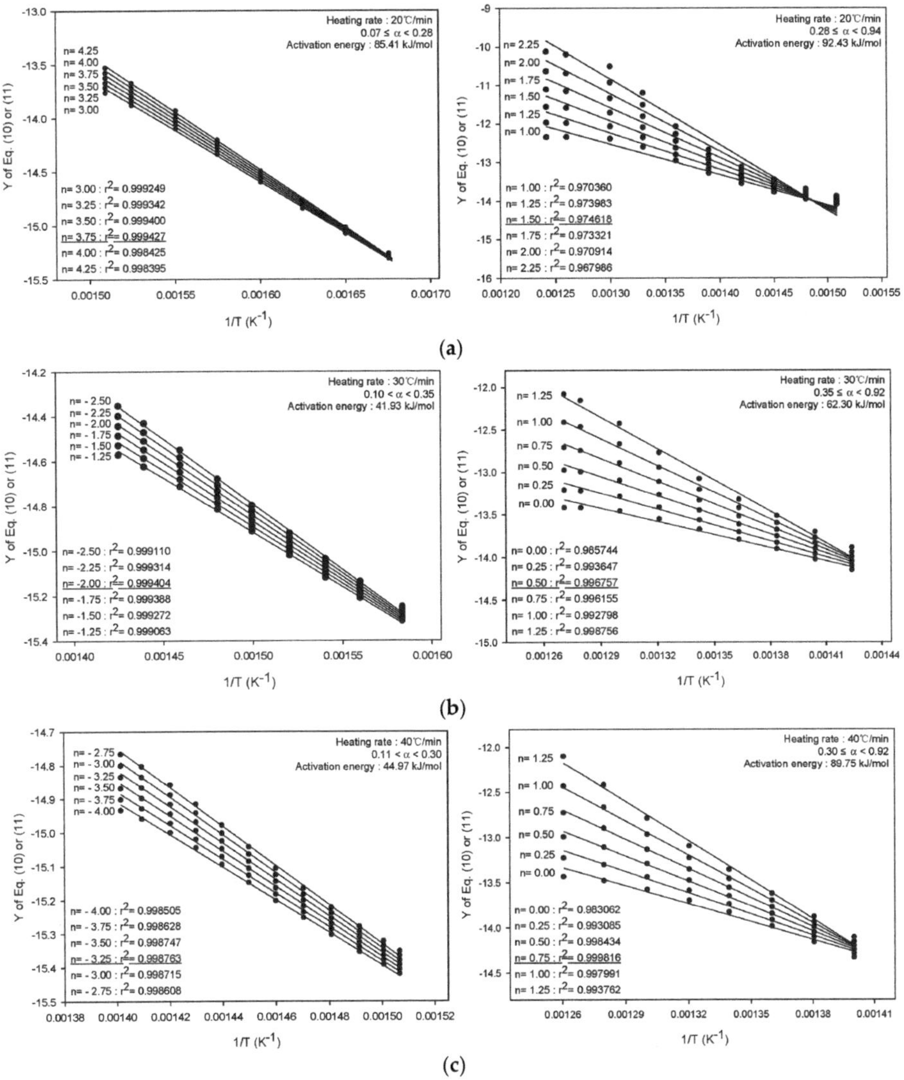

Figure 3. Activation energies at various heating rates of 20 °C/min (**a**), 30 °C/min (**b**), and 40 °C/min (**c**), as determined using the integral method.

Table 1 shows a comparison of the activation energies for the pyrolysis of SRF based on the different kinetic methods applied. We found that in the context of the pyrolysis reactions of PP and PE, the activation energy for the pyrolysis of PP appeared to be lower than that of PE [21]. Therefore, although the different kinetic analysis methods brought forth different activation energies, the universal trend was as follows: PP was the first to decompose, followed by PE in subsequent stages. This confirmed the validity of the kinetic analysis method used, as the activation energy for the initial pyrolysis of PP is lower than that for the later stages of PE decomposition. Based on the analysis results of the kinetic analysis method applied in this study, the differential approach that uses multiple heating rates is judged to be more useful than the integral approach that uses the TGA results of the single heating rate, because it can examine the change in activation energy according to the pyrolysis conversion rate.

Table 1. Activation energy of pyrolysis of SRF used in this work.

Differential Method		Integral Method		
Conversion (α)	**Activation Energy (kJ/mol)**	**Heating Rate**	**Conversion (α)**	**Activation Energy (kJ/mol)**
0.20 ≤ a < 0.40	59.94~116.32	20 °C/min	0.07 ≤ a < 0.28	85.41
			0.28 ≤ a < 0.94	92.43
		30 °C/min	0.10 ≤ a < 0.35	41.93
0.40 ≤ a < 0.85	112.47~131.32		0.35 ≤ a < 0.92	62.30
		40 °C/min	0.11 ≤ a < 0.30	44.97
			0.30 ≤ a < 0.92	89.75

2.2. Product Analysis

2.2.1. Fixed Bed Reactor

Figure 4 shows the changes in yield for each product based on the pyrolysis reaction temperature when SRF was pyrolyzed using a fixed bed reactor. As the pyrolysis reaction temperature increased, the yield of liquid products such as heavy oil and light oil decreased (Figure 4). This is likely because the gasification reactions became more active, increasing the yield of gaseous products. Additionally, even though the yield of the solid residue slightly decreased when the reaction temperature exceeded 550 °C, the temperature-dependent changes it underwent were minimal.

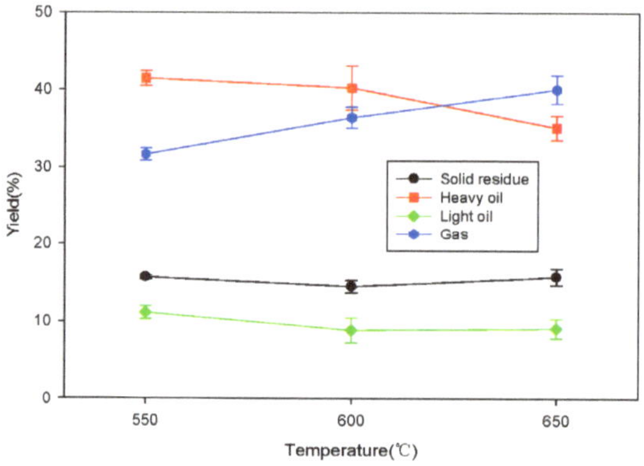

Figure 4. Product yields of SRF pyrolysis using fixed bed reactor across different temperatures.

Figure 5 shows the characteristics of the carbon content of the liquid products obtained from GC-MS analysis of SRF pyrolysis using a fixed bed reactor. Figure 5 shows that the heavy oil mostly consisted of components with carbon numbers greater than C21, culminating in the impact of the pyrolysis reaction temperature on its composition being minimal. The heavy oil also contained components in the C5–C10 range, suggesting that more refined condensation could potentially increase the yield of light oil. However, the light oil mostly consisted of components in the C7 –C8 range. As the reaction temperature increased, there was a noticeable increase in lower-molecular-weight components, especially at 650 °C, culminating in all the components having carbon numbers of C8 or less. This is likely because the pyrolysis becomes more intense as the reaction temperature increases. However, temperatures that are too high can decrease the yield of liquid products due to the corresponding increase in the yield of gaseous products. Therefore, selecting the optimal pyrolysis reaction temperature for SRF should entail considering the yield changes for each product in Figure 4.

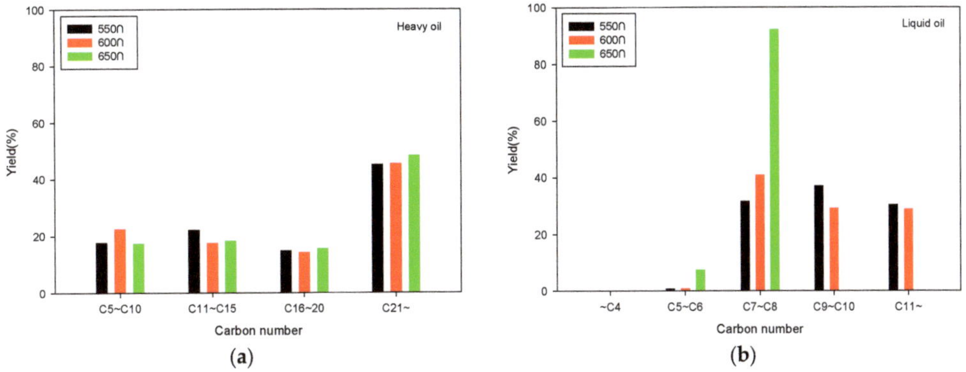

Figure 5. Compounds according to the carbon number of heavy (**a**) and light oil (**b**) obtained from fixed bed reactor.

Figure 6 shows the gas chromatogram for liquid products at a reaction temperature of 600 °C. Heavy oil evidently has components whose heterogeneity is greater than those of light oil.

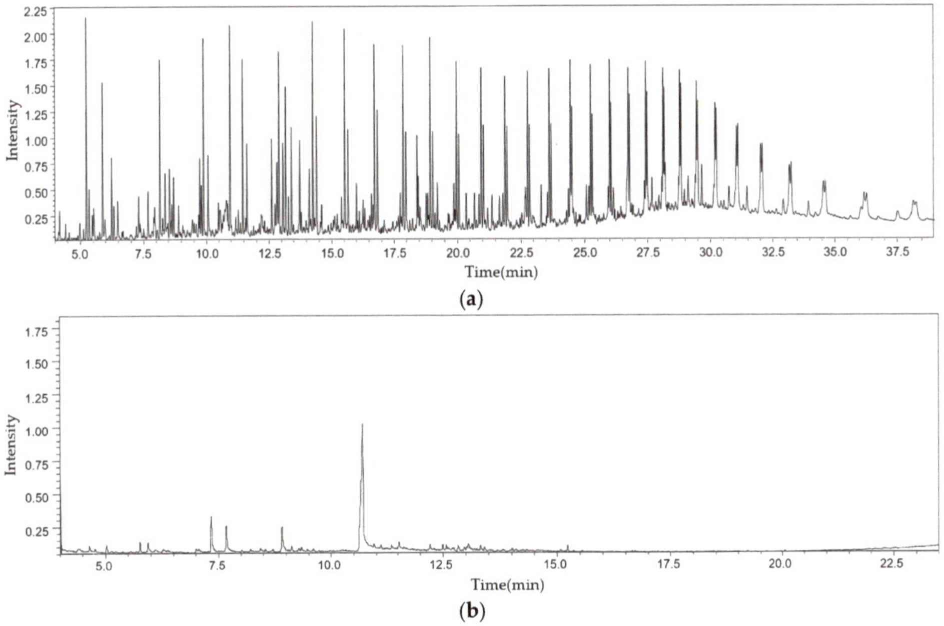

Figure 6. Gas chromatogram of heavy (**a**) and light oil (**b**) obtained from a fixed bed reactor at a reaction temperature of 600 °C.

In addition, the higher heating values of the light oil and heavy oil generated from the fixed bed reactor were found to be 9502.0 to 11,395.5 kcal/kg and 6868.2 to 8164.4 kcal/kg, respectively, indicating that the higher heating value of the light oil was higher than that of the heavy oil. For the gas products released through the pyrolysis of waste plastics, they are hydrocarbon compounds with carbon numbers of C4 or less according to several studies [22–24]. In the case of the characteristics of the solid residues obtained from the pyrolysis of the fixed bed reactor in this study, the higher heating value ranged from 4570.8 to 5421.2 kcal/kg, as shown in Table 2, and the elemental analysis results confirmed that the content of residual carbon was not low.

Table 2. Properties of solid residues obtained from fixed bed reactor.

Elements (wt%)	C	42.17–46.39
	H	1.41–2.39
	N	0.90–1.16
	O	4.18–7.63
	S	0
	Other	45.44–51.46
Higher heating (kcal/kg)	4570.8–5421.2	

2.2.2. Fluidized Bed Reactor

Figure 7 shows the changes in yield for each product based on the pyrolysis reaction temperature and the fluidized gas flow rate when SRF was pyrolyzed using a fluidized bed reactor.

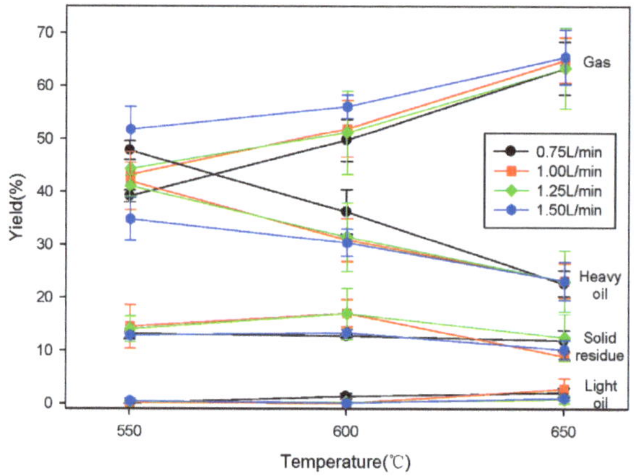

Figure 7. Product yields of SRF pyrolysis using fluidized bed reactor according to temperatures and fluidized gas flow rate.

Figure 7 shows that compared to results obtained via a fixed bed reactor, as the reaction temperature increased, the yield of gaseous products significantly increased, substantially decreasing the yield of heavy oil. Fluidized bed reactions are known to enhance mass and heat transfer compared to fixed bed reactions in the pyrolysis of waste plastics [25]. This trend was further confirmed by the increase in gaseous products as the fluidizing gas velocity increased. Additionally, compared to the fixed bed reactor, the fluidized reactor produced a greater yield of gaseous products; this is because fluidized reactors inherently facilitate decomposition (i.e., into smaller molecular weight components) that is more intensive than that seen in fixed bed reactors. However, in the case of light oil, while a slight decrease was observed as the reaction temperature increased in fixed bed reactions, an increase was noted in fluidized bed reactions. Therefore, as previously mentioned, we confirmed that SRF decomposed into lower-molecular-weight components more readily in fluidized bed reactions compared to fixed bed reactions. For the solid residues, temperature-driven changes were nearly negligible, similar to those in fixed bed reactions.

Figure 8 shows the characteristics of the carbon content of the liquid products obtained from GC-MS analysis of SRF pyrolysis using a fluidized bed reactor. Figure 8 suggests that, similar to fixed bed reactions, components with a carbon number of C21 and above were most prevalent in the heavy oil. However, the proportion of these high-carbon components was higher in fluidized bed reactions, especially at 650 °C, where they were produced in substantial amounts. This could be due to the pyrolysis of SRF in fluidized bed reactors progressing more rapidly than that in fixed bed reactors. As a result, the hydrogen content in the liquid products decreases, leading to the formation of oils with higher molecular weights, and higher C/H ratios. Additionally, in the case of light oil, components ranging from C7–C8 predominated, similar to what we saw in fixed bed reactions. We observed that lower-molecular-weight components increased as the reaction temperature and fluidizing gas velocity increased.

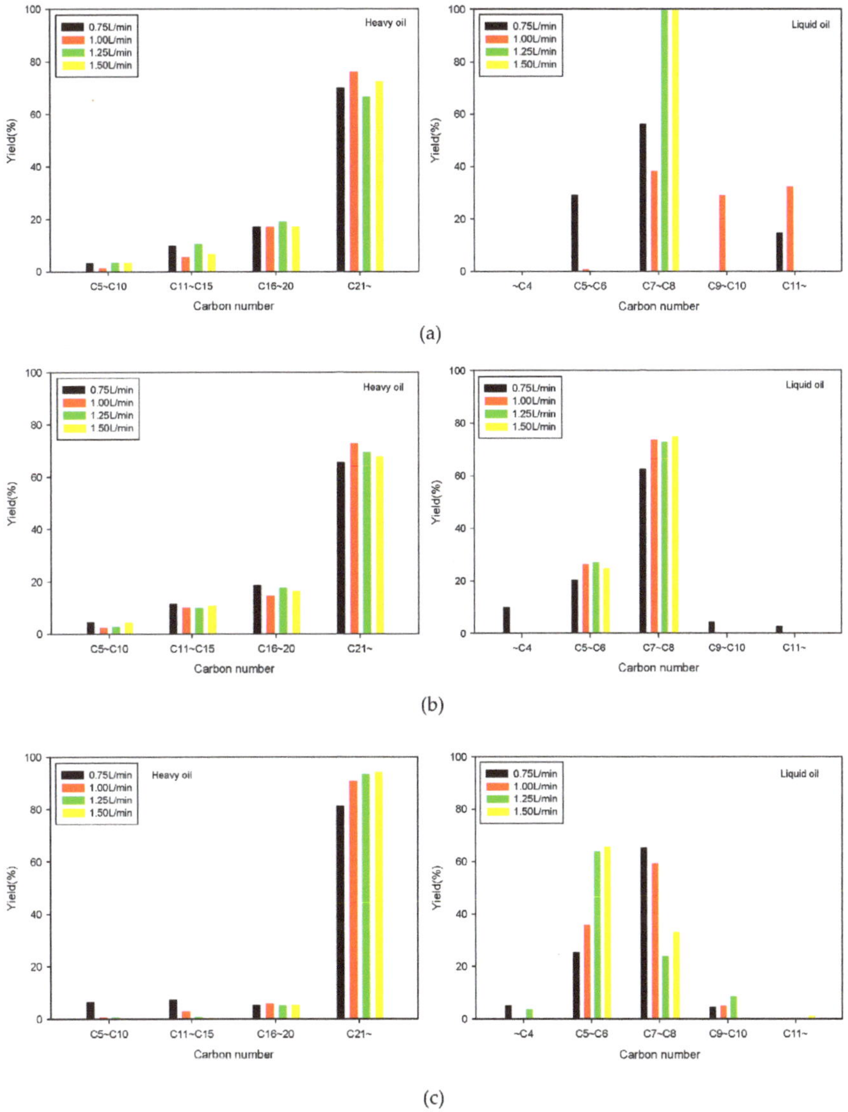

Figure 8. Compounds according to the carbon number of heavy and light oil obtained from fluidized bed reactor at reaction temperatures of 550 (**a**), 600 (**b**), and 650 °C (**c**).

Figure 9 shows the gas chromatograms of heavy and light oil obtained at a reaction temperature of 600 °C and a fluidizing gas flow rate of 1.0 L/min. This trend shows that the composition of heavy oil was more diverse than that of light oil, consistent with the results of the fixed bed reactor.

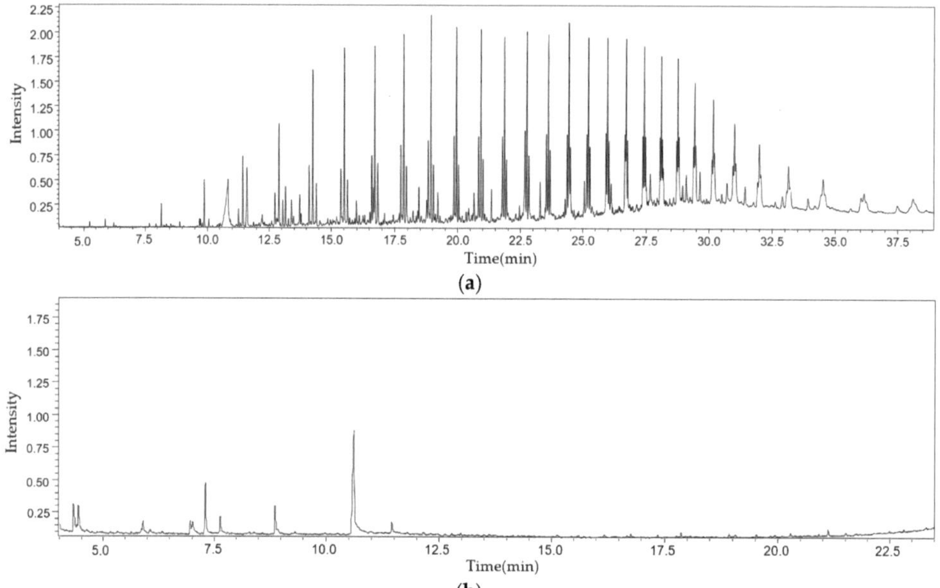

Figure 9. Gas chromatogram of heavy (**a**) and light oil (**b**) obtained from a fluidized bed reactor at a reaction temperature of 600 °C and a gas flow rate of 1.0 L/min.

In addition, the higher heating values of the light oil and heavy oil generated from the fluidized bed reactor were found to range from 9411.5 to 10,271.4 kcal/kg and 6702.4 to 7736.6 kcal/kg, respectively, and the higher heating value of the light oil was also higher than that of the heavy oil, as in the case of the fixed bed reactor. Table 3 shows the characteristics of the solid residues obtained from the pyrolysis of the fluidized bed reactor in this study. It was found that the higher heating value ranged from 4853.6 to 5836.2 kcal/kg, and that the content of residual carbon was not low as in the case of the fixed bed reactor.

Table 3. Properties of solid residues obtained from fluidized bed reactor.

Elements (wt%)	C	40.40–44.16
	H	1.95–3.31
	N	0.94–1.34
	O	6.41–9.75
	S	0
	Other	45.87–46.53
Higher heating (kcal/kg)	4853.6–5836.6	

3. Materials and Methods

3.1. Materials

In this study, SRF produced from plastic packaging waste, such as PP and PE, was mainly used as a pyrolysis sample. The characteristics of the SRF used in this study are shown in Table 4. The composition of SRF was then analyzed using an elemental analyzer (EA) (Flash 2000, UK). As can be seen from the table, the main components of the SRF used in this study were hydrogen and carbon as it was mostly composed of plastics, and it is judged that the content of PET was not low considering the oxygen content. The presence of nitrogen appeared to be due to the ABS (acrylonitrile butadiene styrene) content. The fact that most of the packaging used was made of plastic is reflected in the high amount of volatile matter (i.e., 90.6%) that was detected, while the fixed carbon and ash were recorded at 6.07% and 3.33%, respectively.

Table 4. Properties of SRF used in this work.

Elements (wt%)	C	63.1
	H	9.8
	N	0.6
	O	16.4
	S	0
	Other	10.1
Volatile (%)	90.60	
Fixed carbon (%)	6.07	
Ash (%)	3.33	
Higher heating value (kcal/kg)	8871.7	

3.2. Experimental Methods

The schematic diagrams of the fixed bed and fluidized bed reactors used in this study's pyrolysis experiments are shown in Figure 10. The fixed bed reactor (Figure 10a) consisted of a quartz tube with a diameter and length of 60 and 550 mm, respectively. The pyrolysis products were collected through a two-stage cooler using cooling water and dry ice. For the fixed bed pyrolysis experiment, an initial 15 g sample was loaded into a ceramic boat and inserted into the reactor. Heating, which was performed at a rate of 10 °C/min, was controlled using a PID controller until reaction temperatures of 550, 600, and 650 °C were reached. The reaction time was maintained for 30 min after reaching the predetermined temperature for the sufficient pyrolysis of SRF. The resultant liquid recovered from the cooler and solid residues were then analyzed. The yield of gaseous products was calculated based on the initial sample weight and the amounts of liquid and solid residues. The fluidized bed reactor (Figure 10b) was designed to allow continuous sample injection, unlike the fixed bed reactor. It consisted of a stainless-steel tube, with a lower section that had a diameter and length of 62 and 370 mm, respectively. Its upper section had a diameter and length of 26 and 730 mm, respectively. As in the use of the fixed bed reactor, the pyrolysis products were collected through a two-stage cooler using cooling water and dry ice. In addition, 50 g of sand with a diameter of 125–180 μm was used as the fluidizing medium for the fluidized bed reaction. The fluidized bed pyrolysis experiments were conducted at nitrogen flow rates ranging from 0.75–1.5 L/min, allowing for optimal fluidization. These flow rates were adjusted at an interval of 0.25 L/min. Upon reaching the set reaction temperature, samples were continuously injected at a rate of 0.5–0.7 g/min for 1 h. Like in the case of the fixed bed reactor, pyrolysis was also set to occur at 550, 600, and 650 °C, and these temperatures were controlled using a PID controller. Finally, yields of the liquid products were analyzed once the resultant products were collected from the cooler. The amount of solid residue was analyzed by weighing the solid material remaining in the reactor and the initial sand used as the fluidizing medium after the experiment. The yield of gaseous products was calculated similarly for the fixed bed experiments, based on the initial SRF sample weight and the amounts of liquid and solid residues. In addition, the pyrolysis experiments performed using the fixed and fluidized bed reactors were repeated five times for the reliability of the experiment results.

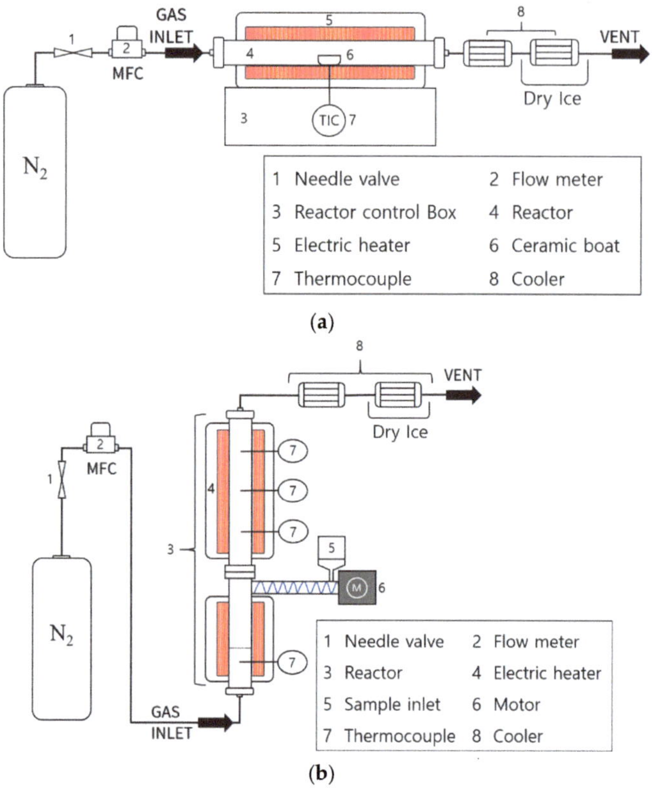

Figure 10. Schematic diagram of fixed (**a**) and fluidized (**b**) bed reactors used in this study.

The composition of the liquid products recovered through the fixed and fluidized bed reactors was analyzed using gas chromatography mass spectrometry (GC-MS) (GCMS-QP2010 Ultra, Dong-il SHIMADZU, Japan). The conditions under which the GC-MS was operated are presented in Table 5. Additionally, for the kinetic analysis of the pyrolysis reactions, thermal mass changes were analyzed using thermogravimetric analysis (TGA) (Pyris 1 TGA, USA) at heating rates of 20, 30, and 40 °C/min under a nitrogen atmosphere. The higher heating value of all samples was determined using a bomb calorimeter (Parr Instrument Co., Model 1672, Moline, IL, USA).

Table 5. Conditions under which the gas chromatography mass spectrometry (GC-MS) was used.

Item	Conditions
Column over temp	40.0 °C
Injection temp	250.0 °C
Injection mode	Split
Flow control mode	Linear Velocity
Pressure	60.1 kPa
Total flow	234.1 mL/min
Column flow (He)	1.15 mL/min
Linear velocity	38.6 cm/s
Purge flow	3.0 mL/min
Split ratio	200.0

3.3. *Kinetic Analysis*

The reaction rate equation for the conversion rate of pyrolysis can be represented via an Arrhenius Equation (1). In the case of kinetic analysis based on TGA, different results are generally derived depending on the analysis method [26]. Therefore, in this study, the differential approach that comprehensively uses the TGA results of multiple heating rates and the integral approach that individually uses the TGA results of the single heating rate were utilized to examine the validity of the kinetic analysis results. Differential and integral approaches are widely used in the kinetic analysis of pyrolytic reactions using TGA results [26].

$$\frac{d\alpha}{dt} = Ae^{-\frac{E}{RT}}(1-\alpha)^n \tag{1}$$

where A: pre-exponential factor (min^{-1})
E: apparent activation energy (kJ/mol)
n: apparent order of reaction
R: gas constant (8.3136 J/mol·K)
T: absolute temperature (K)
t: time (min)
α: degree of conversion.

3.3.1. Differential Method

This method utilizes the following logarithmic differential equation derived from Equation (1).

$$\ln\left(\frac{d\alpha}{dt}\right) = \ln\{A(1-\alpha)^n\} - \frac{E}{RT} \tag{2}$$

The first term on the right side of Equation (2) is constant for a fixed conversion rate, α. Therefore, when ln(da/dt) and 1/T are plotted at heating rates of 20, 30, and 40 °C/min, which were used in TGA of this study, for each fixed conversion rate, activation energy E can be obtained from the slope.

3.3.2. Integral Method

By rearranging Equation (1) using the linear heating rate β (K/min), Equation (3) can be derived; Equation (3) can then be further rearranged to yield Equation (4).

$$\frac{d\alpha}{dT} = \frac{A}{\beta}e^{-\frac{E}{RT}}(1-\alpha)^n \tag{3}$$

$$\frac{d\alpha}{(1-\alpha)^n} = \frac{A}{\beta}e^{-\frac{E}{RT}}dT \tag{4}$$

The application of the integral approximation method [27] to the right side of Equation (4) yields the following [28].

$$\int_0^{\alpha} \frac{d\alpha}{(1-\alpha)^n} = \frac{1-(1-\alpha)^{1-n}}{1-n} \quad n \neq 1 \tag{5}$$

$$= -\ln(1-\alpha) \quad n = 1 \tag{6}$$

and

$$A/\beta \int_0^T e^{-E/RT}dT \approx \frac{ART^2}{\beta E}\left(1-\frac{2RT}{E}\right)e^{-E/RT} \tag{7}$$

Additionally, taking the logarithm of Equations (5)–(7) yields Equations (8) and (9).

$$\ln\left\{\frac{1-(1-\alpha)^{1-n}}{T^2(1-n)}\right\} = \ln\frac{AR}{\beta E}\left(1-\frac{2RT}{E}\right) - \frac{E}{RT} \quad n \neq 1 \tag{8}$$

$$\ln\left\{\frac{-\ln(1-\alpha)^{1-n}}{T^2}\right\} = \ln\frac{AR}{\beta E}\left(1-\frac{2RT}{E}\right) - \frac{E}{RT} \quad n = 1 \tag{9}$$

Therefore, after assuming each reaction order, the activation energy values can be determined from the slopes of the plotted relations.

$$Y = -\ln\left\{\frac{1-(1-\alpha)^{1-n}}{T^2(1-n)}\right\} vs \frac{1}{T} \quad n \neq 1 \tag{10}$$

$$Y = -\ln\left\{-\frac{\ln(1-\alpha)}{T^2}\right\} vs \frac{1}{T} \quad n = 1 \tag{11}$$

Using Equations (10) and (11), the reaction orders n can be determined for the cases that best fit a straight line, and the activation energies can be obtained from the slopes of those lines.

4. Conclusions

In this study, we compared the pyrolytic properties of the SRF made from plastic waste using two reactors, namely, fixed bed and fluidized bed reactors, and kinetic analysis for pyrolysis of SRF was performed. Based on the kinetic analysis of the pyrolytic reactions using differential and integral methods applied to the TGA results, we found that the activation energy was lower in the initial stage of pyrolysis. This trend can be mainly attributed to the initial decomposition of PP components, which was subsequently followed by the decomposition of PE. Based on the analysis results of the kinetic analysis method applied in this study, the differential approach that uses multiple heating rates is judged to be more useful than the integral approach that uses the TGA results of the single heating rate because it can examine the change in activation energy according to the pyrolysis conversion rate. Using a fixed bed reactor, we found that as the reaction temperature increased, gasification became more vigorous, resulting in an increase and decrease in the yields of the gas and liquid, respectively. Via the fluidized bed reactor, we confirmed that, similar to the case of the fixed bed reactor, the gas yield increased as the reaction temperature increased and also rose as the flow rate of fluidizing gas increased. Overall, fluidized bed reactions resulted in a more active decomposition of SRF. The relatively superior performance of this reactor can be attributed to the increased mass and heat transfer effects caused by fluidizing gases, which result in greater gas yields. Regarding the characteristics of liquid products generated during pyrolysis, it was seen that the hydrogen content in the liquid products obtained from the fluidized bed reactor decreased, leading to the formation of oils with higher molecular weights and higher C/H ratios, because the pyrolysis of SRF in fluidized bed reactors progresses more rapidly than that in fixed bed reactors.

Author Contributions: S.O. (corresponding author), M.L. and H.K. were responsible for the formulation of the idea underlying the study, project development, and research preparation. M.L. and H.K. were also responsible for the experimental work. In addition, all the authors contributed equally toward data analysis, manuscript preparation, proofreading, and submission of the manuscript. All authors have read and agreed to the published version of the manuscript.

Funding: This research was funded by the Technology Innovation Program (No. 20015401), which is funded by the Ministry of Trade, Industry, and Energy (MOTIE, Republic of Korea).

Institutional Review Board Statement: Not applicable.

Informed Consent Statement: Not applicable.

Data Availability Statement: The data presented in this study are available on request from the corresponding author. The data are not publicly available due to privacy.

Conflicts of Interest: The authors declare no conflict of interest.

References

1. Horodyska, O.; Vakdes, F.J.; Fullana, A. Plastic flexible films waste management-A state of art review. *Waste Manag.* **2018**, *22*, 413–425. [CrossRef] [PubMed]
2. Meys, R.; Frick, F.; Westhues, S.; Sternberg, A.; Klankermayer, J.; Bardow, A. Towards a circular economy for plastic packing waste-the environmental potential of chemical recycling. *Resour. Conserv. Recycl.* **2020**, *162*, 105010. [CrossRef]
3. Park, K.B.; Jeong, Y.S.; Guzelciftci, B.; Kim, J.S. Two-stage pyrolysis of polystyrene: Pyrolysis oil as a source of fuels or bezene, toluene, ethylbenzene, and xylene. *Appl. Energy* **2020**, *259*, 114240. [CrossRef]
4. Mangesh, V.L.; Padmanabhan, S.; Tamizhdural, P.; Ramesh, A. Experimental investigation to identify the type of waste plastic pyrolysis oil suitable for conversion to diesel engine fuel. *J. Clean. Prod.* **2020**, *246*, 119066. [CrossRef]
5. Sekar, M.; Ponnusamy, V.K.; Pugazhendhi, A.; Nizetic, S.; Praveenkumar, T.R. Production and utilization of pyrolysis oil from solid plastic wastes: A review on pyrolysis ptocess and influence of reactors design. *J. Environ. Manag.* **2022**, *302 Part B*, 114046. [CrossRef]
6. Signh, R.K.; Ruj, B.; Sadhukhan, A.K.; Gupta, P. Thermal degradation of waste plastics under non-sweeping atmosphere: Part 1 Effect of temperature, product optimization, and degradation mechanism. *J. Environ. Manag.* **2019**, *239*, 395–406.
7. Kunwar, B.; Cheng, H.N.; Chandrashekaran, S.R.; Sharma, B.K. Plastics to fuel: A review. *Renew. Sustain. Energy Rev.* **2016**, *54*, 421–428. [CrossRef]
8. Nimmegeers, P.; Billen, P. Quantifying the separation complexity of mixed plastic waste streams with statistical entropy: A plastic waste case study in Belgium. *ACS Sustain. Chem. Eng.* **2021**, *9*, 9813–9822. [CrossRef]
9. Vollmer, I.; Jenks, M.J.F.; Roelands, M.C.P.; White, R.J.; van Harmelen, T.; de Wild, P.; van der Laan, G.P.; Meirer, F.; Keurentjes, J.T.F.; Weckhuysen, B.M. Beyond mechanical recycling: Given new life to plastic waste. *Angew. Chem. Int. Ed.* **2020**, *59*, 15402–15423. [CrossRef]
10. Sharuddin, S.D.A.; Abnisa, F.; Daud, W.M.A.W.; Aroua, M.K. A review on pyrolysis of plastic wastes. *Energy Conserv. Manag.* **2016**, *115*, 308–326. [CrossRef]
11. Jahirul, A.I.; Rasul, M.G.; Schaller, D.; Khan, M.M.K.; Hasan, M.M.; Hazrat, M.A. Transport fuel from waste plastics pyrolysis-A review on technologies challenges and opportunities. *Energy Conserv. Manag.* **2022**, *258*, 115451. [CrossRef]
12. Qureshi, M.S.; Oasmaa, A.; Pihkola, H.; Deviatkin, I.; Tenhunen, A.; Mannila, J.; Minkkinen, H.; Pohjakallio, M.; Laine-Ylijoki, J. Pyrolysis of plastic waste: Opportunities and challenges. *J. Anal. Appl. Pyrolysis* **2020**, *152*, 104804. [CrossRef]
13. Al-Salem, S.M.; Antelava, A.; Constantinou, A.; Manos, G.; Dutta, A. A review on thermal and catalytic pyrolysis of plastic solid waste (PSW). *J. Environ. Manag.* **2017**, *197*, 177–198. [CrossRef] [PubMed]
14. Vellaiyan, S. Energy extraction from waste plastics and its optimization study for effective combustion and cleaner exhaust engaging with water and cetane improve: A response surface methodology approach. *Environ. Res.* **2023**, *231 Part 1*, 116113. [CrossRef]
15. Kaimal, V.K.; Vijayabalan, P. A detailed study of combustion characteristics of a DI diesel engine using waste plastic oil and its blends. *Energy Conserv. Manag.* **2015**, *105*, 951–956. [CrossRef]
16. Wathakit, K.; Sukjit, E.; Kaewbuddee, C.; Maithomklang, S.; Klinkaew, N.; Liplap, P.; Arjharn, W.; Srisertpol, J. Characterization and impact of waste plastic oil in a bariable compression ration diesel engine. *Energies* **2021**, *14*, 2230. [CrossRef]
17. Panda, A.K.; Singh, R.K.; Mishra, D.K. Thermolysis of waste plastics to liquid fuel: A suitable method for plastic waste management and manufacture of value added products-A world prospective. *Renew. Sustain. Energy Rev.* **2010**, *14*, 223–248. [CrossRef]
18. Al-Salem, S.M.; Lettieri, P.; Baeyens, J. Recycling and recovery routes of plastic solid waste (PSW): A review. *Waste Manag.* **2009**, *29*, 2625–2643. [CrossRef]
19. Harussani, M.M.; Sapuan, S.M.; Rashid, U.; Khalina, A.; Ilyas, R.A. Pyrolysis of polypropylene plastic waste into carbonaceous char: Priority of plastic waste management amidst COVID-19 pandemic. *Sci. Total Environ.* **2022**, *803*, 149911. [CrossRef]
20. Marcilla, A.; Garcia-Quesada, J.C.; Sanchez, S.; Ruiz, R. Study of the catalytic pyrolysis behaviour of polyethylene–polypropylene mixtures. *J. Anal. Appl. Pyrolysis* **2005**, *74*, 387–392. [CrossRef]
21. Kim, H.T.; Oh, S.C. Kinetic of thermal degradation of waste polypropylene and high-density polyethylene. *J. Ind. Eng. Chem.* **2005**, *11*, 648–656.
22. López, A.; de Marco, I.; Caballero, B.M.; Laresgoiti, M.F.; Adrados, A. Pyrolysis of municipal plastic wastes: Influence of raw material composition. *Waste Manag.* **2010**, *30*, 620–627. [CrossRef] [PubMed]
23. Jaafar, Y.; Abdelouahed, L.; Hage, R.E.; Samrani, A.E.; Taouk, B. Pyrolysis of common plastics and their mixtures to produce valuable petroleum-like products. *Polym. Degrad. Stab.* **2022**, *195*, 109770. [CrossRef]
24. Tokmurzin, D.; Nam, J.Y.; Lee, T.R.; Park, S.J.; Nam, H.S.; Yoon, S.J.; Mun, T.Y.; Yoon, S.M.; Moon, J.H.; Lee, J.G.; et al. High temperature flash pyrolysis characteristics of waste plastics (SRF) in a bubbling fluidized bed: Effect of temperature and pelletizing. *Fuel* **2022**, *326*, 125022. [CrossRef]
25. Jung, S.-H.; Cho, M.-H.; Kang, B.-S.; Kim, J.-S. Pyrolysis of a fraction of waste polypropylene and polyethylene for the recovery of BTX aromatics using a fluidized bed reactor. *Fuel Process. Technol.* **2010**, *91*, 277–284. [CrossRef]
26. Park, J.W.; Oh, S.C.; Lee, H.P.; Kim, H.T.; Yoo, K.O. Kinetic analysis of thermal decomposition of polymer using a dynamic model. *Korean J. Chem. Eng.* **2000**, *17*, 489–496. [CrossRef]

27. Coats, A.; Redfern, J. Kinetic parameters from thermogravimetric data. *Nature* **1964**, *201*, 68–69. [CrossRef]
28. Cooney, J.D.; Day, M.; Wiles, D.M. Thermal degradation of poly(ethylene terephthalate): A kinetic analysis of thermogravimetric data. *J. Appl. Polym. Sci.* **1983**, *28*, 2887–2902. [CrossRef]

molecules

MDPI

Article

Primary Products from Fast Co-Pyrolysis of Palm Kernel Shell and Sawdust

David O. Usino *, Päivi Ylitervo and Tobias Richards

Swedish Centre for Resource Recovery, University of Borås, 501 90 Borås, Sweden; paivi.ylitervo@hb.se (P.Y.); tobias.richards@hb.se (T.R.)
* Correspondence: david.usino@hb.se

Abstract: Co-pyrolysis is one possible method to handle different biomass leftovers. The success of the implementation depends on several factors, of which the quality of the produced bio-oil is of the highest importance, together with the throughput and constraints of the feedstock. In this study, the fast co-pyrolysis of palm kernel shell (PKS) and woody biomass was conducted in a micro-pyrolyser connected to a Gas Chromatograph–Mass Spectrometer/Flame Ionisation Detector (GC–MS/FID) at 600 °C and 5 s. Different blend ratios were studied to reveal interactions on the primary products formed from the co-pyrolysis, specifically PKS and two woody biomasses. A comparison of the experimental and predicted yields showed that the co-pyrolysis of the binary blends in equal proportions, PKS with mahogany (MAH) or iroko (IRO) sawdust, resulted in a decrease in the relative yield of the phenols by 19%, while HAA was promoted by 43% for the PKS:IRO-1:1 pyrolysis blend, and the saccharides were strongly inhibited for the PKS:MAH-1:1 pyrolysis blend. However, no difference was observed in the yields for the different groups of compounds when the two woody biomasses (MAH:IRO-1:1) were co-pyrolysed. In contrast to the binary blend, the pyrolysis of the ternary blends showed that the yield of the saccharides was promoted to a large extent, while the acids were inhibited for the PKS:MAH:IRO-1:1:1 pyrolysis blend. However, the relative yield of the saccharides was inhibited to a large extent for the PKS:MAH:IRO-1:2:2 pyrolysis blend, while no major difference was observed in the yields across the different groups of compounds when PKS and the woody biomass were blended in equal amounts and pyrolysed (PKS:MAH:IRO-2:1:1). This study showed evidence of a synergistic interaction when co-pyrolysing different biomasses. It also shows that it is possible to enhance the production of a valuable group of compounds with the right biomass composition and blend ratio.

Keywords: fast pyrolysis; primary products; co-pyrolysis; Py-GC-MS/FID; biomass blend

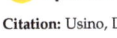

Citation: Usino, D.O.; Ylitervo, P.; Richards, T. Primary Products from Fast Co-Pyrolysis of Palm Kernel Shell and Sawdust. *Molecules* **2023**, *28*, 6809. https://doi.org/10.3390/molecules28196809

Academic Editor: Mohamad Nasir Mohamad Ibrahim

Received: 21 August 2023
Revised: 18 September 2023
Accepted: 20 September 2023
Published: 26 September 2023

1. Introduction

Fast pyrolysis can be used to convert biomass into bio-oil and chemicals. The pyrolysis oil (often denoted as bio-oil when the material is of biomass origin) is composed of compounds such as anhydrosugars, furans, alcohols, ketones, aldehydes, acids, and phenols [1,2]. However, the bio-oil cannot be used directly in fuel engines or mixed with petroleum products due to its high water content, high acidity, and low miscibility, and, therefore, it requires a further upgrade to make it suitable as transportation fuel [1,3]. Several attempts have been employed to improve the quality of the oil. These include the use of catalysts and hydrogen [4,5]. However, these processes are complex and costly due to the equipment requirement and the catalysts needed for the successful upgrade [4,5]. The pretreatment of biomass with dilute acid solutions has also been used to improve the quality and minimise the negative effects of inorganic materials during the fast pyrolysis of biomass [6–8]. Recently, attention has been directed towards the co-pyrolysis of different biomasses [9–12].

The co-pyrolysis of biomasses is described as the pyrolysis of blends, including two or more different biomasses. Previous studies have shown this to improve the overall quality of the pyrolysis oil, such as by increasing the calorific value and promoting the yield of volatile compounds [4,12–15]. Many studies have been carried out on the co-pyrolysis of different biomasses to produce biofuel and chemicals. However, most of the studies found in the literature focused on the co-pyrolysis of biomass and plastic [16–19], and very few studies have been carried out on the co-pyrolysis of different biomass materials [10,20–23]. Moreover, most of the studies that were carried out on the co-pyrolysis of different biomasses were achieved with the use of the TGA and focused on the gas and char yields. For example, El-Sayed et al. [22] showed that the co-pyrolysis of Egyptian olive pomace and wood dust (Kroneiki olive-pomace (KROP), Shamlali olive-pomace (SHOP), and Fine Swedish sawdust (FSSD)) showed an increase in the amount of volatile matter in the blend and had the best synergistic pyrolysis performance at a heating rate of 10 °C/min. Additionally, Nie et al. [10] observed that the co-pyrolysis of wood sawdust (WS) and peanut shell (PS) resulted in an increase in the comprehensive pyrolysis index for the blend ratio W3P7 (WS:PS = 3:7) compared to the single pyrolysis of WS and PS at a heating rate of 10–30 °C/min, while Ge et al. [23] did not observe any clear synergic interaction on the biomass mass loss during the co-pyrolysis of pine wood waste and straw waste. In terms of the bio-oil yield, Biswas et al. [20] showed that the co-pyrolysis of Phumdi (PH) and Para grass (PG) (1:1) resulted in a bio-oil yield of 11.66 wt% and was composed mainly of phenolic compounds, while Hopa et al. [14] observed that the co-pyrolysis of rice husk and sugarcane bagasse resulted in an increased yield of bio-oil with 28.4%. They suggested that this was due to a synergistic interaction between the two biomasses. However, several factors can influence the yield and quality of the pyrolysis product formed from mixing two or more biomasses. These include the biomass type, composition, blending ratio, reactor type, and temperature [4,24]. Tauseef et al. [25] observed a synergistic effect when coal and rice husk were co-pyrolysed. Edmunds et al. [3], in contrast, found that the pyrolysis product distribution was a simple linear combination when switch grass and pine residue were co-pyrolysed. The mixing as well as the blend ratios are important to estimate the final resulting composition and yield of the bio-oil. Furthermore, biomasses differ in their compositions and physical structures, and this can influence the quality of the bio-oil [3]. Palm kernel shell (PKS), for example, has a high lignin content (≈58 wt%) [8,24] compared to woody biomass (15–40%) [26], while woody biomass, such as mahogany (MAH), has a higher carbohydrate content (65 wt%) [27]. During the pyrolysis of raw biomass, cellulose and hemicellulose undergo dehydration, depolymerisation, and rearrangement reactions to form anhydrous sugars, furans, and light oxygenate compounds [28], while lignin undergoes depolymerisation, demethylation, and fragmentation reactions to form mainly phenolic-type compounds [29]. The co-pyrolysis of these biomasses could result in a synergistic effect that could either enhance or decrease the primary products formed. A comparison of the co-pyrolysis of a blend from pure cellulose, hemicellulose, and lignin with the pyrolysis of native birch wood shows a decrease in the product yields of the sugars and phenolic compounds from the native birch wood, while the yields of the hemicellulose-derived products, such as aldehydes and ketones, were promoted [30]. The decreased yield of the sugars, especially levoglucosan, is suggested to depend on the presence of a covalent bond in the morphology of the native biomass and inorganic materials in the native biomass [30].

A review of most of the previous studies shows that the information is scarce about the primary products' characteristics and interactions during the fast co-pyrolysis of different native biomass blends comprising an agricultural residue and a woody biomass residue. Moreover, most of the studies were carried out with the use of a fixed-bed and/or thermogravimetric analyser (TGA) [10,20,22,23], which may result in secondary reactions. With the right mixture of biomasses, it may be possible to control the product distribution and enhance the quality of the volatile compounds formed from the interaction of the different biomass components during pyrolysis. This study aims to investigate the interactions between an agricultural residue and two woody biomass materials and their impact on the primary-product distribution. Additionally, this study investigates the influence of the blending ratio. Overall, this study provides a method for reducing the environmental impact associated with the disposal of these wastes and promoting the co-utilisation of these waste streams for bioenergy production.

2. Results and Discussion

2.1. Characterisation of Biomass

A summary of the proximate analysis in terms of the moisture content, volatile matter, fixed carbon, and ash content, as well as the biomass characterisations for all three biomasses used as blends in this study, are shown in Table 1. The proximate analysis shown in Table 1 provides information about the thermal behaviour and fuel characteristics of the PKS and the two woody biomasses (MAH and iroko (IRO)). It also shows the major components and chemical compositions of the PKS, MAH, and IRO, which are important for characterising the energy content, conversion efficiency, and suitability of each biomass for bioenergy production [14,31]. It can be seen from the proximate analysis that the woody biomass samples had higher volatile matter contents compared to the PKS. Volatile matter is important for understanding the energy content of the biomass. Table 1 shows that the PKS had the highest fixed carbon content and heating value, as well as the lowest moisture and ash content, compared to the woody biomasses. These characteristics make PKS a suitable material for blending and co-pyrolysis. The fixed carbon gives an indication of the energy potential of the biomass, while the high moisture content of a biomass material can negatively influence the energy content and efficiency [32]. The ash content is used to determine the inorganic materials in a biomass fuel, which could act as a catalyst during biomass pyrolysis [32,33]. High ash content has been noted to decrease the yield of bio-oil while, at the same time, increase the char and gas yield [34]. Table 1 also shows that the woody biomass had a higher amount of holocellulose (cellulose and hemicellulose) compared to the PKS, which had a higher content of lignin.

Table 1. Proximate analysis and characterisation of individual biomass feedstocks.

Feedstock	PKS (wt.%)	MAH (wt.%)	IRO (wt.%)
Proximate analysis			
Moisture content	2.2 ± 0.1	3.9 ± 0.1	3.8 ± 0.2
Ash content	0.9 ± 0.1	2.6 ± 0.4	4.8 ± 0.1
Volatile matter	76.7 ± 1.0	82.5 ± 0.6	78.6 ± 0.1
Fixed carbon (by difference)	20.3	11.1	12.9
Calorific value (HHV, MJ/kg)	20.7 ± 0.2	18.9 ± 0.3	19.5 ± 0.2
Component analysis			
Cellulose	8.4 ± 1.3	27.5 ± 0.6	25.0 ± 0.3
Hemicellulose (by difference)	33.5	38.0	31.8
Lignin	57.2 ± 0.7	31.9 ± 1.6	38.4 ± 0.9

2.2. Effect of Blending Two Biomasses on Product Distribution and Yield

The product distribution from the pyrolysis of the individual biomasses and their blends are found in Table S1 in the supplementary material. The pyrolysis of the two biomass blends was achieved by blending them in equal proportions: PKS:MAH_1:1 (0.50:0.50), PKS:IRO_1:1 (0.50:0.50), and MAH:IRO_1:1 (0.50:0.50). It should be noted that the hydroxyacetaldehyde (HAA) and acetic acid (AA) peaks were observed to co-elute for the blended biomasses, as shown in Figure 1. These two peaks (HAA and AA) were observed as the most pronounced peaks in the chromatogram and were resolved by magnifying the chromatographic peak area using the same baseline in order to determine the fraction of the peak area for each chemical compound formed. Hence, the predominant volatile compounds formed when PKS is co-pyrolysed with MAH or IRO (Figure 1a) are as follows: (1) HAA, (2) AA, (3) acetaldehyde, (4) 1-hydroxy-2-propanone, and (5) phenol. Additionally, the main chemical compounds identified when the two woody biomasses (MAH:IRO-1:1) are co-pyrolysed (Figure 1b) are as follows: (1) HAA, (2) acetaldehyde,

(3) AA, (4) 1-hydroxy-2-propanone, and (5) 2,3-pentanedione. It can be observed in Table S1 that the relative yield of AA and phenol was two times larger when PKS was co-pyrolysed with either MAH or IRO compared to when the two woody biomasses were co-pyrolysed. This may be attributed to the high lignin content of the PKS, as shown in Table 1, which was the main source of the phenol formation and its interaction with the holocellulose composition of the woody biomass. Previous studies suggest that the amount of AA formed during pyrolysis is dependent on the degree of the acetylation of the biomass feedstock, especially biomasses with high lignin contents [35,36]. Another study showed that lignin contributed to the increased production of AA during the wet oxidation of lignocellulosic biomass [37]. This clearly shows that the types of biomasses used in co-pyrolysis influence the yield and the chemical compounds formed.

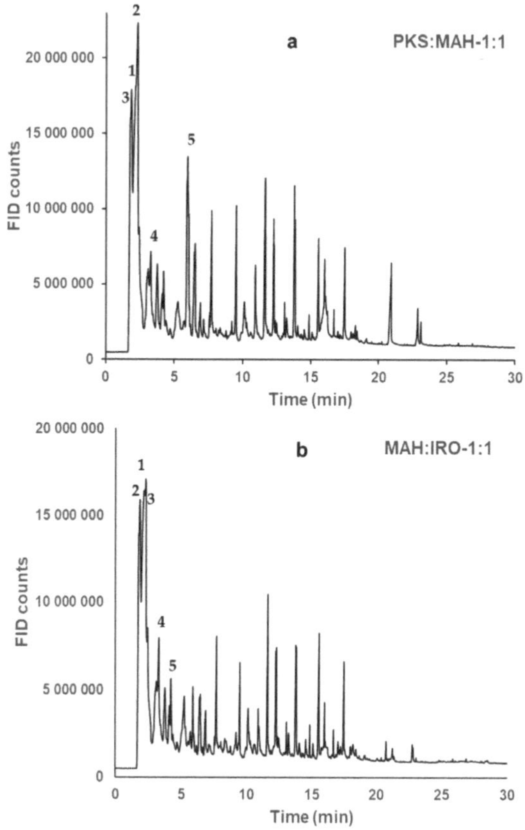

Figure 1. Chromatogram (GC-FID) after co-pyrolysis of two biomass blends at 600 °C and 5 s. The five main components of PKS:MAH-1:1 (**a**) are as follows: (1) hydroxyacetaldehyde, (2) acetic acid, (3) acetaldehyde, (4) 1-hydroxy-2-propanone, and (5) phenol, while those of MAH:IRO-1:1 (**b**) are as follows: (1) hydroxyacetaldehyde, (2) acetaldehyde, (3) acetic acid, (4) 1-hydroxy-2-propanone, and (5) 2,3-pentanedione. NB: (**a**) represents GC-FID chromatogram for PKS:MAH-1:1 and PKS:IRO-1:1, as both blends showed similar primary-product distributions.

To investigate the synergistic effects on the product yield of the co-pyrolysed biomass blends, the yield based on pure biomass according to the blend ratio was added (the so-called predicted value) and compared to the measured value (the experimental value). An analysis of the volatile compounds formed for the two biomass blends during pyrolysis, as shown in Figure 2 and Table S1, indicates that the six main chemical compounds formed when PKS is co-pyrolysed with MAH or IRO for each class of biomass are as follows: phenol (phenols), 1-hydroxy-2-propanone (ketones), hydroxyacetaldehyde (aldehydes), acetic acid (acids), levoglucosan (saccharides), and furfural (furans). However, in contrast to the formation of phenol as the main phenolic compound formed during the co-pyrolysis of PKS with MAH or IRO, 2,6-dimethoxyphenol was observed as the main phenolic compound formed for the pyrolysis of the woody biomass (MAH:IRO-1:1). This can be attributed to the composition and structural differences in the lignin between the PKS and the two woody biomasses. The pyrolysis of PKS produced a higher yield of guaiacyl compounds, while the pyrolysis of either MAH or IRO produced a higher yield of syringyl compounds (see Table S1). For the two component blends, the relative yield of AA was slightly promoted for PKS:MAH-1:1, while HAA was promoted by 43% for PKS:IRO-1:1. AA is formed mainly by the deacetylation of hemicellulose [1,24], or by the thermal cracking of cellulose and depolymerisation of the lignin polymer, while HAA is mainly formed by the primary decomposition and ring cleavage of the glucosidic bond of the cellulose and hemicellulose monomers in the biomass [28]. The increment in the relative yield of the AA may be due to the acetylation of lignin during the pyrolysis of PKS and MAH and the presence of alkali and alkaline earth metals (AAEMs). The presence of inorganic materials, such as AAEMs, in the biomass blend could act as a catalyst that may enhance the production of organic acids, such as AA and other light oxygenated compounds [3]. Pan et al. and Richards et al. [38,39] showed that the yields of AA and HAA were promoted mainly by the presence of potassium and calcium during the pyrolysis of cottonwood. This implies that the AAEMs inherent in the biomass have a strong catalytic effect on the production of AA. Figure 2 also shows that the relative yields of 2,6-dimethoxyphenol and furfural were promoted by 21 and 37%, respectively, for the pyrolysis of the MAH:IRO-1:1 blend, while no difference was observed in the relative yields for 1-hydroxy-2-propanone, HAA, and AA. This may be due to the structural similarity and higher content of holocellulose in the woody biomasses for MAH and IRO, as shown in Table 1, when compared to PKS.

The relative yields of phenol, 1-hydroxy-2-propanone, and furfural for PKS co-pyrolysed with either MAH or IRO in equal proportions showed no difference. However, the relative content of levoglucosan was strongly inhibited for the pyrolysis of PKS:MAH-1:1. The inhibition of levoglucosan for the PKS:MAH-1:1 blend and the promotion of AA, and HAA when PKS is co-pyrolysed with MAH or IRO, as well as the enhancement of 2,6-dimethoxyphenol and furfural for the co-pyrolysis of the woody biomass, are evidence of a synergistic interaction of the biomass components.

The experimental (Exp.) and predicted values (Pred.) for the different classes of compounds are presented in Figure 3. It was found that the relative content of the phenols was inhibited by 19% when PKS was co-pyrolysed with either MAH or IRO in equal proportions (PKS:MAH-1:1 and PKS:IRO-1:1). The suppression of the phenols observed may be due to the interaction between the different biomass components and the presence of inherent metal oxides in the biomass. Zhang et al. [40] observed that the yield of the phenolic compounds formed during the pyrolysis of pure cellulose and lignin impregnated with K_2O and CaO decreased, and the decreased yield was most pronounced for the guaiacol and 4-allyl-2-6-dimethoxyphenol compounds. Chang et al. [41] also observed a decrease in the relative content of the phenols when PKS and *Nannochloropsis* sp. (NC) were co-pyrolysed at 600 °C with a blend ratio of 1:1. The inhibition of the phenols is important for improving the quality of the bio-oil, as they contribute to the instability of the bio-oil during storage and transport [42]. No difference was found in the relative yields for the furans, acids, and aldehydes for the co-pyrolysis of PKS with either MAH or IRO. The relative yield of the saccharides, although very low in relation to the total amount, was strongly inhibited for the pyrolysis of PKS:MAH-1:1. The inhibition of the saccharides, especially levoglucosan, observed for the pyrolysis of PKS:MAH-1:1 may also be attributed to the presence of inorganic materials, such as K and Ca, in both the PKS and the woody biomass, and the interaction of holocellulose with lignin [39]. The major metal ions found in PKS are Si, K, and Ca, while those of the woody biomass (iroko) are K and Ca [43,44].

Richards et al. [39] observed that the yield of levoglucosan was negatively affected during the pyrolysis of cotton wood, and they attributed this to the presence of metal ions, such as K, Li, and Ca, and the interaction of levoglucosan formation with lignin [39]. Their result is similar to that presented by Usino et al. [30], who investigated the co-pyrolysis of pure cellulose, hemicellulose, and lignin to mimic a native birch wood. They observed that the yields of saccharides and phenols were inhibited to a large extent; moreover, they attributed this to the presence of inorganic materials and the formation of active compounds from the hemicellulose unit, which may have reacted with compounds from the cellulose and lignin and subsequently inhibited their reaction. The inhibition of the phenols and sugars observed in this study may be attributed to the inherent inorganics present in these biomasses and the higher content of lignin present in the PKS that may have reacted with the holocellulose in the woody biomass during pyrolysis. The metal ions present in the biomass are known to induce the hemolytic cleavage of the pyranose ring during the decomposition of cellulose. They are also known to decrease the stability of the glycosidic bonds and the hydroxyl group during pyrolysis, thereby resulting in dehydration, ring scission, and cracking reactions that favour the formation of low-molecular-weight compounds and the inhibition of levoglucosan formation [15].

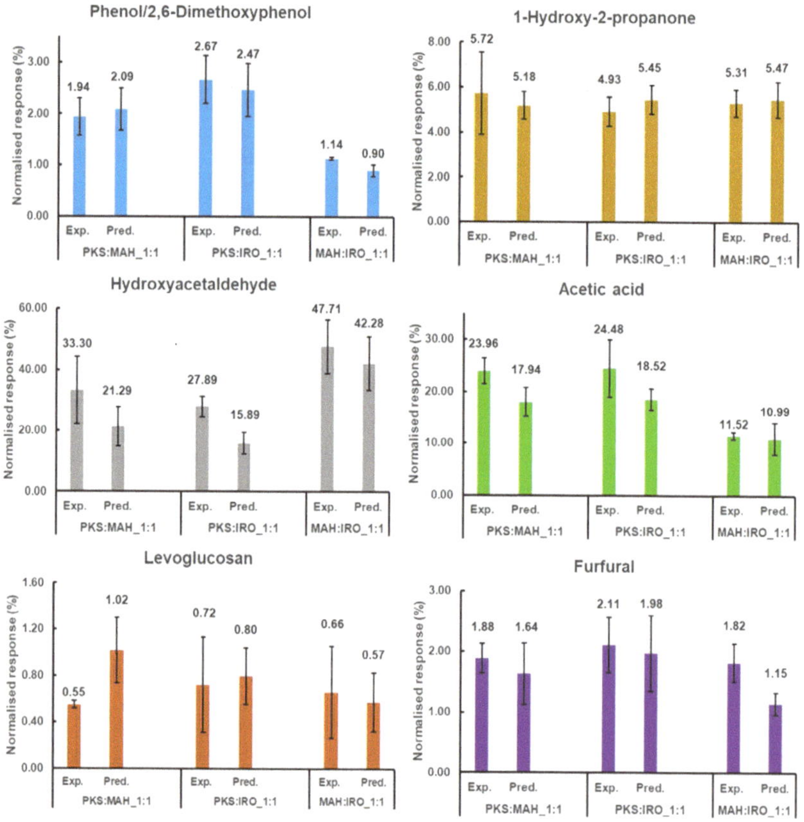

Figure 2. Comparison of the experimental (Exp.) and predicted (Pred.) results of the six main chemical compounds formed for the different classes of compounds after the co-pyrolysis of two different biomass blends. Product distribution is given in normalised weight (%) based on the calculated response (count/μg sample). Phenol is the main compound formed among the phenols during the pyrolysis of PKS:MAH-1:1 and PKS:IRO-1:1, while 2,6-Dimethoxyphenol is the main compound for MAH:IRO-1:1.

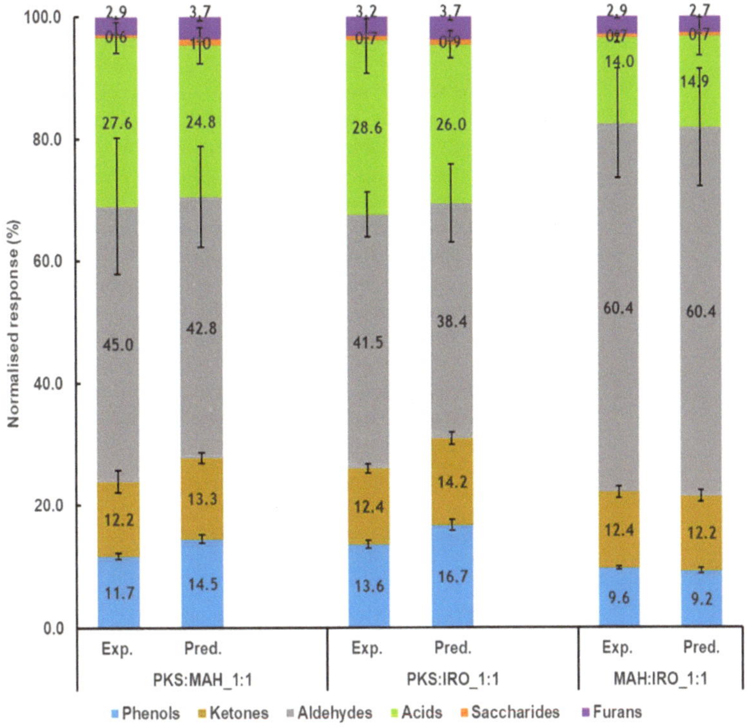

Figure 3. Comparison of the experimental (Exp.) and predicted (Pred.) results of the different classes of compounds after co-pyrolysis of two biomasses.

Additionally, a comparison of the experimental and predicted values for the woody biomasses (MAH and IRO), co-pyrolysed in equal proportions, indicates that no difference was observed in the relative yield across the different classes of compounds. This could be due to the woody biomasses being composed of similar polysaccharide structures. The result is similar to that presented by Edmunds et al. [3], who observed that the pyrolysis products obtained from the co-pyrolysis of switch grass and pine residue were a simple linear combination of the two biomasses investigated. Generally, Figure 2 shows that whenever PKS is co-pyrolysed with either MAH or IRO, a decrease in the relative contents, especially of the phenols and the saccharides, was observed. This indicates the occurrence of a synergistic interaction between these biomass components.

Considering the total yields of the volatile compounds formed (count/µg sample) from the pyrolysis of the biomass blends PKS:MAH-1:1, PKS:IRO-1:1, and MAH:IRO-1:1, it can be seen that the experimental values were about two times higher than the predicted values (Table S2). The yield (count/µg sample) increased from 4.48×10^4 (predicted value) to 7.53×10^4 (experimental value) for PKS:MAH-1:1, from 3.57×10^4 (predicted value) to 7.36×10^4 (experimental value) for PKS:IRO 1:1, and, finally, from 3.60×10^4 (predicted value) to 6.59×10^4 (experimental value) for MAH:IRO 1:1. This shows that co-pyrolysis could lead to an increase in the amount of volatile compounds formed.

2.3. Effect of Blending Three Biomasses on Product Distribution and Yield

The pyrolysis of the three biomass blends was achieved by first blending them in equal proportions (PKS:MAH:IRO-1:1:1 (0.33:0.33:0.33)), and then, secondly, mixing one part of the PKS with two parts each of the woody biomass (PKS:MAH:IRO-1:2:2

(0.20:0.40:0.40)), and thirdly, using two parts of the PKS and one part each of the woody biomass (PKS:MAH:IRO-2:1:1 (0.50:0.25:0.25)).

Similar to the two biomass blends, the main chemical compounds formed for each class of the three-biomass blend were as follows: phenol (phenols), 1-hydroxy-2-propanone (ketones), HAA (aldehyde), AA (acids), levoglucosan (saccharides), and furfural (furans) (see Figure 4 and Table S1). A comparison of the experimental and predicted values for the main chemical compounds indicates that the relative contents of HAA and levoglucosan were promoted by 34 and 24%, while AA was slightly inhibited for the PKS:MAH:IRO_1:1:1 pyrolysis blend. However, the relative content of AA was promoted for the PKS:MAH:IRO-1:2:2 and PKS:MAH:IRO-2:1:1 pyrolysis blends. It can be observed in Figure 4 that the relative yield of levoglucosan was inhibited when the proportion of the woody biomass was increased in the biomass blend for the co-pyrolysed feedstock (PKS:MAH:IRO-1:2:2). No difference was observed in the relative yield for 1-hydroxy-2-propanone and furfural for any of the three biomass blends, nor for phenol for the PKS:MAH:IRO_1:1:1 and PKS:MAH:IRO-2:1:1 blends. These results show that the composition and blending ratio of the biomasses influence the primary-product distribution when different types of native biomasses are co-pyrolysed.

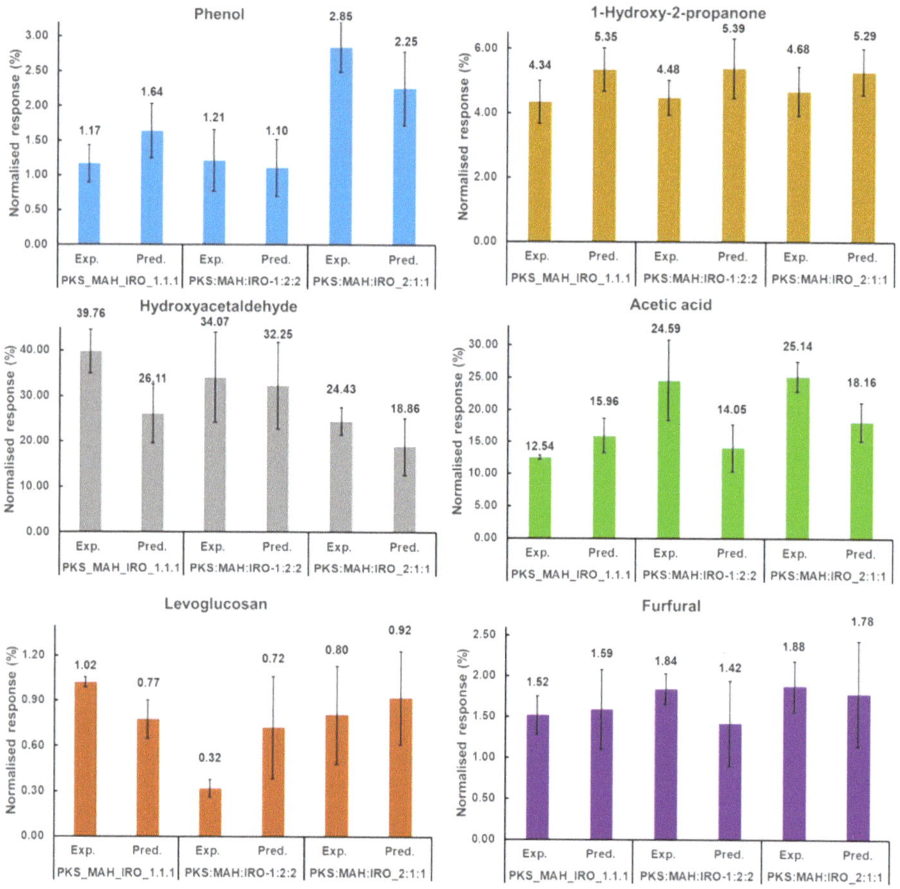

Figure 4. Comparison of the experimental (Exp.) and predicted (Pred.) results of the six main chemical compounds formed after the co-pyrolysis of three different biomass blends. Product distribution is given in normalised weight (%), based on the calculated response (count/μg sample).

Figure 5 shows the effect of blending on the various groups of compounds obtained from the co-pyrolysis of PKS, MAH, and IRO. It was found that the relative content of the saccharides was promoted to a large extent when the three biomasses were blended in equal proportions (PKS:MAH:IRO_1:1:1), which can be attributed to the increased yield of levoglucosan. The promotion may be due to the interaction of free radicals from the PKS and the woody biomass during their co-pyrolysis and the deoxygenation of the PKS-derived oxygenated compounds via the depolymerisation reaction [42]. It should be noted that the total relative yield of the saccharides was low (i.e., less than 2%) in comparison to the total yield of the volatile compounds. However, Table S1 shows that the amount of saccharides (especially levoglucosan) was almost as high for the PKS as it was for the MAH, and it was twice the amount of the IRO. An increased yield of sugars was also observed by Ojha et al. [45], who investigated the fast co-pyrolysis of cellulose and polypropylene at different mass ratios and temperatures. They observed that the yield of anhydrosugars increased with the temperature when cellulose and polypropylene were blended in equal proportions (50:50). However, the relative yield of the saccharides was inhibited to a large extent for the PKS:MAH:IRO-1:2:2 pyrolysis blend. This decrease can be attributed mainly to the inhibition of levoglucosan formation (see Figure 4). Additionally, the relative yield of the acids was promoted for the PKS:MAH:IRO-1:2:2 and PKS:MAH:IRO-2:1:1 pyrolysis blends, while that of the PKS:MAH:IRO-1:1:1 pyrolysis blend was slightly inhibited. The inhibition of the acids observed for the PKS:MAH:IRO-1:1:1 pyrolysis blend may be due to the deoxygenation of the acid via the decarboxylation reaction [42]. It has previously been reported that the strong presence of carboxylic acids in the bio-oil, such as AA, could lead to corrosion in the pipes and burners [46,47]. Moreover, the relative yield of phenols was inhibited by 25% for the PKS:MAH:IRO-1:2:2 pyrolysis blend, while only a slight inhibition was observed for the PKS:MAH:IRO_1:1:1 pyrolysis blend. However, no difference in the yield was observed for the ketones, aldehydes, and furans for any of the three biomass pyrolysis blends (PKS:MAH:IRO_1:1:1, PKS:MAH:IRO-1:2:2, and PKS:MAH:IRO-2:1:1). The increased and decreased yields for the different biomass pyrolysis blends are indications of a synergistic interaction between the different biomass components and could be due to the presence of inorganic materials in the biomasses, which may have acted as a catalyst during the pyrolysis of the blended biomasses [30]. The inhibition of the phenols observed in this study is similar to that reported in a previous study [30], in which it was observed that the relative yield of the phenolic compounds was inhibited when the individual biomass components of cellulose and hemicellulose were co-pyrolysed with lignin. The inhibition of the phenolic compounds was attributed to the presence of inorganic materials and the formation of active sites from the hemicellulose unit inhibiting the decomposition of the cellulose and lignin pyrolysis [30]. Vasu et al. [24] also reported that the amount of lignin-derived compounds was inhibited when the proportion of PKS was reduced in the co-pyrolysis of PKS and palm oil sludge. Nonetheless, the palm oil sludge was reported to have low volatile matter (48 wt.%) and high ash content (24 wt.%) compared to the high volatile matter (≈81%) and low ash content (≈3.7%) of the woody biomasses used in the current study. The blending and co-pyrolysis of the PKS, with a very low ash content (0.9 wt.%), and the woody biomass, with a higher ash content (2.6–4.8%), implies that there was a higher proportion of inorganic materials in the co-pyrolysed biomasses. This increase may have induced catalytic reactions and negatively affected the total yield of the phenolic compounds. The differences in the results among the different blends may be due to interactions as a result of the chemical reactions of the different components and physical action, such as the blend ratio [15]. The results show that the blending ratio plays a vital role in the product yield, and the PKS:MAH:IRO-1:1:1 blend produced volatile compounds with better fuel qualities than the PKS:MAH:IRO-1:2:2 and PKS:MAH:IRO-2:1:1 blends.

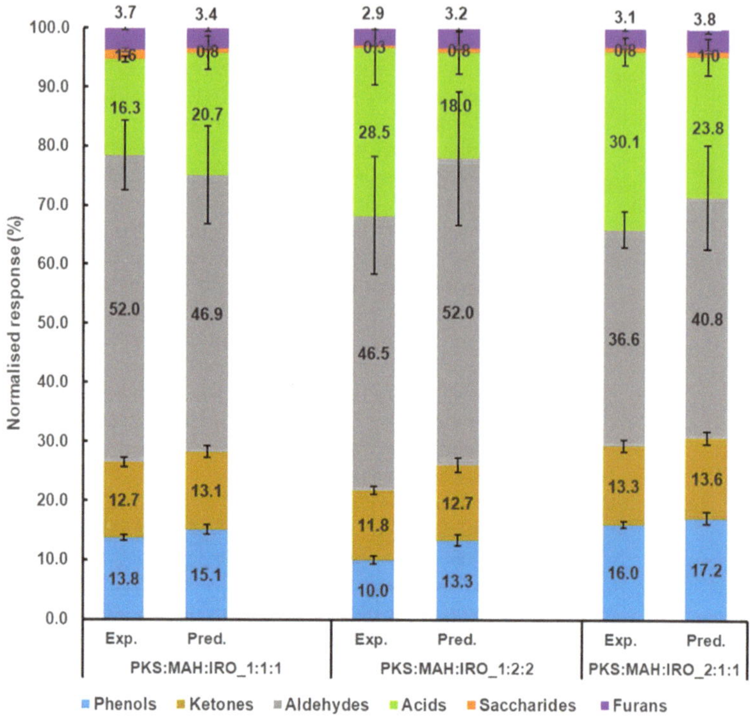

Figure 5. Comparison of the experimental (Exp.) and predicted (Pred.) volatile compositions after co-pyrolysis of three different biomass blends.

The results from the three biomass blends showed an impact on the primary pyrolysis product. While the yield of the saccharides was promoted when the three biomasses were blended in equal promotion and co-pyrolysed, an increase in the woody biomass composition resulted in a decreased yield of levoglucosan. Additionally, the synergistic interaction was more pronounced when the blend ratio of PKS was lower in the blend, as seen in the PKS:MAH:IRO_1:1:1 and PKS:MAH:IRO-1:2:2 (i.e., PKS:woody biomass_1:2 and 1:4) blends, than when they were blended and co-pyrolysed in equal proportions, as seen in the PKS:MAH:IRO-2:1:1 (i.e., PKS:woody biomass_2:2) for the ternary blend. An analysis of the volatile compounds formed in response (count/µg sample), shown in Table S3, shows that the total yield of the volatile compounds formed for the PKS:MAH:IRO_1:1:1 blend was similar for both the experimental and predicted values compared to the PKS:MAH:IRO-1:2:2 and PKS:MAH:IRO-2:1:1 blends, in which the total yield of the volatile compounds formed (experimental value) was promoted by almost two times the predicted value. This may have been the reason for the improved quality of the products from the PKS:MAH:IRO_1:1:1 blend in comparison to the other blends. This study thus shows that there is an interaction during the co-pyrolysis of different biomasses, and the interaction is dependent on the biomass type/composition and blend ratio. This makes it possible to optimise the pyrolysis process with the selective production of specific valuable chemicals.

3. Materials and Methods

3.1. Materials and Sample Preparation

The materials used for this study were palm kernel shell (*Elaeis guineensis*), mahogany (*Khaya ivorensis*) sawdust, and iroko (*Chlorophora excelsa*) sawdust. These biomasses were selected because of the environmental concerns associated with their disposal, the volume of the produced waste, and their suitability as feedstock for thermochemical conversion.

The woody biomasses (MAH and iroko (IRO) sawdust) were collected from sawmills, while the PKS was obtained from a palm oil site, all in the southern parts of Nigeria. The three biomass samples were milled in a cutting mill (Retsch SM 100) with a screen size of 6 mm and sieved to a particle size of 125–250 μm by a Haver (EML 200 pure) test sieve shaker. The samples were then dried in an oven at 105 °C overnight to a constant weight and stored for further use. The biomasses were then blended into binary and tertiary blends. The binary blends were made from equal amounts of two samples, while the tertiary blends were made from all three samples, but in different proportions. The binary blends correspond to the following weight proportions: PKS:MAH_1:1 (0.50:0.50), PKS:IRO_1:1 (0.50:0.50), and MAH:IRO_1:1 (0.50:0.50), while the tertiary blends correspond to the following weight ratios: PKS:MAH:IRO_1:1:1 (0.33:0.33:0.33), PKS:MAH:IRO_1:2:2 (0.20:0.40:0.40), and PKS:MAH:IRO_2:1:1 (0.50:0.25:0.25).

3.2. Proximate Analysis

The proximate analysis included the moisture content and the volatile matter of the biomass, and these were determined in reference to the ASTM standard test methods: ASTM E 871-82 and ASTM E872-82 [48,49]. For the moisture content, 3.0 g of each biomass sample was dried in an oven at 105 °C for 16 h, and the weight difference between the raw and dried biomass samples was used to calculate the weight loss of the biomass. Additionally, 1.0 g of each biomass sample was heated in a furnace at a constant temperature of 950 ± 20 °C for 7 min and cooled in a desiccator to determine the volatile matter of the raw biomass. The Nabertherm B 150 furnace was used to determine the ash content of the raw biomass by heating it at 550 °C for 4 h, according to the E1755-01 standard [50]. Prior to heating the biomass, 4.0 g of raw biomass was dried in an oven overnight at 105 °C. The residue left after the combustion of the raw biomass in the furnace was used to calculate the ash content of the biomass. Finally, the fixed carbon was calculated by subtracting the moisture content, volatile matter, and ash content from 100%.

3.3. Compositional Analysis of Biomass

Carbohydrate and lignin contents of PKS, MAH, and IRO were determined according to the NREL protocol (NREL/TP-510-42618) [51]. In this procedure, 0.3 g of the biomass sample was mixed with 3 mL 72% sulfuric acid solution. The mixture was stirred every 10 min for 1 h in a water bath at a temperature of 35 °C. Thereafter, 84 mL of Milli-Q water was added to the mixture before it was placed in an autoclave at 121 °C for 60 min. Filtration of the sample was carried out after cooling at room temperature. The collected filtrate was stored in a 50 mL bottle before analysing the acid-soluble lignin and cellulose contents. A spectrophotometer was used to determine the acid-soluble lignin. This was determined by drying the residue overnight at 100 °C and weighing the lignin content after cooling. High-performance liquid chromatography (HPLC) was used to determine the glucose content, while the hemicellulose content was determined by calculating the difference. The column (Aminex HPX-87P, Bio-Rad, Hercules, CA, USA) used was set at 85 °C, and the eluent flow was 0.6 mL/min. The experiments were carried out in duplicates, and the values are reported as average values.

3.4. Calorific Analysis

The calorific values of the biomass samples were determined with a bomb calorimeter (IKA C200 (IKA-Werke GmbH & Co. KG, Staufen Breisgau, Germany)) by placing 1.0 g of each biomass sample inside a steel vessel containing 500 μL of deionised water. Then, the vessel was filled with oxygen at 30 bar pressure before it was placed inside the equipment in which the sample was ignited.

3.5. Py-GC-MS/FID

Pyrolysis and analysis of the gaseous product from the raw and blended biomass samples were obtained with a micro-pyrolyser (Pyrola2000 (Pyrolab, Lund, Sweden)) connected to a Gas Chromatograph (GC), Mass Spectrometer (MS), and Flame Ionisation Detector (FID), also known as a Py-GC-MS/FID. The biomass samples were blended in different proportions, and close to 500 µg was weighed on an analytical balance (KERN ABT 320-4M) and placed on the platinum filament in the micro-pyrolyser. The platinum filament enables the investigation of primary reactions under isothermal conditions due to its high heating ramp of about 100,000 °C/s and good temperature control. Fast pyrolysis of the raw and blended biomasses was performed at a temperature of 600 °C with a residence time of 5 s. The operating condition was selected in order to obtain the maximum product yield of the released volatile compounds, as reported in previous studies [28,29].

Both the micro-pyrolyser chamber temperature and the transfer tube were set to 150 °C. The low chamber temperature minimises secondary reactions, while it is high enough to keep many of the products in the gas phase. The released pyrolysis vapour was directly transferred to the GC (Trace GC Ultra, Thermo Scientific, Waltham, MA, USA) via a continuous flow of helium gas with a purity of 99.9%. The volatile compounds were identified via the MS (ISQTM, Thermo Scientific) and quantified via an FID (Thermo Scientific). A Zebron™ ZB-5MS (30 m × 0.25 mm × 0.25 µm) capillary column was used with helium carrier gas (1.5 mL min^{-1}) and a split ratio of 1:7, while the GC oven programme was set at a temperature ramp of 1 min at 60 °C, followed by a ramp of 8 °C/min to 265 °C, and was then held at 265 °C for 20 min. The MS was operated with an ionisation energy of 70 eV, while the MS transfer line and the FID base temperatures were set to 250 °C. Furthermore, the ion-source temperature was kept at 200 °C, and the MS scan was obtained from 25 to 250 m/z. More than 50 chromatography peaks were extracted by the GC-MS Xcalibur software (Version 2.1.0–2.3.0) and identified by the NIST library (NIST MS Search 2.0) and based on previous research [28,30]. A minimum of three replicates were performed to ensure reproducibility and to minimise errors. Quantification of the released volatile compounds was based on their response in count/µg sample for each of the evolved volatile compounds, as described in a previous study [30]. Moreover, the data presented in this study are based on the released volatile compounds that did not condense in the glass cell of the pyrola. This set-up, thus, shows the potential to reduce secondary reactions due to its high heating ramp and precise temperature control compared to non-isothermal reactors mostly used to investigate the co-pyrolysis of different biomasses.

4. Conclusions

The co-pyrolysis of palm kernel shell and sawdust from two different woody biomasses was investigated to understand their influence on the primary-product yield and synergistic interactions.

1. The binary blends show that the co-pyrolysis of PKS with MAH or IRO in equal proportions (PKS:MAH-1:1 and PKS:IRO-1:1) decreased the relative yield of phenolic compounds by 19% compared to the pyrolysis of each material individually;
2. The saccharides, mainly levoglucosan, were inhibited to a large extent, while HAA was promoted by 43% for the PKS:IRO-1:1 pyrolysis blend;
3. The relative yields of 2,6-dimethoxyphenol and furfural were also promoted by 21 and 37%, respectively, for the pyrolysis of the MAH:IRO-1:1 blend;
4. No major difference in the relative yield was observed across the different classes of compounds when the woody biomasses were co-pyrolysed together, which is due to their similar chemical structures;
5. The ternary blends showed that the pyrolysis of PKS, MAH, and IRO in equal proportions (PKS:MAH:IRO_1:1:1) led to an increase in the relative yield of the saccharides to a large extent, while an increase in the proportion of the woody biomass in the pyrolysis blend (PKS:MAH:IRO-1:2:2) led to a strong inhibition in the relative yield of the saccharides;
6. Analysis of the individual volatile compounds formed shows that the pyrolysis of PKS:MAH:IRO-1:2:2 resulted in a decreased yield of phenols by 25%, while the relative

yields of HAA and levoglucosan were promoted by 34 and 24%, respectively, for PKS:MAH:IRO_1:1:1.

Supplementary Materials: The following supporting information can be downloaded at https://www.mdpi.com/article/10.3390/molecules28196809/s1, Tables S1–S3: Product distribution of fast pyrolysis of individual and co-pyrolysed biomass blends at 600 °C and 5 s, comparison of the experimental and predicted results of the two-biomass pyrolysis blend, and comparison of the experimental and predicted results of the three-biomass pyrolysis blend.

Author Contributions: Conceptualisation, D.O.U.; conceptualisation developed by D.O.U., P.Y. and T.R.; methodology, D.O.U.; investigation and analysis of results, D.O.U.; writing—original draft preparation, D.O.U.; writing—review and editing, D.O.U., P.Y. and T.R.; supervision, P.Y. and T.R. All authors have read and agreed to the published version of the manuscript.

Funding: This research has been supported by the University of Borås.

Informed Consent Statement: Not applicable.

Data Availability Statement: Data can be obtained on request from the corresponding author.

Conflicts of Interest: The authors declare no conflict of interest.

Sample Availability: Not applicable.

References

1. Kim, J.-S.; Choi, G.-G. Chapter 11—Pyrolysis of Lignocellulosic Biomass for Biochemical Production. In *Waste Biorefinery*; Bhaskar, T., Pandey, A., Mohan, S.V., Lee, D.-J., Khanal, S.K., Eds.; Elsevier: Amsterdam, The Netherlands, 2018; pp. 323–348.
2. Mahadevan, R.; Shakya, R.; Adhikari, S.; Fasina, O.; Taylor, S.E. Fast Pyrolysis of Biomass: Effect of Blending Southern Pine and Switchgrass. *Trans. ASABE* **2016**, *59*, 5–10. [CrossRef]
3. Edmunds, C.W.; Reyes Molina, E.A.; André, N.; Hamilton, C.; Park, S.; Fasina, O.; Adhikari, S.; Kelley, S.S.; Tumuluru, J.S.; Rials, T.G.; et al. Blended Feedstocks for Thermochemical Conversion: Biomass Characterization and Bio-Oil Production from Switchgrass-Pine Residues Blends. *Front. Energy Res.* **2018**, *6*, 79. [CrossRef]
4. Abnisa, F.; Wan Daud, W.M.A. A review on co-pyrolysis of biomass: An optional technique to obtain a high-grade pyrolysis oil. *Energy Convers. Manag.* **2014**, *87*, 71–85. [CrossRef]
5. Pinto, F.; Paradela, F.; Carvalheiro, F.; Duarte, L.C.; Costa, P.; André, R.N. Co-pyrolysis of pre-treated biomass and wastes to produce added value liquid compounds. *Chem. Eng. Trans.* **2018**, *65*, 211–216.
6. Chen, D.; Gao, D.; Huang, S.; Capareda, S.C.; Liu, X.; Wang, Y.; Zhang, T.; Liu, Y.; Niu, W. Influence of acid-washed pretreatment on the pyrolysis of corn straw: A study on characteristics, kinetics and bio-oil composition. *J. Anal. Appl. Pyrolysis* **2021**, *155*, 105027. [CrossRef]
7. Dong, Q.; Zhang, S.; Zhang, L.; Ding, K.; Xiong, Y. Effects of four types of dilute acid washing on moso bamboo pyrolysis using Py–GC/MS. *Bioresour. Technol.* **2015**, *185*, 62–69. [CrossRef] [PubMed]
8. Usino, D.O.; Sar, T.; Ylitervo, P.; Richards, T. Effect of Acid Pretreatment on the Primary Products of Biomass Fast Pyrolysis. *Energies* **2023**, *16*, 2377. [CrossRef]
9. Khodaparasti, M.S.; Khorasani, R.; Tavakoli, O.; Khodadadi, A.A. Optimal Co-pyrolysis of municipal sewage sludge and microalgae Chlorella Vulgaris: Products characterization, synergistic effects, mechanism, and reaction pathways. *J. Clean. Prod.* **2023**, *390*, 135991. [CrossRef]
10. Nie, Y.; Deng, M.; Shan, M.; Yang, X. Is there interaction between forestry residue and crop residue in co-pyrolysis? Evidence from wood sawdust and peanut shell. *J. Therm. Anal. Calorim.* **2023**, *148*, 2467–2481. [CrossRef]
11. Chen, C.; Qiu, S.; Ling, H.; Zhao, J.; Fan, D.; Zhu, J. Effect of transition metal oxide on microwave co-pyrolysis of sugarcane bagasse and Chlorella vulgaris for producing bio-oil. *Ind. Crops Prod.* **2023**, *199*, 116756. [CrossRef]
12. Muniyappan, D.; Pereira Junior, A.O.; Ramanathan, A. Synergistic recovery of renewable hydrocarbon resources via microwave co-pyrolysis of biomass residue and plastic waste over spent toner catalyst towards sustainable solid waste management. *Energy* **2023**, *278*, 127652. [CrossRef]
13. Park, D.K.; Kim, S.D.; Lee, S.H.; Lee, J.G. Co-pyrolysis characteristics of sawdust and coal blend in TGA and a fixed bed reactor. *Bioresour. Technol.* **2010**, *101*, 6151–6156. [CrossRef]
14. Hopa, D.Y.; Alagöz, O.; Yılmaz, N.; Dilek, M.; Arabacı, G.; Mutlu, T. Biomass co-pyrolysis: Effects of blending three different biomasses on oil yield and quality. *Waste Manag. Res.* **2019**, *37*, 925–933. [CrossRef]
15. Wang, W.; Lemaire, R.; Bensakhria, A.; Luart, D. Review on the catalytic effects of alkali and alkaline earth metals (AAEMs) including sodium, potassium, calcium and magnesium on the pyrolysis of lignocellulosic biomass and on the co-pyrolysis of coal with biomass. *J. Anal. Appl. Pyrolysis* **2022**, *163*, 105479. [CrossRef]

16. Vibhakar, C.; Sabeenian, R.S.; Kaliappan, S.; Patil, P.Y.; Patil, P.P.; Madhu, P.; Sowmya Dhanalakshmi, C.; Ababu Birhanu, H. Production and Optimization of Energy Rich Biofuel through Co-Pyrolysis by Utilizing Mixed Agricultural Residues and Mixed Waste Plastics. *Adv. Mater. Sci. Eng.* **2022**, *2022*, 8175552. [CrossRef]
17. Du, J.; Zhang, F.; Hu, J.; Yang, S.; Liu, H.; Wang, H. Co-pyrolysis of industrial hemp stems and waste plastics into biochar-based briquette: Product characteristics and reaction mechanisms. *Fuel Process. Technol.* **2023**, *247*, 107793. [CrossRef]
18. Xu, D.; Zhang, Z.; He, Z.; Wang, S. Machine learning-driven prediction and optimization of monoaromatic oil production from catalytic co-pyrolysis of biomass and plastic wastes. *Fuel* **2023**, *350*, 128819. [CrossRef]
19. Chen, W.-H.; Naveen, C.; Ghodke, P.K.; Sharma, A.K.; Bobde, P. Co-pyrolysis of lignocellulosic biomass with other carbonaceous materials: A review on advance technologies, synergistic effect, and future prospectus. *Fuel* **2023**, *345*, 128177. [CrossRef]
20. Biswas, B.; Sahoo, D.; Sukumaran, R.K.; Krishna, B.B.; Kumar, J.; Reddy, Y.S.; Adarsh, V.P.; Puthiyamadam, A.; Mallapureddy, K.K.; Ummalyma, S.B.; et al. Co-hydrothermal liquefaction of phumdi and paragrass an aquatic biomass: Characterization of bio-oil, aqueous fraction and solid residue. *J. Energy Inst.* **2022**, *102*, 247–255. [CrossRef]
21. Fricler, V.Y.; Nyashina, G.S.; Vershinina, K.Y.; Vinogrodskiy, K.V.; Shvets, A.S.; Strizhak, P.A. Microwave pyrolysis of agricultural waste: Influence of catalysts, absorbers, particle size and blending components. *J. Anal. Appl. Pyrolysis* **2023**, *171*, 105962. [CrossRef]
22. El-Sayed, S.A.; Mostafa, M.E. Pyrolysis and co-pyrolysis of Egyptian olive pomace, sawdust, and their blends: Thermal decomposition, kinetics, synergistic effect, and thermodynamic analysis. *J. Clean. Prod.* **2023**, *401*, 136772. [CrossRef]
23. Ge, Y.; Ding, S.; Kong, X.; Kantarelis, E.; Engvall, K.; Pettersson, J.B.C. Online monitoring of alkali release during co-pyrolysis/gasification of forest and agricultural waste: Element migration and synergistic effects. *Biomass Bioenergy* **2023**, *172*, 106745. [CrossRef]
24. Vasu, H.; Wong, C.F.; Vijiaretnam, N.R.; Chong, Y.Y.; Thangalazhy-Gopakumar, S.; Gan, S.; Lee, L.Y.; Ng, H.K. Insight into Co-pyrolysis of Palm Kernel Shell (PKS) with Palm Oil Sludge (POS): Effect on Bio-oil Yield and Properties. *Waste Biomass Valorization* **2020**, *11*, 5877–5889. [CrossRef]
25. Tauseef, M.; Ansari, A.A.; Khoja, A.H.; Naqvi, S.R.; Liaquat, R.; Nimmo, W.; Daood, S.S. Thermokinetics synergistic effects on co-pyrolysis of coal and rice husk blends for bioenergy production. *Fuel* **2022**, *318*, 123685. [CrossRef]
26. Donaldson, L.; Nanayakkara, B.; Harrington, J. Wood Growth and Development. In *Encyclopedia of Applied Plant Sciences*, 2nd ed.; Thomas, B., Murray, B.G., Murphy, D.J., Eds.; Academic Press: Oxford, UK, 2017; pp. 203–210.
27. Tekpetey, S.L.; Essien, C.; Appiah-Kubi, E.; Opuni-Frimpong, E.; Korang, J. Evaluation of the chemical composition and natural durability of natural and plantation grown African Mahogany Khaya ivorensis A. Chev. in Ghana. *J. Indian Acad. Wood Sci.* **2016**, *13*, 152–155. [CrossRef]
28. Usino, D.O.; Supriyanto; Ylitervo, P.; Pettersson, A.; Richards, T. Influence of Temperature and Time on Initial Pyrolysis of Cellulose and Xylan. *J. Anal. Appl. Pyrolysis* **2020**, *147*, 104782. [CrossRef]
29. Supriyanto; Usino, D.O.; Ylitervo, P.; Dou, J.; Sipponen, M.H.; Richards, T. Identifying the primary reactions and products of fast pyrolysis of alkali lignin. *J. Anal. Appl. Pyrolysis* **2020**, *151*, 104917. [CrossRef]
30. Usino, D.O.; Ylitervo, P.; Moreno, A.; Sipponen, M.H.; Richards, T. Primary interactions of biomass components during fast pyrolysis. *J. Anal. Appl. Pyrolysis* **2021**, *159*, 105297. [CrossRef]
31. Sukarni; Sudjito; Hamidi, N.; Yanuhar, U.; Wardana, I.N.G. Potential and properties of marine microalgae *Nannochloropsis* oculata as biomass fuel feedstock. *Int. J. Energy Environ. Eng.* **2014**, *5*, 279–290. [CrossRef]
32. Mitchual, S.J.; Frimpong-Mensah, K.; Darkwa, N.A. Evaluation of fuel properties of six tropical hardwood timber species for briquettes. *J. Sustain. Bioenergy Syst.* **2014**, *2014*, 44225. [CrossRef]
33. Woodyard, D. Chapter Four—Fuels and Lubes: Chemistry and Treatment. In *Pounder's Marine Diesel Engines and Gas Turbines*, 9th ed.; Woodyard, D., Ed.; Butterworth-Heinemann: Oxford, UK, 2009; pp. 87–142.
34. Abnisa, F.; Wan Daud, W.M.A. Optimization of fuel recovery through the stepwise co-pyrolysis of palm shell and scrap tire. *Energy Convers. Manag.* **2015**, *99*, 334–345. [CrossRef]
35. Mante, O.D.; Babu, S.P.; Amidon, T.E. A comprehensive study on relating cell-wall components of lignocellulosic biomass to oxygenated species formed during pyrolysis. *J. Anal. Appl. Pyrolysis* **2014**, *108*, 56–67. [CrossRef]
36. Zhou, S.; Xue, Y.; Sharma, A.; Bai, X. Lignin Valorization through Thermochemical Conversion: Comparison of Hardwood, Softwood and Herbaceous Lignin. *ACS Sustain. Chem. Eng.* **2016**, *4*, 6608–6617. [CrossRef]
37. Jin, F.; Cao, J.; Zhou, Z.; Moriya, T.; Enomoto, H. Effect of Lignin on Acetic Acid Production in Wet Oxidation of Lignocellulosic Wastes. *Chem. Lett.* **2004**, *33*, 910–911. [CrossRef]
38. Pan, W.-P.; Richards, G.N. Influence of metal ions on volatile products of pyrolysis of wood. *J. Anal. Appl. Pyrolysis* **1989**, *16*, 117–126. [CrossRef]
39. Richards, G.N.; Zheng, G. Influence of metal ions and of salts on products from pyrolysis of wood: Applications to thermochemical processing of newsprint and biomass. *J. Anal. Appl. Pyrolysis* **1991**, *21*, 133–146. [CrossRef]
40. Zhang, C.; Hu, X.; Guo, H.; Wei, T.; Dong, D.; Hu, G.; Hu, S.; Xiang, J.; Liu, Q.; Wang, Y. Pyrolysis of poplar, cellulose and lignin: Effects of acidity and alkalinity of the metal oxide catalysts. *J. Anal. Appl. Pyrolysis* **2018**, *134*, 590–605. [CrossRef]
41. Chang, G.; Miao, P.; Wang, H.; Wang, L.; Hu, X.; Guo, Q. A synergistic effect during the co-pyrolysis of *Nannochloropsis* sp. and palm kernel shell for aromatic hydrocarbon production. *Energy Convers. Manag.* **2018**, *173*, 545–554. [CrossRef]

42. Zulkafli, A.H.; Hassan, H.; Ahmad, M.A.; Din, A.T.M. Co-pyrolysis of palm kernel shell and polypropylene for the production of high-quality bio-oil: Product distribution and synergistic effect. *Biomass Convers. Biorefinery* **2022**. [CrossRef]
43. Azeez, A.M.; Meier, D.; Odermatt, J.; Willner, T. Fast Pyrolysis of African and European Lignocellulosic Biomasses Using Py-GC/MS and Fluidized Bed Reactor. *Energy Fuels* **2010**, *24*, 2078–2085. [CrossRef]
44. Uchegbulam, I.; Momoh, E.O.; Agan, S.A. Potentials of palm kernel shell derivatives: A critical review on waste recovery for environmental sustainability. *Clean. Mater.* **2022**, *6*, 100154. [CrossRef]
45. Ojha, D.K.; Vinu, R. Fast co-pyrolysis of cellulose and polypropylene using Py-GC/MS and Py-FT-IR. *RSC Adv.* **2015**, *5*, 66861–66870. [CrossRef]
46. Shurong, W. High-Efficiency Separation of Bio-Oil. In *Biomass Now*; Miodrag Darko, M., Ed.; IntechOpen: Rijeka, Croatia, 2013.
47. Panwar, N.L.; Paul, A.S. An overview of recent development in bio-oil upgrading and separation techniques. *Environ. Eng. Res.* **2021**, *26*, 200382. [CrossRef]
48. *ASTM E871-82*; Standard Test Method for Moisture Analysis of Particulate Wood Fuels. ASTM International: West Conshohocken, PA, USA, 2019. [CrossRef]
49. *ASTM Standard E872-82*; Standard Test Method for Volatile Matter in the Analysis of Particulate Wood Fuels. ASTM International: West Conshohocken, PA, USA, 2019.
50. *ASTM E1755-01*; Standard Test Method for Ash in Biomass. ASTM International: West Conshohocken, PA, USA, 2020. [CrossRef]
51. Sluiter, A. *Determination of Structural Carbohydrates and Lignin in Biomass: Laboratory Analytical Procedure (LAP): Issue Date, April 2008, Revision Date: August 2012 (Version 08-03-2012)*; National Renewable Energy Laboratory: Cole Boulevard, CO, USA, 2012.

molecules

MDPI

Article

A Novel Kinetic Modeling of Enzymatic Hydrolysis of Sugarcane Bagasse Pretreated by Hydrothermal and Organosolv Processes

João Moreira Neto [1,*], Josiel Martins Costa [2,*], Antonio Bonomi [3] and Aline Carvalho Costa [4]

1 Department of Engineering, Federal University of Lavras, Lavras 37200-000, MG, Brazil
2 School of Food Engineering, University of Campinas, Campinas 13083-862, SP, Brazil
3 Brazilian Biorenewables National Laboratory (LNBR), Brazilian Center for Research in Energy and Materials (CNPEM), Campinas 13083-100, SP, Brazil; antonio.bonomi@lnbr.cnpem.br
4 Laboratory of Fermentative and Enzymatic Process Engineering, School of Chemical Engineering, University of Campinas, Campinas 13083-852, SP, Brazil; al286899@unicamp.br
* Correspondence: joao.neto@ufla.br (J.M.N.); josiel@unicamp.br (J.M.C.)

Abstract: Lignocellulosic biomasses have a complex and compact structure, requiring physical and/or chemical pretreatments to produce glucose before hydrolysis. Mathematical modeling of enzymatic hydrolysis highlights the interactions between cellulases and cellulose, evaluating the factors contributing to reactor scale-up and conversion rates. Furthermore, this study evaluated the influence of two pretreatments (hydrothermal and organosolv) on the kinetics of enzymatic hydrolysis of sugarcane bagasse. The kinetic parameters of the model were estimated using the Pikaia genetic algorithm with data from the experimental profiles of cellulose, cellobiose, glucose, and xylose. The model considered the phenomenon of non-productive adsorption of cellulase on lignin and inhibition of cellulase by xylose. Moreover, it included the behavior of cellulase adsorption on the substrate throughout hydrolysis and kinetic equations for obtaining xylose from xylanase-catalyzed hydrolysis of xylan. The model for both pretreatments was experimentally validated with bagasse concentration at 10% *w*/*v*. The Plackett–Burman design identified 17 kinetic parameters as significant in the behavior of process variables. In this way, the modeling and parameter estimation methodology obtained a good fit from the experimental data and a more comprehensive model.

Keywords: biomass; lignocellulose; parameter estimation; pretreatment; simulation

Citation: Moreira Neto, J.; Costa, J.M.; Bonomi, A.; Costa, A.C. A Novel Kinetic Modeling of Enzymatic Hydrolysis of Sugarcane Bagasse Pretreated by Hydrothermal and Organosolv Processes. *Molecules* **2023**, *28*, 5617. https://doi.org/10.3390/molecules28145617

Academic Editor: Mohamad Nasir Mohamad Ibrahim

Received: 28 June 2023
Revised: 20 July 2023
Accepted: 21 July 2023
Published: 24 July 2023

1. Introduction

The situation of world food production is linked to the problem of gases generated by the greenhouse effect, which creates pressure for the development of promising technologies for converting lignocellulosic biomass [1]. Its reuse aims to cogenerate energy and value-added products [2,3]. In addition, its use and recycling reduce environmental problems regarding the disposal of agro-industrial waste. One of the ways to process this residue is through enzymatic hydrolysis. The cellulase and xylanase enzymes hydrolyze cellulose and xylan into fermentable sugars such as glucose and xylose, respectively [4]. The technique is considered sustainable due to its low cost, ease of acquisition, and great potential for obtaining sugars from agro-industrial residues [5,6].

The structure and the composition of lignocellulosic biomass vary according to its source. Cellulose stores the energy conserved by photosynthesis and is one of the main components in lignocellulosic biomass. It is entangled in the hemicellulose structure and covered by lignin [7]. In biomass, cellulose is also found in the form of cellobiose, which has two glucose molecules [8]. For efficient conversion to obtain biofuels, cellulose polymers must be decomposed into small molecules, allowing microbial assimilation. Therefore, the outer layer of lignin needs to be broken down [9].

Sugarcane bagasse is an important byproduct with a high cellulose content [10,11]. It consists of cellulose (45%), hemicellulose (28%), and lignin (18%) [12]. Its pretreatment facilitates subsequent operations, which can result in second-generation ethanol [13]. Pretreatment can occur by various methods, such as physical–chemical, chemical, and biological [14]. The organosolv chemical method is a process of extracting lignin from lignocellulosic materials using organic solvents or an aqueous-organic solution. It can swell the plant cell walls and disrupt the lignin structure. The physicochemical hydrothermal method consists of keeping the lignocellulosic material in the presence of water over a wide range of 160–240 °C, facilitating enzymatic digestibility. It can efficiently convert hemicelluloses into soluble compounds that are mainly composed of mono- and oligosaccharides. However, lignin is not removed effectively [15,16].

Approximate simulations of the hydrolysis operation model the process to improve, optimize, and maximize the value derived from the biomass used as substrate [17]. A reliable hydrolysis model must reflect the actual process, considering the reaction rates, mechanisms, and activities of the reactants and products [18]. Substrate heterogeneity, enzyme inhibition, and operating conditions, such as pH and temperature control, make system modeling difficult.

Enzymatic hydrolysis models are classified according to the number of solubilization activities and substrate state variables [19]. They are known as non-mechanistic, semi-mechanistic, functional, and structural models. In semi-mechanistic models, equations for reaction rates are generally described using Michaelis–Menten enzymatic kinetics with or without incorporating enzyme adsorption effects, temperature, substrate characteristics (accessible surface area, crystallinity, etc.), and product-inhibitory effects. Although this model encompasses many industrial projects, it may be limited in the effect of substrate characteristics. Godoy et al. [20] considered a semi-mechanistic approach to modeling batch and semi-batch enzymatic hydrolysis of hydrothermal sugarcane bagasse. The fit demonstrated that the model performed well compared to the experimental data. For reactor design and process simulation and optimization, these models are crucial as they describe structural information.

In this regard, this study aimed to validate and develop the mathematical modeling of the kinetic parameters in the enzymatic hydrolysis of sugarcane bagasse subjected to hydrothermal and organosolv pretreatment. Relevant experimental data were systematically accumulated. The proposed semi-mechanistic model included the non-productive adsorption of cellulase on lignin and cellulase inhibition by xylose. Moreover, Plackett–Burman sensitive analysis was used to identify the significant parameters in the kinetic model of enzymatic hydrolysis.

2. Results and Discussion

2.1. Estimation of Kinetic Parameters for Sugarcane Bagasse under Hydrothermal and Organosolv Pretreatment

The model considered as input data the enzyme concentrations and the initial concentrations of cellulose, lignin, hemicelluloses (xylan), cellobiose, glucose, and xylose of each assay. The initial concentration of the resulting sugars in the hydrolysis (cellobiose, glucose, and xylose) was 0 mg/mL in all assays. Table 1 displays the cellulose, lignin, and xylan concentration in hydrothermal bagasse (HB) and organosolv bagasse (OB), while endoglycanase/cellobiohydrolase (EG/CBH), β-glucosidase (BG) concentration, and xylanase activity used in the model simulation are shown in Table 2. Supplementary Materials: Table S1 displays the main genetic parameters selected in the Pikaia genetic algorithm. As reported previously, the default values and options were adopted [21–23].

Table 3 shows the estimated kinetic parameters for HB and OB data compared to the other studies.

Table 1. Concentration of cellulose, lignin, and xylan in HB and OB.

Pretreatment	Bagasse Concentration % (m/v)	Cellulose (mg/mL)	Lignin (mg/mL)	Xylan (mg/mL)
HB	4	24.4 ± 3.9	12.79 ± 0.02	0.84 ± 0.02
HB	6	36.6 ± 5.8	19.18 ± 0.03	1.26 ± 0.04
HB	8	48.9 ± 7.8	25.58 ± 0.04	1.68 ± 0.05
HB	10	61.1 ± 9.7	31.97 ± 0.05	2.10 ± 0.06
HB	12	73.3 ± 11.6	38.36 ± 0.06	2.52 ± 0.07
OB	4	34.8 ± 1.6	2.7 ± 1.1	1.8 ± 0.1
OB	6	52.2 ± 2.4	4 ± 2	2.65 ± 0.15
OB	8	69.5 ± 3.2	5.3 ± 2.2	3.5 ± 0.2
OB	10	87 ± 4	6.6 ± 2.7	4.42 ± 0.25
OB	12	104.3 ± 4.8	8 ± 3	5.3 ± 0.3

Table 2. EG/CBH, BG concentration, and xylanase activity for HB and OB.

Bagasse Concentration % (m/v)	EG/CBH (mg/mL)	BG (mg/mL)	Xylanase (U/mL)
4	0.420	0.1017	5.853
6	0.629	0.1526	8.779
8	0.839	0.2034	11.706
10	1.049	0.2543	14.632
12	1.259	0.3052	17.559

Table 3. Estimated kinetic parameters of the Pikaia genetic algorithm for HB and OB data.

	Current Study		[24]	[25]	[26]	[27]	[28]
Parameter	HB	OB	Acid Treatment—Wild Ryegrass	Cotton Pretreated with N-Oxide-N-Methylmorpholine	Sugarcane Straw	Corncob Stock	Wheat Straw
k_{1r} (mL/mg h)	19.178	20.289	16.5	32.10	0.509	94.72	1.224
k_{2r} (h^{-1})	196.56	230.82	267.6	263.89	165.7	432.16	252
k_{3r} (mL/mg h)	8.576	7.236	7.1	13.56	12.75	958.3	19.08
K_{1iG2} (mg/mL)	0.769	0.11	0.02	7.52	0.016	1.00×10^{-5}	0.0014
K_{1iG} (mg/mL)	0.03	4.875	0.1	0.34	0.710	7.33	0.073
K_{1iX} (mg/mL)	31.92	285.15	-	-	0.559	8.92	0.1007
K_{2iG} (mg/mL)	14.853	10.02	2.1	3.19	0.011	1.45×10^{-5}	3.9
K_{2m} (mg/mL)	22.48	11.295	25.5	11.63	47.20	0.022	24.3
K_{2iX} (mg/mL)	278.2	51.48	-	-	110.0	39.19	201
K_{3iG2} (mg/mL)	0.913	2.574	132.5	38.41	89.18	7.33	132
K_{3iG} (mg/mL)	0.853	0.167	0.01	1.58	0.551	1.15×10^{-3}	0.34
K_{3iX} (mg/mL)	86.38	180.22	-	-	0.581	6.13	0.029
k_4 (h^{-1})	18.066	37.56	-	-	13.46 [b]	167.27 [b]	9.72 [b]
Keq (U/mL)	0.0786	0.0066	-	-	-	-	-
K_{4iX} (mg/mL)	0.0111	0.0262	-	-	134.1	23.12	201
ks (mg/mL)	0.0354	45.8	-	-	-	-	-
[a] λ (h^{-1})	$\frac{0.1817L^{9.45}}{18.3549^{9.45}+L^{9.45}}$	0.2004	-	-	-	-	-

[a] Function of λ dependence on lignin concentration at concentrations from 0 to 12% w/v. [b] k_4 in: (mL/mg h).

In addition to the different experimental conditions and substrates, the parameter difference may be related to assumptions in the development of the models. The hydrolysis phenomena not considered in the model may affect the parameter estimation step. For example, Zheng et al. [24] considered the effect of non-productive adsorption of lignin. However, the effect of xylose inhibition on enzymes was neglected. Flores-Sánchez et al. [27]

added the formation and inhibition of arabinose in the model, while Prunescu and Sin [28] considered acetic acid formation and furfural inhibition. The model developed in this study, in addition to the phenomenon of non-productive adsorption of cellulase on lignin and inhibition of cellulase by xylose, included the behavior of cellulase adsorption on the substrate throughout the hydrolysis and kinetic equations for the formation of xylose from the hydrolysis of xylan catalyzed by xylanase.

Important differences in the order of magnitude, such as k_{3iX} and k_{4iX} in Table 3, demonstrate the contradictions and difficulty of direct comparison of model parameters. The comparisons can become inconvenient due to different experimental conditions, especially substrate and enzyme type, and operational factors, such as pH, temperature, and operation mode [29]. Differences in reaction rate constants can indicate biomass types and/or pretreatment method variations. These constants describe the susceptibility of the corresponding substrate in each of the reactions (r_1, r_2, and r_3). Analyzing the constants k_{1r}, k_{2r}, and k_{3r} of HB and OB, the order of magnitude followed the same pattern, $k_{2r} > k_{1r} > k_{3r}$. Therefore, in the hydrolysis of HB and OB, the homogeneous cellobiose hydrolysis reaction was faster than the heterogeneous reactions r_1 and r_3. The reaction rate constants k_{1r} and k_{3r} kept practically the same value in the hydrolysis of HB and OB. However, the reaction rate constant of cellobiose to glucose conversion (k_{2r}) was higher in OB hydrolysis than in HB hydrolysis.

The enzymatic inhibition by each sugar is inversely proportional to its inhibition constant. The inhibition of the r_1 reaction by cellobiose should be stronger than that by glucose because cellobiose is the product of the r_1 reaction. On the other hand, since glucose is the direct product of the r_3 reaction, this reaction is more strongly inhibited by glucose than by cellobiose. The xylose present in the r_1, r_2, and r_3 reactions is not a direct product of the cellulase and β-glucosidase catalyzed reactions. Therefore, it acts as a weaker inhibitor than glucose and cellobiose, respectively. In general, the comparison between the inhibition constants is consistent in that $K_{1iX} > K_{1iG} > K_{1iG_2}$ and $K_{3iX} > K_{2iG_2} > K_{3iG}$ (Table 3).

The enzyme inhibition constants were different for the two pretreatments. The studied biomasses also presented different values of enzyme inhibition constant [24,25]. The oscillation observed in the behavior of the HB and OB inhibition constants made it difficult to establish a relationship between these effects and the type of biomass pretreatment. In r_1, K_{1iG_2} decreased from 0.769 to 0.11 comparing HB with OB, indicating a greater inhibitory effect of cellobiose in converting cellulose to cellobiose in the hydrolysis of OB. However, in r_3, K_{3iG_2} exhibited a stronger inhibitory effect on HB than on OB. The cellulase inhibition constant by glucose in r_1 (K_{1iG}) demonstrated a greater effect on HB than on OB. However, in r_3, K_{3iG} exhibited a greater inhibitory effect on OB than on HB.

The parameter λ was dependent on the rate of decrease in cellulose surface area with the concentration of lignin (L) in HB, as illustrated in Figure 1. The increase in λ with increased lignin concentration was related to the decrease in glucose yield observed with increasing solids concentrations in enzymatic hydrolysis reactions. Lignin acted as a barrier that limited the availability of cellulose area accessible to enzyme action.

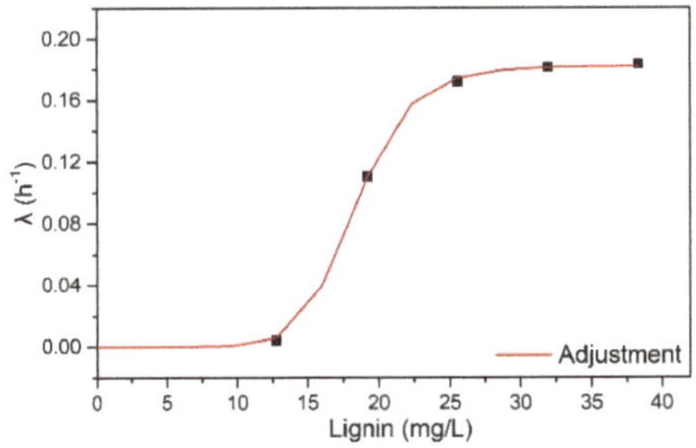

Figure 1. λ dependence with lignin concentration.

According to Figure 1, lignin exhibited a non-linear effect on λ, indicating two behaviors: (1) at low lignin concentrations (from 0 to 4% *w*/*v* HB), the λ profile was close to 0 and constant; (2) between concentrations of 12.8 to 38.4 mg/mL of lignin (from 4 to 12% m/v of HB), there was an exponential profile with a constant λ value from 25.6 mg/mL of lignin (8% m/v of HB).

The dependence of λ on L was fitted by a simple exponential function (Equation (1)), valid up to 40 mg/mL of lignin. The OriginPro 8.0 Software Levenberg–Marquardt algorithm (OriginLab, Northampton, MA, USA) estimated the following parameters: λ_{max}—0.1817 h^{-1}, n—9.45, k—18.354 h^{-1}, and R^2—0.999.

$$\lambda = \frac{\lambda_{max} L^n}{K^n + L^n} \quad (1)$$

As shown in Table 3, λ was not dependent on lignin concentration for OB. As a delignifying pretreatment, OB exhibited a lower lignin content (4.42%) than HB (31.97%), leading to a low lignin concentration in the enzymatic hydrolysis. The highest lignin concentration in the OB hydrolysis was 5.3 mg/mL (OB assay at 12% *w*/*v*).

Figure 2 shows the simulation of enzymatic hydrolysis of HB and OB using the kinetic parameters of Table 3. The model represented the behavior of the experimental profiles of cellulose and glucose for the assays used in the parameter estimation procedure (Figure 2a–f,i,j), model validation (Figure 2g,h), and xylose temporal profiles (Figure 3). A semi-mechanistic model was a suitable choice for the enzymatic hydrolysis model. It presents the smallest possible number of parameters, reducing the amount of experimental data for estimating the values [30].

In Figure 2a, although the model represented the trend of the experimental profiles for the assay with 4% *w*/*v* HB in the initial hours of hydrolysis, it exhibited a simulated glucose profile slightly below the experimental profile. In addition, the simulated cellulose profile was slightly overestimated compared to the experimental.

In Figure 2c,e,g,i, similarly to Figure 2a, the model accurately described the experimental profiles of cellulose and glucose after 12 h of reaction. However, between the reaction times of 1 and 12 h, the simulated glucose profiles were slightly overestimated compared to the experimental. In addition, the cellulose profiles showed lower values than the experimental. Experimental/simulated cellulose and glucose profiles of HB and OB demonstrated rates of cellulose consumption and formation of similar glucose. All glucose was formed in the initial 15 h of hydrolysis. As previously reported, this observation was compatible since the reaction rate constants k_{1r} and k_{3r} for HB were close to the constants for OB.

Figure 3 displays the xylose temporal profiles for HB and OB. For a concentration of 12% *w*/*v* of HB, a simulated lower xylose profile and an overestimated profile were observed compared to the experimental data. The difference between the experimental and simulated data was due to the experimental profile of xylose corresponding to the 12% *w*/*v* HB assay presenting a conversion concerning xylan of 107%, due to an imprecision in the stage of quantification of the experimental data or even analysis of the composition of HB hemicelluloses. Furthermore, the content of hemicelluloses in HB was 2.1%, increasing the uncertainty in quantification. The conversion of xylan to xylose in HB was close to 100%, while the average conversion of xylan to xylose in OB was around 65%. The difference between the conversion values for each biomass was related to the xylan/xylanase ratio used in the assays. The xylan concentration in the OB was twice that of the HB, and as the enzyme dosage was the same in each biomass, the xylan/xylanase ratio in the OB was half that of the HB.

The residual standard deviation (RSD) of the kinetic model for the cellulose, glucose, and xylose concentrations of the assays for HB and OB are shown in Table S2. The cellulose, glucose, and xylose profiles for HB and OB showed RSD values close to just over 20%, demonstrating the excellent fit of the model to the experimental data. Although the glucose and xylose profiles exhibited RSD values of 22.98 and 21.44%, respectively, Figure 3 shows that the model followed the trend of the experimental data even without perfectly overlapping them. Therefore, the semi-mechanistic model described and predicted the enzyme–substrate conversion kinetics for engineering applications. Within the range of its validity (substrate loading range of 4 to 12% w/v), it assists in designing technologies for pilot- and industrial-scale and optimization studies [31,32].

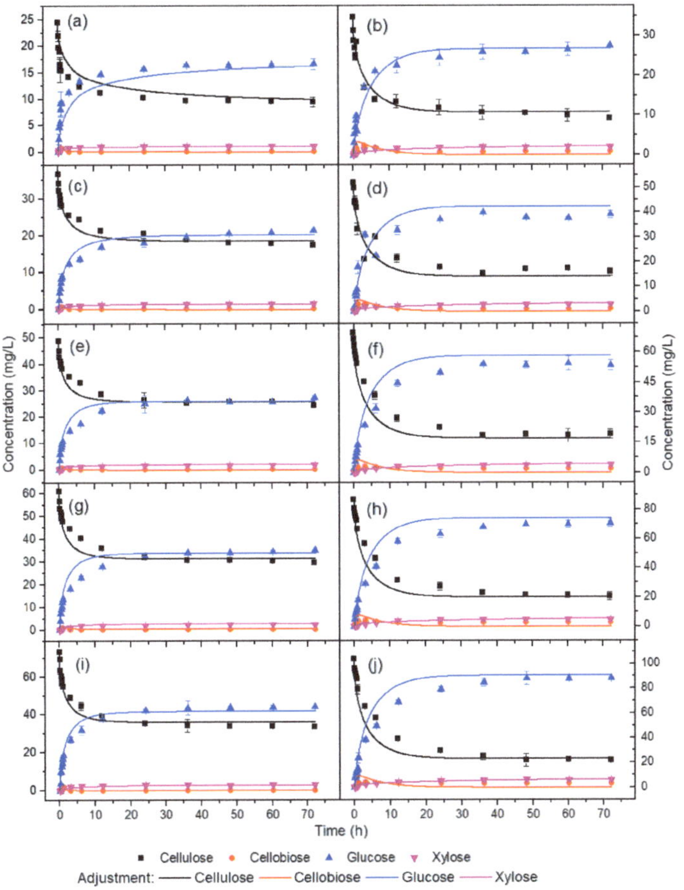

Figure 2. Temporal profiles of enzymatic hydrolysis with the following pretreatments and concentrations (% w/v): (**a**) HB 4, (**b**) OB 4, (**c**) HB 6, (**d**) OB 6, (**e**) HB 8, (**f**) OB 8, (**g**) HB 10, (**h**) OB 10, (**i**) HB 12, (**j**) OB 12.

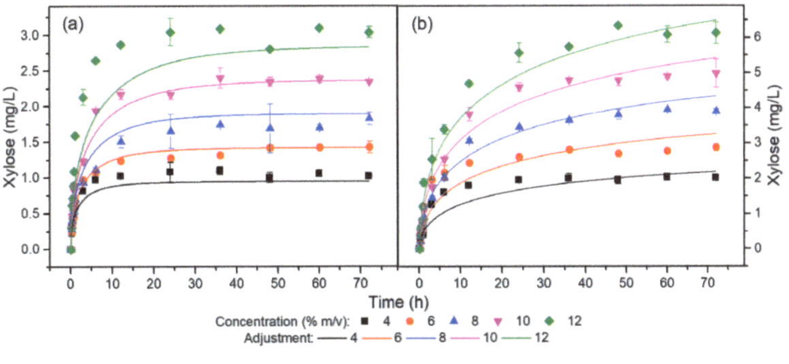

Figure 3. Xylose temporal profiles for (**a**) HB and (**b**) OB at different concentrations.

2.2. Plackett–Burman Design for the Selection of Significant Parameters

A Plackett–Burman design was used to analyze the dynamic behavior of the 17 parameters of the kinetic model of enzymatic hydrolysis of HB. The responses to represent the batch experiment were the concentrations of C, G_2, G, X_n, and X at intervals from 10 min to 1 h, and at times 3, 6, 12, 24, 36, 48, 60, and 72 h. The kinetic parameters of the model analyzed were k_{1r}, k_{2r}, k_{3r}, K_{1iG_2}, K_{1iG}, K_{1iX}, K_{2iG}, K_{2m}, K_{2iX}, K_{3iG_2}, K_{3iG}, K_{3iX}, k_4, K_{eq}, K_{4iX}, k_{4s}, and λ. Values were evaluated at low (−) and high (+) levels. The difference between these levels and the nominal values of the parameters was 10%. The initial conditions of C, G_2, G, X_n, and X were set at 61.07, 0, 0, 2.1, and 0 mg/mL. The enzyme loads of cellulase, β-glucosidase, and xylanase enzymes were fixed at 15 FPU/g of HB (1.049 mg protein/mL), 25 CBU/g of HB (0.2543 mg of protein/mL), and 14.632 U/mL, respectively. The effects of kinetic parameters on responses were determined by BP factorial design using Statistica 7.0 Software (Statsoft).

Figure 4 illustrates the effects of the kinetic parameters over the time of hydrolysis. The parameter λ at the beginning of hydrolysis had a low effect on cellulose and glucose concentration, as shown in Figure 4a,c. However, over time the influence of λ became significant. The effects of all kinetic parameters for cellobiose concentration decreased with hydrolysis time (Figure 4b), demonstrating greater influence in the initial hours of reaction. Therefore, the analysis in a time interval of the hydrolysis can eliminate the procedure of re-estimating important parameters for describing the enzymatic hydrolysis. According to Figure 4d,e, k_4 and k_{4iX} effects decreased along with the hydrolysis, tending to close to zero for the xylan and xylose concentration.

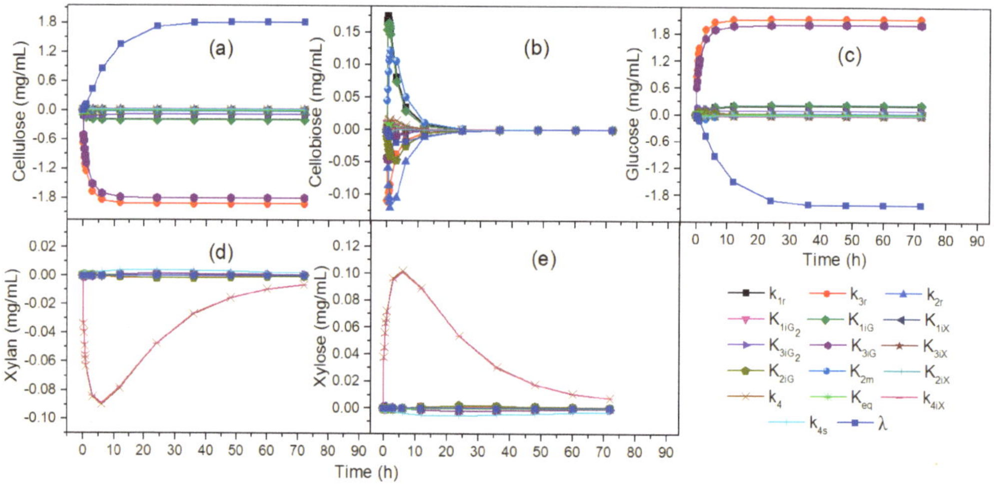

Figure 4. Effects of the kinetic parameters: (**a**) cellulose, (**b**) cellobiose, (**c**) glucose, (**d**) xylan, (**e**) xylose concentration.

Table 4 lists the effects of kinetic parameters on C, G_2, G, X_n, and X concentrations. The black areas indicate that the parameter had a great influence on the response, the gray areas indicate low influence, and the white areas indicate practically negligible influence. The significant parameters for cellulose concentration were the same for glucose concentration. Therefore, the parameters with the greatest influence on cellulose and glucose concentration were k_{3r}, K_{3iG}, and λ. The parameters of low influence were k_{1r} and K_{1iG}, while the parameters k_{2r}, K_{1iG_2}, K_{1iX}, K_{2iG}, K_{2m}, K_{2iX}, K_{3iG_2}, K_{3iX}, k_4, K_{eq}, K_{4iX}, and k_{4s} showed no influence on the cellulose and glucose concentration in the hydrolysis.

Table 4. Effect of kinetic parameters on C, G_2, G, X_n, and X concentrations during hydrolysis.

Parameter	Cellulose	Cellobiose	Glucose	Xylan	Xylose
k_{1r}	░	■	░		
k_{2r}		■			
k_{3r}	■		■		
K_{1iG2}					
K_{1iG}	░	■	░		
K_{1iX}					
K_{2iG}					
K_{2m}		■			
K_{2iX}					
K_{3iG2}					
K_{3iG}	■		■		
K_{3iX}					
k_4				■	■
K_{eq}					
K_{4iX}				■	■
k_{4s}				░	░
λ	■		■		

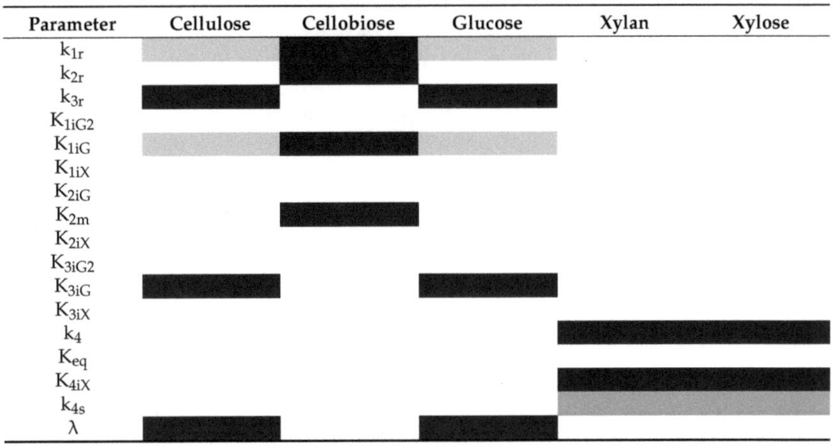

For cellobiose concentration, the significant parameters were k_{1r}, k_{2r}, K_{1iG}, and K_{2m}, while k_{3r}, K_{1iG_2}, K_{1iX}, K_{2iG}, K_{2iX}, K_{3iG_2}, K_{3iX}, k_4, K_{eq}, K_{4iX}, k_{4s}, and λ did not influence hydrolysis. The gradual reduction of all the effects of the parameters until null values at the end of the hydrolysis occurred due to the consumption of glucose to obtain glucose after the initial hours of hydrolysis, justifying the non-influence of the parameters in the final hours of reaction.

Significant parameters for xylan concentration were the same as for xylose concentration. The parameters k_4 and K_{4iX} showed significant effects under xylan and xylose concentrations. The parameter k_{4s} indicated low influence, while the parameters k_{1r}, k_{2r}, k_{3r}, K_{1iG_2}, K_{1iG}, K_{1iX}, K_{2iG}, K_{2m}, K_{2iX}, K_{3iG_2}, K_{3iG}, K_{3iX}, λ, and K_{eq} did not influence the concentration of xylan and xylose.

3. Materials and Methods

3.1. Biomass Preparation and Pretreatment

Sugarcane bagasse (*Saccharum officinarum*) was dried at room temperature for 4 days, ground in a cutting mill (Pulverisette 19, Fritsch), and sieved with a 0.5 mm sieve and stored. Two pretreatments (hydrothermal and organosolv) were applied separately to the biomass. For the hydrothermal pretreatment, 300 g of dried bagasse and 3 L of distilled water were added to a 7.5 L reactor. The reaction occurred for 10 min at 190 °C. After pretreatment, the HB was washed with water until the pH remained constant to remove soluble compounds in the hydrolyzate.

For the organosolv pretreatment, 300 g of dried bagasse and 3 L of a water/ethanol solution (1:2 *v*/*v*) were added to a 7.5 L reactor. The reaction occurred for 150 min at 190 °C. The pretreated bagasse was washed with a 1% (*m*/*v*) sodium hydroxide solution to solubilize residual lignin from the fibers. After pretreatment, the OB was washed with water until the pH stabilized. Both HB and OB were dried at room temperature and stored.

3.2. Enzymatic Activity

The enzymes used were cellulase from *Trichoderma reesei* (Celluclast 1.5 L from Novozyme) and β-glycosidase from *Aspergillus niger* (Novozym 188). The cellulolytic activity was quantified in filter paper units per milliliter (FPU/mL). In addition, a 15 mmol/L cellobiose solution was used to determine β-glucosidase activity. Its unit was expressed in units per milliliter (CBU/mL). Cellulase indicated an enzymatic activity of 75.69 FPU/mL and β-glucosidase 491.71 CBU/mL.

Xylanase activity was determined by assays in 96-well conical-bottomed plates as here described: 10 μL of culture supernatant with a suitable dilution was added to 40 μL of 0.05 M sodium citrate buffer (pH 4.8) and 50 μL of 0.05% beechwood xylan substrate. The mixture was incubated in a thermocycler at 50 °C for 10 min. Then, the reducing sugar was measured by the 3,5-dinitrosalicylic acid method by adding 100 μL of reagent and incubated again in a thermocycler at 99 °C for 5 min with immediate cooling. After cooling, 100 μL was transferred to a 96-well Elisa plate where the absorbance reading at 540 nm was performed using a Spectra Plate Reader Max 384 (Molecular Device). The xylanase activity was 738.34 U/mL. One unit (U) of xylanase was defined as the amount of enzyme required to release 1 μmol of xylose per minute under the test conditions.

3.3. Enzymatic Hydrolysis

The enzymatic hydrolysis of the pretreated bagasse was performed in a 1 L reactor containing a mixture of 250 mL of citrate buffer with pH adjusted to 4.8 and supplemented with 0.02% sodium azide per gram of biomass. Different concentrations of pretreated bagasse were added to each assay (4, 6, 8, 10, and 12% *w*/*v*), and cellulase and β-glucosidase loads were fixed at 15 FPU/g of bagasse and 25 CBU/g of bagasse. The hydrolysis reaction occurred in a jacketed reactor at 50 °C with 150 rpm stirring. Aliquots were collected in duplicate at intervals of 10 min to 1 h. After 1 h, aliquots in duplicate were collected at times 3, 6, 12, 24, 36, 48, 60, and 72 h. All samples were boiled to deactivate the enzymes.

3.4. Quantification of Sugars

The boiled samples were filtered through a GS membrane filter in cellulose ester with 0.22 μm pores (Millipore), and the monosaccharide content (glucose, cellobiose, xylose, and arabinose) was quantified in an HPLC system (model 1260 Infinity Agilent Technologies HPLC) equipped with a refractive index detector. Separation was performed on an Aminex HPX-87H column at 35 °C using a 0.01 mol/L H_2SO_4 solution prepared with filtered and degassed ultra-pure water (Milli-Q) as mobile phase with a flow rate of 0.6 mL/min [33]. The compound separated in the stationary phase was monitored with a refractive index detector at 30 °C for a run time of 20 min.

3.5. Kinetic Model

As shown in Figure 5, the hydrolysis of cellulose to glucose occurs by three reaction rates: r_1, r_2, and r_3. r_1 is the heterogeneous reaction rate for the production of cellobiose from cellulose catalyzed by the enzymes endoglycanase (EG) and cellobiohydrolase (CBH) adsorbed on cellulose; r_2 is the homogeneous reaction rate for the production of glucose from cellobiose catalyzed by the enzyme β-glucosidase (BG) in solution; and r_3 is the heterogeneous reaction rate for the production of glucose from cellulose catalyzed by the enzymes EG and CBH adsorbed on cellulose.

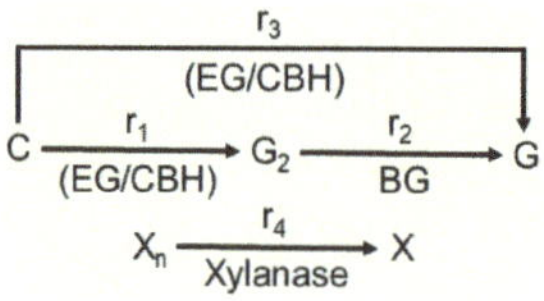

Figure 5. Mechanisms involved in the enzymatic hydrolysis of lignocellulosic biomass. r_1 is the heterogeneous reaction rate for the production of cellobiose from cellulose catalyzed by the enzymes endoglycanase (EG) and cellobiohydrolase (CBH) adsorbed on cellulose; r_2 is the homogeneous reaction rate for the production of glucose from cellobiose catalyzed by the enzyme β-glucosidase (BG) in solution; r_3 is the heterogeneous reaction rate for the production of glucose from cellulose catalyzed by the enzymes EG and CBH adsorbed on cellulose; and r_4 is the heterogeneous reaction rate for xylose production from xylan catalyzed by the enzyme xylanase adsorbed on xylan.

Each enzymatic reaction is potentially inhibited by the generated sugars (cellobiose and/or glucose) or other sugars present in the system, such as xylose. Although the enzymatic hydrolysis of cellulose does not result in xylose, it is present in the reaction medium in greater quantity, and its inhibitory effect on the catalytic action of cellulases has been reported [34]. Therefore, the r_4 reaction, referring to the hydrolysis of xylan in xylose, was coupled to the model. r_4 is the heterogeneous reaction rate for xylose production from xylan catalyzed by the enzyme xylanase adsorbed on xylan. Therefore, the model considered only the sugars generated by the reactions r_1 to r_4 (cellobiose, glucose, and xylose).

For the development of the model, the following considerations were addressed:

(1) The enzymatic adsorption follows the Langmuir adsorption isotherm, where the r_1 and r_3 reactions occur on the cellulose surface;
(2) Enzymatic deactivation by thermal and mechanical effects was negligible;
(3) The cellulosic matrix was uniform in terms of enzyme accessibility in the substrate, without distinction between the amorphous and crystalline fractions of cellulose;
(4) The cellulose consisted of EG, CBH, and low β-glucosidase activity. The model did not distinguish EG from CBH. Due to the low amount of β-glucosidase, the model considered the enzyme only from *Aspergillus niger*;
(5) The xylanase from the reaction medium was present in the cellulase used in the experimental assays;
(6) The hemicelluloses of HB and OB were composed solely of xylan;
(7) The conversion of cellobiose into glucose represented by r_2 occurred in solution and followed the Michaelis–Menten kinetics;
(8) The conversion of xylan into xylose represented by r_4 occurred in a single reaction, absent intermediate compounds such as xylobiose;
(9) The proportion of lignin exposed to the enzyme of the total lignin present in the pretreated bagasse was equal to 1, demonstrating that cellulose did not block the adsorption of enzymes on lignin [24];
(10) β-glucosidase did not adsorb to cellulose and lignin;
(11) The lignin of the pretreated biomass was not degraded during enzymatic hydrolysis.

Equations (2)–(8) describe the mass balance for the enzymatic hydrolysis model of HB and OB.

$$\frac{dC}{dt} = -r_1 - r_3 \quad (2)$$

$$\frac{dG_2}{dt} = 1.056r_1 - r_2 \quad (3)$$

$$\frac{dG}{dt} = 1.053r_2 + 1.111r_3 \quad (4)$$

$$\frac{dX_n}{dt} = -r_4 \quad (5)$$

$$\frac{dX}{dt} = 1.136r_4 \quad (6)$$

$$\text{Cellulase}: E_{1T} = E_{1b} + E_{1f} \quad (7)$$

$$\beta - \text{glucosidase}: E_{2T} = E_{2b} + E_{2f} \quad (8)$$

where C, G_2, G, X_n, and X are the concentrations (mg/mL), t is the reaction time (h), and r_1, r_2, r_3, and r_4 are the reaction rates (mg/mL h). In Equations (3), (4) and (6), 1.056, 1.111, 1.053, and 1.136 are the stoichiometric factors of the reactions; 1.056 is the cellulose to cellobiose conversion factor, 1.111 is the cellulose to glucose conversion factor, 1.053 is the cellobiose to glucose conversion factor, and 1.136 is the xylan to xylose conversion factor.

The Langmuir isotherm for the adsorption of EG/CBH on HB and OB, which contains cellulose and lignin, is described by Equation (9) [33].

$$E_{1b} = \frac{E_{max}K_pE_{1f}}{1+K_pE_{1f}}S \quad (9)$$

where E_{1b} is the concentration of EG/CBH adsorbed on pretreated bagasse (mg protein/mL), E_{max} is the maximum amount of EG/CBH adsorbed per unit mass of pretreated bagasse (mg protein/g substrate), E_{1f} is the concentration of free EG/CBH in solution considering the pretreated bagasse as substrate (mg protein/mL), K_p is the dissociation constant for the adsorption/desorption reaction of EG/CBH with pretreated bagasse (mL/mg protein), and S is the pretreated bagasse concentration (mg/mL).

During the enzymatic hydrolysis of HB and OB, the adsorption of EG/CBH in the biomass decreases as the hydrolysis reaction proceeds. This behavior led to different profiles for the Langmuir isotherms and, therefore, different parameters in each isotherm. In this way, the E_{max} and K_p parameters were correlated with the reaction time through Equations (10) and (11).

$$E_{max} = e^{a+bX+cX^2} \quad (10)$$

$$K_p = e^{a+bX+cX^2} \quad (11)$$

where a, b, and c are constants and X is the conversion (%). Values of a, b, and c are displayed in Table 5, as adjusted in a previous study [35].

Table 5. Values of constants and determination coefficient (R^2) for Langmuir isotherm.

Parameters	a	b	C	R^2
E_{max} (HB)	3.607	−0.00719	−0.0000772	0.980
K_p (HB)	0.2501	0.00134	−0.000146	0.979
E_{max} (OB)	3.383	−0.0027	0.000003	0.996
K_p (OB)	1.008	−0.014	0.00008	0.889

Equation (12) describes the Langmuir adsorption isotherm of cellulase on lignin.

$$E_{1bL} = \frac{E_{maxL}K_{pL}E_{1fL}}{1+K_{pL}E_{1fL}}L \quad (12)$$

where E_{1bL} is the concentration of EG/CBH adsorbed on lignin (mg protein/mL), E_{maxL} is the maximum amount of EG/CBH adsorbed per unit mass of lignin (mg protein/g lignin), E_{1fL} is the concentration of EG/CBH free in solution considering lignin as substrate (mg protein/mL), K_{pL} is the dissociation constant for the adsorption/desorption reaction of EG/CBH in lignin (mL/mg protein), and L is the lignin concentration (mg/mL).

As the adsorption of EG/CBH occurs on both cellulose and lignin, the amount of EG/CBH adsorbed on cellulose, E_{1bC}, is calculated by Equation (13).

$$E_{1bC} = E_{1b} - E_{1bL} \quad (13)$$

where E_{1bC} is the concentration of EG/CBH adsorbed on pretreated bagasse cellulose (mg protein/mL).

The kinetic rates described in Figure 5 are estimated according to Equations (14)–(17).

$$r_1 = \frac{k_{1r}E_{1bc}C}{1+\frac{G_2}{K_{1iG_2}}+\frac{G}{K_{1iG}}+\frac{X}{K_{1iX}}}e^{-\lambda t} \quad (14)$$

$$r_2 = \frac{k_{2r}E_{2f}G_2}{K_{2m}\left(1+\frac{G}{K_{2iG}}+\frac{X}{K_{2iX}}\right)+G_2} \quad (15)$$

$$r_3 = \frac{k_{3r}E_{1bc}C}{1+\frac{G_2}{K_{3iG_2}}+\frac{G}{K_{3iG}}+\frac{X}{K_{3iX}}}e^{-\lambda t} \quad (16)$$

where k_{ir} are reaction rate constants (i = 1 for reaction of cellulose to cellobiose (mL/mg h); i = 2 for reaction of cellobiose to glucose (mL/mg h); and i = 3 for reaction of cellulose to glucose (h^{-1}). K_{1iG_2}, K_{1iG}, and K_{1iX} are competitive inhibition constants of EG/CBH by cellobiose, glucose, and xylose in r_1 (mg/mL). K_{2iG} and K_{2iX} are competitive inhibition constants of β-glucosidase by glucose and xylose in r_2 (mg/mL). K_{3iG_2}, K_{3iG}, and K_{3iX} are competitive inhibition constants of EG/CBH by cellobiose, glucose, and xylose in r_3 (mg/mL). K_{2m} is the cellobiose saturation constant for β-glucosidase (mg/mL), E_{2f} is the concentration of free β-glucosidase in solution (mg/mL), and λ is the rate of decrease in cellulose surface area (h^{-1}).

The xylan reaction rate equation, r_4, considered xylan (X_n) a limiting term for xylose formation.

$$r_4 = \frac{K_4X_n}{1+\frac{X}{K_{4iX}}}\frac{X_n}{k_S+X_n} \quad (17)$$

where K_4 is a concentrated constant of the reaction rate of xylan in xylose with dependence on the xylanase dosage (h^{-1}), K_{4iX} is the competitive inhibition constant of xylanase per xylose (mg/mL), and k_S is the saturation constant for the term of substrate limitation (mg/mL). An analogous Langmuir-type dependence between the xylanase activity in the reaction medium and the K_4 constant was described by Equation (18), as reported previously [36].

$$r_4 = \frac{k_4E_3}{K_{eq}+E_3} \quad (18)$$

where E_3 is the activity of the xylanase enzyme (U/mL), k_4 is the maximum specific rate of hydrolysis of xylan to xylose (h^{-1}), and K_{eq} is the saturation constant for adsorption of the xylanase enzyme (U/mL).

3.6. Parameter Estimation

Compaq Visual Fortran software version 6.6 estimated and modeled the parameters. The model's resolution (Equations (2)–(6)) was performed in Fortran language with an integration algorithm based on the fourth-order Runge–Kutta method (IVPRK routine from the IMSL Math Library Fortran-90). The kinetic parameters of the model (k_{1r}, k_{2r}, k_{3r}, K_{1iG_2}, K_{1iG}, K_{1iX}, K_{2iG}, K_{2iX}, K_{2m}, K_{3iG_2}, K_{3iG}, K_{3iX}, λ, k_4, K_{eq}, K_{4iX}, and k_S) were estimated using the Pikaia genetic algorithm. This optimization subroutine implemented in Fortran estimated the parameters through the minimization of an objective function E(θ), according to Equation (19). θ is the vector that contains all the kinetic parameters to be optimized. The objective of the optimization is to find θ by minimizing the objective function using the experimental profiles of cellulose, cellobiose, and glucose in the bagasse concentration range defined in Section 3.3.

$$E(\theta) = \sum_{i=1}^{np}\sum_{i=1}^{m}\left[\left(\frac{C_{i,j}-Ce_{i,j}}{Ce_{i\,max}}\right)^2+\left(\frac{G_{2i,j}-G_2e_{i,j}}{G_2e_{i\,max}}\right)^2+\left(\frac{G_{i,j}-Ge_{i,j}}{Ge_{i\,max}}\right)^2+\left(\frac{X_{i,j}-Xe_{i,j}}{Xe_{i\,max}}\right)^2\right] \quad (19)$$

In Equation (19), np is the number of points referring to batch enzymatic hydrolysis samples, and m is the number of experimental profiles. $C_{ei,j}$, $G_2e_{i,j}$, $Ge_{i,j}$, and $Xe_{i,j}$ are the concentrations of cellulose, cellobiose, glucose, and xylose measured at sampling time i for profile j. $C_{i,j}$, $G_{2i,j}$, $G_{i,j}$, and $X_{i,j}$ are the concentrations of cellulose, cellobiose, glucose, and xylose predicted by the model at sampling time i for profile j. Ce_{imax}, G_2e_{imax}, Ge_{imax}, and Xe_{imax} are the maximum cellulose, cellobiose, glucose, and xylose concentrations measured at sampling time i for profile j. Four profiles of cellulose, cellobiose, glucose, and xylose obtained from the enzymatic hydrolysis of HB and OB were used to fit the kinetic model. The validation of the model considered the profiles at a concentration of 10% m/v.

4. Conclusions

A semi-mechanistic model for the enzymatic hydrolysis processes of HB and OB was successfully developed and validated. The model structure considered the representation of the cellulase/cellulose system, the inhibition by the sugars released by the biomass (cellobiose, glucose, and xylose), and enzymatic adsorption (productive and non-productive adsorption of enzymes on cellulose and lignin). The model adequately described the concentration of cellulose, glucose, and xylose and initial cellulase and β-glucosidase. However, the kinetic parameters estimated for HB and OB were different, demonstrating the influence of pretreatment on the morphological characteristics and composition of the biomass in enzymatic hydrolysis. This model differed from previous models. It included the behavior of cellulase adsorption on the substrate and kinetic equations for xylose formation. The Plackett–Burman design indicated that some parameters influenced the beginning of hydrolysis while others influenced the end. Therefore, this tool was important for determining the significant parameters of the model and thus eliminating the re-estimation procedure.

Supplementary Materials: The following supporting information can be downloaded at: https://www.mdpi.com/article/10.3390/molecules28145617/s1, Table S1. Technical settings selected in the Pikaia genetic algorithm; Table S2. Residual standard deviation for the kinetic model prediction of both pretreated bagasse.

Author Contributions: Conceptualization, J.M.N., A.B. and A.C.C.; methodology, J.M.N.; software, J.M.N.; validation, J.M.N.; formal analysis, J.M.N.; investigation, J.M.N.; resources, A.C.C.; data curation, J.M.N.; writing—original draft, preparation, J.M.N. and J.M.C.; writing—review and editing, A.C.C.; visualization, J.M.N. and J.M.C.; supervision, A.B. and A.C.C.; project administration, A.B. and A.C.C.; funding acquisition, A.C.C. All authors have read and agreed to the published version of the manuscript.

Funding: This study was supported by the Fundação de Amparo à Pesquisa do Estado de São Paulo (FAPESP, Brazil)—process number 2011/02743-5.

Institutional Review Board Statement: Not applicable.

Informed Consent Statement: Not applicable.

Data Availability Statement: The data presented in this study are available in the article.

Acknowledgments: The authors thank the sugar plant "Usina Tarumã" of the Raizen group (located in Tarumã, São Paulo, Brazil) for the supply of sugarcane bagasse.

Conflicts of Interest: The authors declare no conflict of interest.

Sample Availability: Not applicable.

References

1. Qi, F.; Wright, M. A novel optimization approach to estimating kinetic parameters of the enzymatic hydrolysis of corn stover. *AIMS Energy* **2016**, *4*, 52–67. [CrossRef]
2. Preethi, M.G.; Kumar, G.; Karthikeyan, O.P.; Varjani, S.; Banu, J.R. Lignocellulosic biomass as an optimistic feedstock for the production of biofuels as valuable energy source: Techno-economic analysis, Environmental Impact Analysis, Breakthrough and Perspectives. *Environ. Technol. Innov.* **2021**, *24*, 102080. [CrossRef]
3. Solarte-Toro, J.C.; Chacón-Pérez, Y.; Cardona-Alzate, C.A. Evaluation of biogas and syngas as energy vectors for heat and power generation using lignocellulosic biomass as raw material. *Electron. J. Biotechnol.* **2018**, *33*, 52–62. [CrossRef]
4. Gao, W.; Li, Z.; Liu, T.; Wang, Y. Production of high-concentration fermentable sugars from lignocellulosic biomass by using high solids fed-batch enzymatic hydrolysis. *Biochem. Eng. J.* **2021**, *176*, 108186. [CrossRef]
5. Suarez, C.A.G.; Cavalcanti-Montaño, I.D.; da Costa Marques, R.G.; Furlan, F.F.; da Mota e Aquino, P.L.; de Campos Giordano, R.; de Sousa, R. Modeling the Kinetics of Complex Systems: Enzymatic Hydrolysis of Lignocellulosic Substrates. *Appl. Biochem. Biotechnol.* **2014**, *173*, 1083–1096. [CrossRef]

6. Ashraf, M.T.; Thomsen, M.; Schmidt, J. Hydrothermal Pretreatment and Enzymatic Hydrolysis of Mixed Green and Woody Lignocellulosics from Arid Regions. *Bioresour. Technol.* **2017**, *238*, 369–378. [CrossRef]
7. Lee, H.V.; Hamid, S.B.A.; Zain, S.K. Conversion of Lignocellulosic Biomass to Nanocellulose: Structure and Chemical Process. *Sci. World J.* **2014**, *2014*, 631013. [CrossRef]
8. Keller, R.G.; Weyand, J.; Vennekoetter, J.B.; Kamp, J.; Wessling, M. An electro-Fenton process coupled with nanofiltration for enhanced conversion of cellobiose to glucose. *Catal. Today* **2021**, *364*, 230–241. [CrossRef]
9. Nadar, D.; Naicker, K.; Lokhat, D. Ultrasonically-Assisted Dissolution of Sugarcane Bagasse during Dilute Acid Pretreatment: Experiments and Kinetic Modeling. *Energies* **2020**, *13*, 5627. [CrossRef]
10. Mpatani, F.M.; Aryee, A.A.; Kani, A.N.; Wen, K.; Dovi, E.; Qu, L.; Li, Z.; Han, R. Removal of methylene blue from aqueous medium by citrate modified bagasse: Kinetic, Equilibrium and Thermodynamic study. *Bioresour. Technol. Rep.* **2020**, *11*, 100463. [CrossRef]
11. Silveira, M.H.L.; Chandel, A.K.; Vanelli, B.A.; Sacilotto, K.S.; Cardoso, E.B. Production of hemicellulosic sugars from sugarcane bagasse via steam explosion employing industrially feasible conditions: Pilot scale study. *Bioresour. Technol. Rep.* **2018**, *3*, 138–146. [CrossRef]
12. Nikodinovic-Runic, J.; Guzik, M.; Kenny, S.T.; Babu, R.; Werker, A.; Connor, K.E.O. Chapter Four-Carbon-Rich Wastes as Feedstocks for Biodegradable Polymer (Polyhydroxyalkanoate) Production Using Bacteria. In *Advances in Applied Microbiology*; Sariaslani, S., Gadd, G.M., Eds.; Academic Press: New York, NY, USA, 2013; Volume 84, pp. 139–200.
13. da Siva Martins, L.H.; Komesu, A.; Moreira Neto, J.; de Oliveira, J.A.R.; Rabelo, S.C.; da Costa, A.C. Pretreatment of sugarcane bagasse by OX-B to enhancing the enzymatic hydrolysis for ethanol fermentation. *J. Food Process Eng.* **2021**, *44*, e13579. [CrossRef]
14. Sun, S.; Sun, S.; Cao, X.; Sun, R. The role of pretreatment in improving the enzymatic hydrolysis of lignocellulosic materials. *Bioresour. Technol.* **2016**, *199*, 49–58. [CrossRef] [PubMed]
15. Cao, X.; Peng, X.; Sun, S.; Zhong, L.; Sun, R. Hydrothermal Conversion of Bamboo: Identification and Distribution of the Components in Solid Residue, Water-Soluble and Acetone-Soluble Fractions. *J. Agric. Food Chem.* **2014**, *62*, 12360–12365. [CrossRef]
16. Sun, S.; Cao, X.; Sun, S.; Xu, F.; Song, X.; Sun, R.-C.; Jones, G.L. Improving the enzymatic hydrolysis of thermo-mechanical fiber from Eucalyptus urophylla by a combination of hydrothermal pretreatment and alkali fractionation. *Biotechnol. Biofuels* **2014**, *7*, 116. [CrossRef] [PubMed]
17. Jiménez-Villota, D.S.; Acosta-Pavas, J.C.; Betancur-Ramírez, K.J.; Ruiz-Colorado, A.A. Modeling and Kinetic Parameter Estimation of the Enzymatic Hydrolysis Process of Lignocellulosic Materials for Glucose Production. *Ind. Eng. Chem. Res.* **2020**, *59*, 16851–16867. [CrossRef]
18. Huron, M.; Hudebine, D.; Lopes Ferreira, N.; Lachenal, D. Mechanistic modeling of enzymatic hydrolysis of cellulose integrating substrate morphology and cocktail composition. *Biotechnol. Bioeng.* **2016**, *113*, 1011–1023. [CrossRef] [PubMed]
19. Zhang, Y.-H.P.; Lynd, L.R. Toward an aggregated understanding of enzymatic hydrolysis of cellulose: Noncomplexed cellulase systems. *Biotechnol. Bioeng.* **2004**, *88*, 797–824. [CrossRef]
20. Godoy, C.M.d.; Machado, D.L.; Costa, A.C.d. Batch and fed-batch enzymatic hydrolysis of pretreated sugarcane bagasse–Assays and modeling. *Fuel* **2019**, *253*, 392–399. [CrossRef]
21. Ponce, G.H.S.F.; Moreira Neto, J.; De Jesus, S.S.; Miranda, J.C.d.C.; Maciel Filho, R.; Andrade, R.R.d.; Wolf Maciel, M.R. Sugarcane molasses fermentation with in situ gas stripping using low and moderate sugar concentrations for ethanol production: Experimental data and modeling. *Biochem. Eng. J.* **2016**, *110*, 152–161. [CrossRef]
22. Neto, J.M.; dos Reis Garcia, D.; Rueda, S.M.G.; da Costa, A.C. Study of kinetic parameters in a mechanistic model for enzymatic hydrolysis of sugarcane bagasse subjected to different pretreatments. *Bioprocess Biosyst. Eng.* **2013**, *36*, 1579–1590. [CrossRef] [PubMed]
23. Charbonneau, P.; Knapp, B. *User's Guide to Pikaia 1.0-Ncar Technical Note 418+IA*; National Center for Atmospheric Research: Boulder, CO, USA, 1995.
24. Zheng, Y.; Pan, Z.; Zhang, R.; Jenkins, B. Kinetic Modeling for Enzymatic Hydrolysis of Pretreated Creeping Wild Ryegrass. *Biotechnol. Bioeng.* **2009**, *102*, 1558–1569. [CrossRef]
25. Khodaverdi, M.; Jeihanipour, A.; Karimi, K.; Taherzadeh, M.J. Kinetic modeling of rapid enzymatic hydrolysis of crystalline cellulose after pretreatment by NMMO. *J. Ind. Microbiol. Biotechnol.* **2012**, *39*, 429–438. [CrossRef] [PubMed]
26. Angarita, J.D.; Souza, R.B.A.; Cruz, A.J.G.; Biscaia, E.C.; Secchi, A.R. Kinetic modeling for enzymatic hydrolysis of pretreated sugarcane straw. *Biochem. Eng. J.* **2015**, *104*, 10–19. [CrossRef]
27. Flores-Sánchez, A.; Flores-Tlacuahuac, A.; Pedraza-Segura, L.L. Model-Based Experimental Design to Estimate Kinetic Parameters of the Enzymatic Hydrolysis of Lignocellulose. *Ind. Eng. Chem. Res.* **2013**, *52*, 4834–4850. [CrossRef]
28. Prunescu, R.M.; Sin, G. Dynamic modeling and validation of a lignocellulosic enzymatic hydrolysis process–A demonstration scale study. *Bioresour. Technol.* **2013**, *150*, 393–403. [CrossRef]
29. Kadam, K.L.; Rydholm, E.C.; McMillan, J.D. Development and Validation of a Kinetic Model for Enzymatic Saccharification of Lignocellulosic Biomass. *Biotechnol. Prog.* **2004**, *20*, 698–705. [CrossRef]
30. Scott, F.; Li, M.; Williams, D.L.; Conejeros, R.; Hodge, D.B.; Aroca, G. Corn stover semi-mechanistic enzymatic hydrolysis model with tight parameter confidence intervals for model-based process design and optimization. *Bioresour. Technol.* **2015**, *177*, 255–265. [CrossRef]

31. Morales-Rodriguez, R.; Meyer, A.S.; Gernaey, K.V.; Sin, G. Dynamic model-based evaluation of process configurations for integrated operation of hydrolysis and co-fermentation for bioethanol production from lignocellulose. *Bioresour. Technol.* **2011**, *102*, 1174–1184. [CrossRef]
32. Morales-Rodriguez, R.; Meyer, A.S.; Gernaey, K.V.; Sin, G. A framework for model-based optimization of bioprocesses under uncertainty: Lignocellulosic ethanol production case. *Comput. Chem. Eng.* **2012**, *42*, 115–129. [CrossRef]
33. Machado, D.L.; Moreira Neto, J.; da Cruz Pradella, J.G.; Bonomi, A.; Rabelo, S.C.; da Costa, A.C. Adsorption characteristics of cellulase and β-glucosidase on Avicel, pretreated sugarcane bagasse, and lignin. *Biotechnol. Appl. Biochem.* **2015**, *62*, 681–689. [CrossRef] [PubMed]
34. Qing, Q.; Yang, B.; Wyman, C.E. Xylooligomers are strong inhibitors of cellulose hydrolysis by enzymes. *Bioresour. Technol.* **2010**, *101*, 9624–9630. [CrossRef] [PubMed]
35. Moreira Neto, J.; Machado, D.L.; Bonomi, A.; Gonçalves, V.O.O.; Martins, L.H.S.; Costa, J.M.; Costa, A.C. Cellulase adsorption on pretreated sugarcane bagasse during enzymatic hydrolysis. *Sugar Tech.* **2023**.
36. Drissen, R.E.T.; Maas, R.H.W.; Van Der Maarel, M.J.E.C.; Kabel, M.A.; Schols, H.A.; Tramper, J.; Beeftink, H.H. A generic model for glucose production from various cellulose sources by a commercial cellulase complex. *Biocatal. Biotransform.* **2007**, *25*, 419–429. [CrossRef]

molecules

MDPI

Article

Improving the Cellulose Enzymatic Digestibility of Sugarcane Bagasse by Atmospheric Acetic Acid Pretreatment and Peracetic Acid Post-Treatment

Yuchen Bai [1,2], Mingke Tian [1,2], Zhiwei Dai [3,4] and Xuebing Zhao [3,4,*]

1 China Food Flavor and Nutrition Health Innovation Center, Beijing Technology and Business University, Beijing 100048, China; baiyuchen@btbu.edu.cn (Y.B.)
2 Beijing Key Laboratory of Flavor Chemistry, Beijing Technology and Business University, Beijing 100048, China
3 Key Laboratory of Industrial Biocatalysis, Ministry of Education, Tsinghua University, Beijing 100084, China
4 Institute of Applied Chemistry, Department of Chemical Engineering, Tsinghua University, Beijing 100084, China
* Correspondence: zhaoxb@mail.tsinghua.edu.cn

Abstract: Pretreatment of sugarcane bagasse (SCB) by aqueous acetic acid (AA), with the addition of sulfuric acid (SA) as a catalyst under mild condition (<110 °C), was investigated. A response surface methodology (central composite design) was employed to study the effects of temperature, AA concentration, time, and SA concentration, as well as their interactive effects, on several response variables. Kinetic modeling was further investigated for AA pretreatment using both Saeman's model and the Potential Degree of Reaction (PDR) model. It was found that Saeman's model showed a great deviation from the experimental results, while the PDR model fitted the experimental data very well, with determination coefficients of 0.95–0.99. However, poor enzymatic digestibility of the AA-pretreated substrates was observed, mainly due to the relatively low degree of delignification and acetylation of cellulose. Post-treatment of the pretreated cellulosic solid well improved the cellulose digestibly by further selectively removing 50–60% of the residual linin and acetyl group. The enzymatic polysaccharide conversion increased from <30% for AA-pretreatment to about 70% for PAA post-treatment.

Keywords: sugarcane bagasse; lignocellulosic biomass; pretreatment; acetic acid delignification; kinetic modeling; peracetic acid post-treatment

Citation: Bai, Y.; Tian, M.; Dai, Z.; Zhao, X. Improving the Cellulose Enzymatic Digestibility of Sugarcane Bagasse by Atmospheric Acetic Acid Pretreatment and Peracetic Acid Post-Treatment. *Molecules* **2023**, *28*, 4689. https://doi.org/10.3390/molecules28124689

Academic Editor: Mohamad Nasir Mohamad Ibrahim

Received: 16 May 2023
Revised: 4 June 2023
Accepted: 8 June 2023
Published: 10 June 2023

1. Introduction

Sugarcane bagasse (SCB) is a fibrous matter that remains after sugarcane is crushed to extract juice. It is an abundant lignocellulosic biomass with a global production of more than 100 metric tons [1]. Various applications of SCB have been developed for producing chemicals, fuels, and materials [2,3]. One of the most promising utilizations of SCB is to produce second-generation bioethanol, because SCB has a relatively high cellulose content, having low ash content [4]. However, being similar to other lignocellulosic feedstock, SCB has to be pretreated prior to converting the cellulose to ethanol in order to increase the accessibility of cellulose towards cellulase enzymes for improving saccharafication efficiency. Various pretreatments have been employed to increase the cellulose digestibility of SCB, such as dilute acid [5], steam explosion [6], alkaline pretreatment [7], and ionic liquid pretreatment [8,9]. After pretreatment, the saccharafication efficacy of SCB can be increased to 70–90%. Organosolv pretreatment is a promising pretreatment technique to improve the digestibility of SCB [10,11]. This process employs various organic solvents, with or without the addition of exogenous catalysts at elevated temperatures, to remove a considerable part of lignin, as well as hemicelluloses, thus exposing cellulose [12]. Among the frequently used organic solvents, organic acids, such as acetic acid (AA), show some merits in biomass pretreatment [13]. First, AA has a strong hydrogen bonding ability with a Hildebrand solubility parameter (δ_1) of 25.8 $(J/cm^3)^{-1/2}$ close to that of lignin (δ_2 = 22.5 $(J/cm^3)^{-1/2}$), and, therefore, it is a good solvent to lignin [14]. Second, the delignification process can be performed under atmospheric pressure when additional mineral acids (H_2SO_4 or HCl) are used as catalysts [15]. Third, the AA formed by hydrolysis of the acetyl group in hemicelluloses can be a supplement of the loss AA during pretreatment. Fourth, the H^+ dissociated from AA can facilitate hemicellulose hydrolysis and lignin fragmentation, thus achieving a fractionation of lignocellulosic biomass. Actually, AA has been used for delignification of lignocellulosic biomass with or without addition of mineral acids for producing pulps from various lignocellulosic biomass feedstocks since the middle of last century [16,17]. However, much less attention has been paid to AA pretreatment of biomass, mainly because acetylation of cellulose takes place during AA delignification, which limits the recognition of cellulose substrates by cellulases [18]. Post-treatment of AA delignified substrates with a small amount of alkali has been found to well remove acetyl group and recover cellulose digestibility [19]. Nevertheless, alkaline deacetylation requires a washing step after AA pretreatment, which would lead to high water consumption. Moreover, the AA pretreatment parameters have to be re-optimized with consideration to acetylation and cellulose enzymatic digestibility. On the other hand, oxidative pretreatment with various oxidants pertains to be an effective method to remove lignin to achieve significantly increased cellulose digestibility. Various oxidative pretreatments have been developed, including wet oxidation, alkaline hydrogen peroxide, organic peracids, Fenton oxidation, and ozone oxidation [20]. After oxidative pretreatment, a considerable part of lignin could be removed with change in cell wall structure, thus greatly exposing cellulose to cellulase enzyme. The oxidative modification of lignin also can change its surface properties, including hydrophobility and surface charges, which might reduce the non-productive adsorption of enzymes on the lignin matrix. However, the direct use of oxidants for pretreatment might be too costly because a relatively large amount of oxidants is needed. Thus, a combination of a delignification process prior to oxidative treatment would reduce the amount of oxidants. Therefore, the objective of the present work is to optimize the AA pretreatment process, considering several response variables, including solid recovery degree of delignification, hemicelluloses (xylan removal), glucan recovery, acetyl group and cellulose digestibility, and investigation of the pretreatment kinetics in terms of total solid solubilization, delignification, and carbohydrate (holocellulose) solubilization. Peracetic acid (PAA) post-treatment was further employed to remove the residual lignin and the acetyl group, which avoids using a washing step after AA delignification, which can reduce the amount of PAA for delignification.

2. Results and Discussion

2.1. Effects of AA Pretreatment on Chemical Compositions

The CCD experimental and model predicted results are shown in Table 1. Corresponding three-dimensionsal response surface plots are shown in Supplementary Figures S1–S4, and the regressed quadratic polynomial models are shown as Equations (1)–(4). Statistical analysis (ANOVA) (Supplementary Tables S1–S4) indicated that Equation (1) was very significant (p = 0.0004 < 0.01) to predict the experimental data. The p-values for T, C_{AA}, C_{SA}, and t were very small (<0.01), indicating that these variables showed very significant effects on SY. The p-value for $X_1{}^2$ was less than 0.05, indicating that temperature also had a significant non-linear influence on SY. As illustrated in Figure S1, SY decreased with an increase in the levels of these variables. The decrease in SY was mainly due to the removal of lignin and carbohydrate (holocellulose). For DD, statistical analysis results (Supplementary Table S2) showed that T, C_{AA}, and C_{SA} had very significant effects, while t had no significant effect. This was mainly because most of the lignin was removed within the first two hours, while the residual lignin demonstrated much lower reactivity [15]. Therefore, prolonging pretreatment time showed no significant increase in DD (Supplementary Figure S2). Similarly, temperature also showed a significant non-linear influence on *DD*, while strong interactive effects were found between *T* and C_{AA}, as well as *T* and C_{SA}. As shown in Figure S3 and Table S3, all of these four variables showed very significant effects on *HS*, and *HS* increased with the levels of the variables. The solubilization of holocellulose was mainly attributed to solubilization of hemicelluloses (xylan) because hemicelluloses are much more susceptible than cellulose. It was found that cellulose solubilization during AA pretreatment was generally less than 20%, while xylan solubilization could be higher than 80%, depending on pretreatment conditions. Temperature had a significant impact on *HS* because the hydrolysis of the glucosidic bond was greatly temperature-dependent. The hydrolysis of polysaccharide was catalyzed by H^+, and, therefore, sulfuric acid demonstrated a very significant effect. The effects of AA concentration on *HS* can be attributed to two aspects. First, AA can dissociate H^+ as a supplemental catalyst for carbohydrate hydrolysis. Second, AA works as a solvent to dissolve the lignin fragments and facilitate the delignification reaction, and removing lignin can expose more carbohydrate. Therefore, apparently, C_{AA} demonstrated very significant effects on *HS*.

$$Y_{SY}\,(\%) = 79.28 - 6.77X_1 - 3.77X_2 - 2.48X_3 - 6.05X_4 - 1.25X_1X_2 - 0.77X_1X_3 - 1.28X_1X_4 + 0.013X_2X_3 - 1.36X_2X_4 - 0.16X_3X_4 - 1.71X_1^2 - 0.48X_2^2 + 0.34X_3^2 - 0.29X_4^2 \quad (1)$$

$$Y_{DD}\,(\%) = 36.74 + 8.67X_1 + 5.54X_2 + 0.84X_3 + 7.84X_4 + 2.60X_1X_2 - 0.14X_1X_3 + 4.36X_1X_4 + 0.87X_2X_3 + 1.37X_2X_4 + 1.80X_3X_4 + 2.80X_1^2 + 0.52X_2^2 + 0.40X_3^2 - 0.32X_4^2 \quad (2)$$

$$Y_{HS}\,(\%) = 22.61 + 7.30X_1 + 3.81X_2 + 3.03X_3 + 6.54X_4 + 0.85X_1X_2 + 0.84X_1X_3 + 0.76X_1X_4 - 0.57X_2X_3 + 0.73X_2X_4 + 0.39X_3X_4 + 0.10X_1^2 - 0.14X_2^2 - 1.35X_3^2 - 1.53X_4^2 \quad (3)$$

$$Y_{AGC}\,(\%) = 4.91 - 0.18X_1 + 0.25X_2 - 0.074X_3 - 0.043X_4 + 0.024X_1X_2 - 0.045X_1X_3 - 0.15X_1X_4 + 0.040X_2X_3 + 0.061X_2X_4 - 0.062X_3X_4 - 0.093X_1^2 + 0.030X_2^2 - 0.038X_3^2 - 0.19X_4 \quad (4)$$

Table 1. CCD experimental design and results of AA pretreatment of bagasse.

Run	Variables				*SY* (%)		*DD* (%)		*HS* (%)		*AGC* (%)		*EPC* (%)
	X_1	X_2	X_3	X_4	Exp.	Pred.	Exp.	Pred.	Exp.	Pred.	Exp.	Pred.	Exp.
1	−1	−1	−1	−1	93.30	91.40	24.77	28.11	3.01	2.01	4.60	4.53	7.08
2	1	−1	−1	−1	82.13	84.46	34.88	31.81	14.10	11.71	4.44	4.52	14.43
3	−1	1	−1	−1	92.10	89.06	25.70	29.51	3.44	7.61	4.80	4.78	6.76
4	1	1	−1	−1	78.50	77.12	40.88	43.61	19.82	20.71	4.74	4.86	18.34
5	−1	−1	1	−1	91.35	88.28	23.23	24.73	4.57	6.75	4.40	4.52	8.95
6	1	−1	1	−1	78.00	78.26	21.02	27.87	20.31	19.81	4.54	4.32	21.44
7	−1	1	1	−1	88.20	85.98	28.06	29.61	5.75	10.07	5.05	4.93	9.05
8	1	1	1	−1	75.80	70.96	40.65	43.15	24.29	26.53	4.76	4.83	21.16
9	−1	−1	−1	1	81.00	84.90	31.04	28.73	12.89	11.33	4.70	4.75	10.73
10	1	−1	−1	1	71.85	72.84	47.17	49.87	27.34	24.07	4.25	4.13	23.43
11	−1	1	−1	1	78.60	77.12	38.17	35.61	18.28	19.85	5.25	5.24	17.07
12	1	1	−1	1	57.88	60.06	68.43	67.15	37.48	35.99	4.74	4.72	23.36
13	−1	−1	1	1	81.00	81.14	30.99	32.55	17.45	17.63	4.83	4.49	14.72
14	1	−1	1	1	63.90	66.00	56.75	53.13	37.18	33.73	3.57	3.69	22.47
15	−1	1	1	1	76.65	73.40	39.62	42.91	20.80	23.87	5.10	5.14	16.8
16	1	1	1	1	52.60	53.26	72.96	73.89	41.31	43.37	4.61	4.44	21.68
17	−2	0	0	0	81.60	85.98	33.45	30.60	14.03	8.41	4.79	4.90	9.95
18	2	0	0	0	61.10	58.90	66.95	65.28	33.76	37.61	4.15	4.18	21.31
19	0	−2	0	0	88.35	84.90	28.99	27.74	8.65	14.43	4.36	4.53	10.87
20	0	2	0	0	64.20	69.82	53.14	49.90	37.23	29.67	5.56	5.53	20.67
21	0	0	−2	0	87.48	85.60	36.14	36.66	8.74	11.15	4.97	4.91	14.33
22	0	0	2	0	71.64	75.68	45.06	40.02	27.41	23.27	4.41	4.61	22.00
23	0	0	0	−2	84.40	90.22	27.14	19.78	7.51	3.41	4.28	4.24	12.10
24	0	0	0	2	69.70	66.02	48.30	51.14	27.21	29.57	3.90	4.06	21.44
25	0	0	0	0	77.40	79.28	37.09	36.74	23.36	22.61	4.56	4.91	17.89

Table 1. *Cont.*

Run	Variables				SY (%)		DD (%)		HS (%)		AGC (%)		EPC (%)
	X_1	X_2	X_3	X_4	Exp.	Pred.	Exp.	Pred.	Exp.	Pred.	Exp.	Pred.	Exp.
26	0	0	0	0	80.88	79.28	36.55	36.74	21.09	22.61	5.28	4.91	20.43
27	0	0	0	0	79.85	79.28	36.40	36.74	22.56	22.61	4.79	4.91	21.05
28	0	0	0	0	78.92	79.28	37.01	36.74	23.12	22.61	4.86	4.91	18.49
29	0	0	0	0	80.10	79.28	36.42	36.74	21.55	22.61	4.92	4.91	17.01
30	0	0	0	0	78.54	79.28	36.98	36.74	22.98	22.61	5.02	4.91	20.51

Acetylation of cellulosic solid took place during *AA* pretreatment. This was mainly caused by the esterification reaction between cellulose hydroxyl groups and *AA*. However, the tendency of effects of *T*, C_{AA}, C_{SA}, and *t* on *AGC* were somewhat different from those on *SY*, *DD* and *HS*. There was a maximal value for *AGC* depending on the pretreatment condition, and only temperature and *AA* concentration showed very significant impacts (Supplementary Table S4). It can be known, from Figure S4, that *AGC* increased with *AA* monotonically, while parabolic tendency was observed for *T*, C_{SA}, and *t*. This is because hemicellulose contains an acetyl group, and deacetylation also takes place with solubilization of hemicelluloses during *AA* pretreatment. As the removal of hemicelluloses and lignin increased, more cellulose was exposed, and acetylation of cellulose became significant. Therefore, the maximal *AGC* was dependent on the kinetic rates of hemicellulose deacetylation and cellulose acetylation.

The determination coefficients (R^2) of the quadratic polynomial models for *SY*, *DD*, *HS*, and *AGC* (Equations (1)–(4) were in the range of 0.9173, 0.9484, 0.9142, and 0.8614, respectively. Plots of actual data with the model predicted data shown in Figure 1 indicated that these models were accurate enough to predict most of the experimental data. Further analysis showed that linear relationships were found between *SY* and *DD*, as well as *SY* and *HS*, with R^2 of 0.8733 and 0.9240, respectively (Figure 2). Thus, *DD* and *HS* can be roughly estimated by *SY*, since *SY* was much easier to determine. However, no apparent mathematical relationship was found between *SY* and *AGC*.

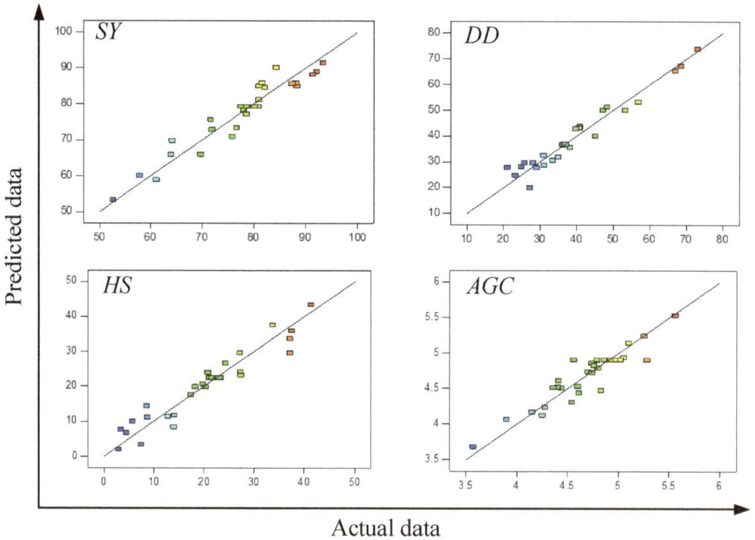

Figure 1. Comparison of actual experimental data and quadratic polynomial model-predicted data for *SY*, *DD*, *HS*, and *AGC*. The red color indicates a high level of the response variable, while the blue color indicates a low level.

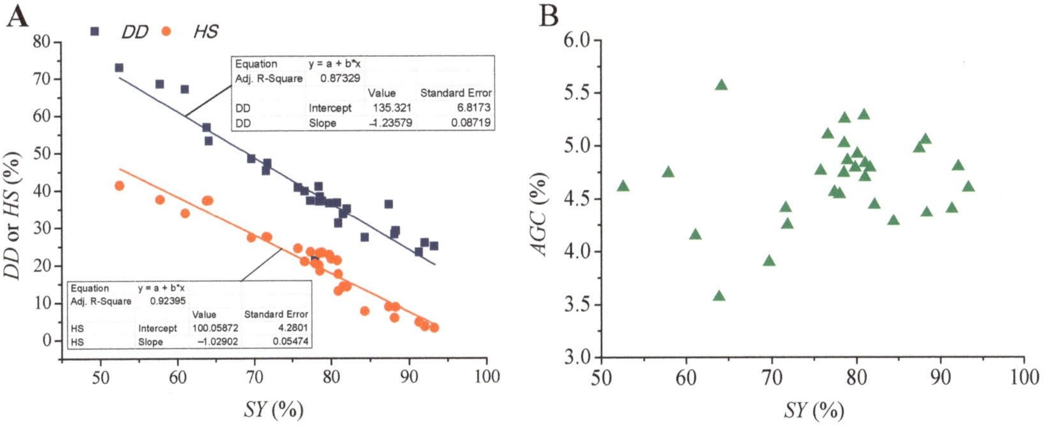

Figure 2. Plots of *SY* with *DD*, *HS*, or *AGC*. (**A**): *SY* with *DD* and *HS*; (**B**): *SY* with *AGC*.

2.2. Kinetics of Delignification and Solubilization of Holocellulose

In order to further understand the kinetic behaviours of AA pretreatment, the kinetics of total solid solubilization (S_S), degree of delignification (D_d), and holocellulose solubilization (H_S) were investigated using 75% AA with an addition of 0.1–0.4% SA at different temperatures. Both Saeman's model and the "potential degree of reaction (PDR)" model were considered to kinetically simulate the AA pretreatment process.

2.2.1. Saeman's Model

Saeman's model is the simplest kinetic model to describe biomass hydrolysis [21]. Taking delignification as an example, this model considers the rate of lignin solublization is a pseudo-homogeneous first-order reaction, with the residual (unreacted) lignin fraction defined as follows:

$$-\frac{\mathrm{d}C_L(t)}{\mathrm{d}t} = k_L C_L(t) \quad (5)$$

where $C_L(t)$ is the lignin concentration (g/L) in the pseudo-homogenous reaction system at time t. One can define the degree of delignification (D_d) at time t as the following equation:

$$D_d(t) = \frac{C_{L0} - C_L(t)}{C_{L0}} \quad (6)$$

where C_{L0} is initial lignin concentration (g/L) in the pseudo-homogenous reaction system, and we, thus, also can describe the rate of xylan solubilization as:

$$\frac{\mathrm{d}D_d}{\mathrm{d}t} = k_L(1 - D_d) \quad (7)$$

with $D_d(0) = 0$. The integral form of Equation (7) is:

$$D_d = 1 - \exp(-k_D t) \quad (8)$$

where the value of D_d is in the range of 0–1, and k_L is the rate constant. Therefore, the rate constant can be determined by ploting $\ln(1 - D_d)$ with t. Similarly, the integral forms of Saeman's models for total solid and holocellulose solubilizations are:

$$S_S = 1 - \exp(-k_S t) \quad (9)$$

$$H_S = 1 - \exp(-k_H t) \tag{10}$$

where S_S and H_S are ratio of total solid solubilization and degree of delignification, respectively; k_S and k_H are corresponding rate constants, which can be determined by plotting $\ln(1 - S_S)$ and $\ln(1 - H_S)$ with t, respectively. According to experimental data, the plots of $\ln(1 - S_S)$, $\ln(1 - D_d)$ and $\ln(1 - H_S)$ with t are shown in Supplementary Figures S5–S7. However, as shown in these figures, $\ln(1 - S_S)$, $\ln(1 - D_d)$, and $\ln(1 - H_S)$ actually had apparent deviation from linear relation. The determination coefficients (R^2) were in the range of 0.46–0.98, with most of them being less than 0.6. It indicated that AA-pretreatment did not follow the kinetics described by Saeman's model.

2.2.2. The "Potential Degree of Reaction (PDR)" Model

The "Potential degree of reaction (PDR)" model was developed by Zhao et al. to describe the kinetic behaviors of dilute acid [22] and organosolv pretreatments [15,23]. This model has been found to fit well with different chemical pretreatments of various biomass feedstocks [24]. The PDR model was developed based on the multilayered structure of the plant cell wall and heterogeneity of the reaction system. The biomass component distributed in different layers of the cell wall showed different reactivity, depending on the reaction severity. Thus, a parameter representing the potential degree of reaction, such as potential degree of delignification, was introduced into the kinetic model. In the present work, potential degree of total solid solubilization (δ_{SS}), delignification (δ_{DD}), and holocellulose solubilization (δ_{HS}) were proposed. The PRD models for total solid solubilization, delignification, and holocellulose solubilization, thus, can be expressed as Equations (11)–(13), with integral forms shown as Equations (14)–(16), respectively.

$$\frac{dS_S}{dt} = k_S(\delta_{SS} - S_S) \tag{11}$$

$$\frac{dD_d}{dt} = k_L(\delta_{DD} - D_d) \tag{12}$$

$$\frac{dH_S}{dt} = k_H(\delta_{SS} - H_S) \tag{13}$$

$$S_S = \delta_{SS}[1 - \exp(-k_S t)] \tag{14}$$

$$D_d = \delta_{DD}[1 - \exp(-k_D t)] \tag{15}$$

$$H_S = \delta_{HS}[1 - \exp(-k_H t)] \tag{16}$$

The rate constants and parameters for PDR can thus be determined according to experimental results. The experimental and model predicted data are shown in Figures 3–5, and the corresponding regressed rate constants (k, including k_S, $k_{L,}$ and k_H) and parameters for potential degree of reaction (δ, including δ_{SS}, δ_{DD}, and δ_{HS}) are shown in Table 2. It is clear that the PDR model demonstrated much higher accuracy than Saeman's model to describe the kinetics of *SS*, *DD*, and *HS* during 75% AA pretreatment. The determination coefficients of the models are in the range of 0.95–0.99, indicating that the model fitted the experimental results very well.

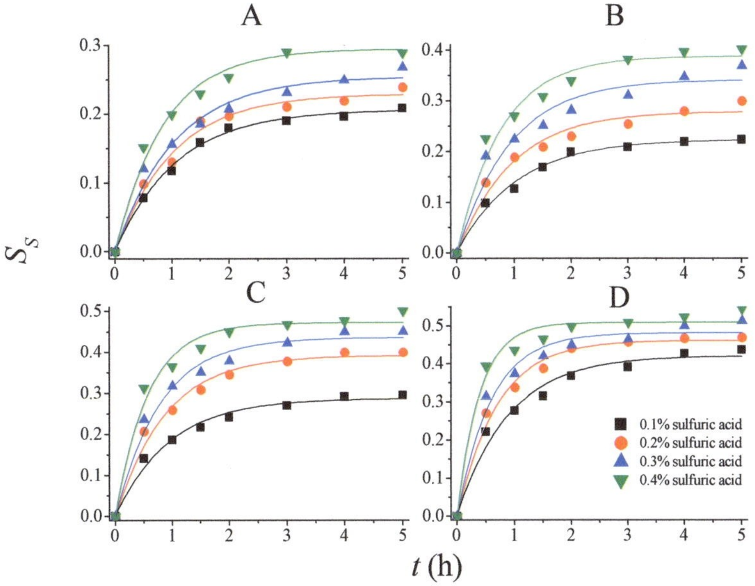

Figure 3. Experimental and model-predicted kinetic data for total solid solubilization at different temperatures. (**A**): 80 °C; (**B**): 90 °C; (**C**): 100 °C; and (**D**): 110 °C.

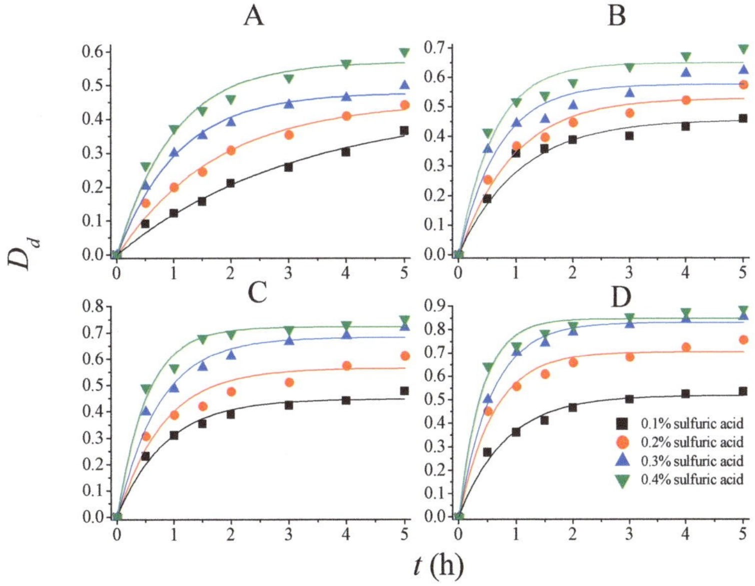

Figure 4. Experimental and model-predicted kinetic data for degree of delignification at different temperatures. (**A**): 80 °C; (**B**): 90 °C; (**C**): 100 °C; and (**D**): 110 °C.

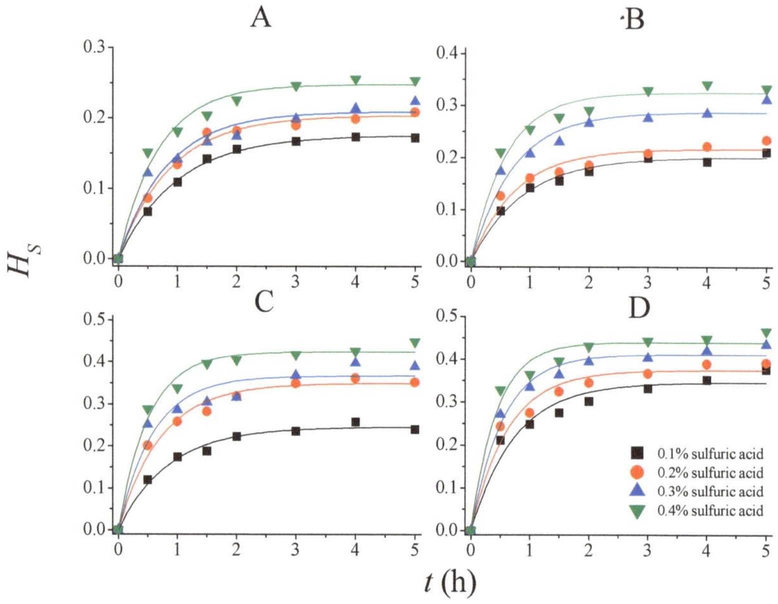

Figure 5. Experimental and model-predicted kinetic data for holocellulose solubilization at different temperatures. (**A**): 80 °C; (**B**): 90 °C; (**C**): 100 °C; and (**D**): 110 °C.

Table 2. Determined parameters for *SS*, *DD*, and *HS* by the PDR model, according to experimental data, with 75% AA.

T (°C)	C_{SA} (mol/L)							
	0.0102		0.0204		0.0306		0.0408	
	δ	k (h^{-1})	δ	k (h^{-1})	δ	k (h^{-1})	δ	k (h^{-1})
				For *SS*				
80	0.2063	0.9280	0.2297	0.9860	0.2543	0.9598	0.2951	1.1617
90	0.2239	0.9684	0.2790	1.0650	0.3413	1.0901	0.3884	1.2900
100	0.2876	1.0570	0.3924	1.1641	0.4367	1.2756	0.4726	1.7496
110	0.4207	1.1189	0.4625	1.4572	0.4819	1.7093	0.5092	2.5670
				For *DD*				
80	0.4437	0.3116	0.4548	0.5714	0.4794	0.9436	0.5716	1.0047
90	0.4567	0.9447	0.5297	1.0737	0.5760	1.4615	0.6493	1.6422
100	0.4485	1.1850	0.5663	1.2515	0.6844	1.3721	0.7232	1.9464
110	0.5190	1.2391	0.7063	1.6942	0.8315	1.7779	0.8473	2.5142
				For *HS*				
80	0.1759	1.0153	0.2039	1.1438	0.2093	1.2217	0.2477	1.4665
90	0.2007	1.1562	0.2170	1.3381	0.2874	1.4000	0.3246	1.6903
100	0.2460	1.1801	0.3499	1.3930	0.3674	1.6862	0.4250	1.9468
110	0.3462	1.3510	0.3749	1.6038	0.4118	1.8492	0.4400	2.2970

Both δ and k increased with C_{SA} and temperature. This is because the fragmentation of lignin and degradation of polysaccharides can be facilitated by acid-catalysts. The fragmentation of lignin during organosolv pretreatment is mainly attributed to the cleavage of ether linkages. In acidic systems, easily hydrolysable α-ether linkages are most readily broken. However, cleavage of β-aryl ether bonds also takes place, and this may be more important than cleavage of α-ether linkages for lignin fragmentation, especially under strong acid system [25]. The effect of T on the reaction rate can be described by the

Arrhenius equation. In order to correlate k or δ with C_{SA} and T, an extended Arrhenius equation or modified logistic equation is used as follows, respectively, according to the work of Dong et al. [24].

$$k = k_0 \exp(-\frac{E_a}{RT})C_{SA}^{\alpha} \tag{17}$$

$$\delta = 1 - \frac{1}{1 + AC_{SA}^{m}R_0^{n}} \tag{18}$$

where; k_0 and E_a are pre-exponential factor and activation energy, respectively; R_0 is the temperature-dependent severity factor ($R_0 = \exp(\frac{T' - 100}{14.75})$, where T' is reaction temperature in unit of °C), A is an adjustable parameter, and m and n are corresponding order parameters for C_{SA} and R_0. Equations (17) and (18) can be expressed as:

$$\ln k = \ln k_0 - \frac{E_a}{RT} + \alpha \ln C_{SA} \tag{19}$$

$$\ln\left(\frac{1}{1-\delta} - 1\right) = \ln A + m \ln C_{SA} + n \ln R_0 \tag{20}$$

Therefore, based on the data listed in Table 2, corresponding kinetic constants could be determined by multiple linear regression, and the results are shown in Table 3. Corresponding comparison of actual values (shown in Table 2) and regressed model-predicted values for k and δ are shown in Figure 6. The results demonstrated that Equations (17) or (18) could well correlate the relation of k or δ with C_{SA} and T. Therefore, the S_S, D_d and H_S can be calculated by the following integral kinetic models:

$$S_S = (1 - \frac{1}{1 + 3.2135C_{SA}^{0.4284}R_0^{0.5011}})\{1 - \exp\left[-343.52\exp\left(-\frac{14573}{RT}\right)C_{SA}^{0.2348}t\right]\} \tag{21}$$

$$D_d = (1 - \frac{1}{1 + 30.3386C_{SA}^{0.7421}R_0^{0.5281}})\{1 - \exp\left[-9.3887 \times 10^5\exp\left(-\frac{35533}{RT}\right)C_{SA}^{0.5176}t\right]\} \tag{22}$$

$$H_S = (1 - \frac{1}{1 + 2.2262C_{SA}^{0.4003}R_0^{0.4619}})\{1 - \exp\left[-393.52\exp\left(-\frac{13636}{RT}\right)C_{SA}^{0.3021}t\right]\} \tag{23}$$

Table 3. Determination of k and δ by multivariate linear regression.

k	k_0	E_a (kJ/mol)	α	R^2	F	P
k_S	343.52	14.573	0.2348	0.8272	28.7148	0.0000
k_L	9.3887×10^5	35.533	0.5176	0.8448	35.3855	0.0000
k_H	393.52	13.636	0.3021	0.9386	99.3466	0.0000
δ	A	m	n	R^2	F	P
δ_{SS}	3.2135	0.4284	0.5011	0.9627	167.9312	0.0000
δ_{DD}	30.3386	0.7421	0.5281	0.8595	39.7707	0.0000
δ_{HS}	2.2262	0.4003	0.4619	0.9434	108.4232	0.0000

The results demonstrated that the activation energy for delignification under the experimental condition (75% AA) was 35.533 kJ/mol, which was much lower than that reported by other researchers [15,26,27]. However, it should be noted that the activation energy determined in the present work was the apparent activation energy using 75% AA for delignification. If the contribution of AA was excluded, the determined activation energy (intrinsic activation energy) should be higher. The experimental results also indicated that the S_S and H_S had smaller activation energies, indicating that they are less sensitive to temperature. Actually, holocellulose contains cellulose and hemicellulose, and hemicelluloses are much easier than cellulose to degrade during AA pretreatment. It has been found that, under the used severest pretreatment conditions (0.0408 mol/L SA and

90% AA), S_S reached about 57%, with both D_d and xylan solubilization being higher than 90%, while cellulose solubilization was less than 25%. Therefore, because cellulose was much inerter than lignin, the k and δ for H_S were apparently smaller than those of D_d.

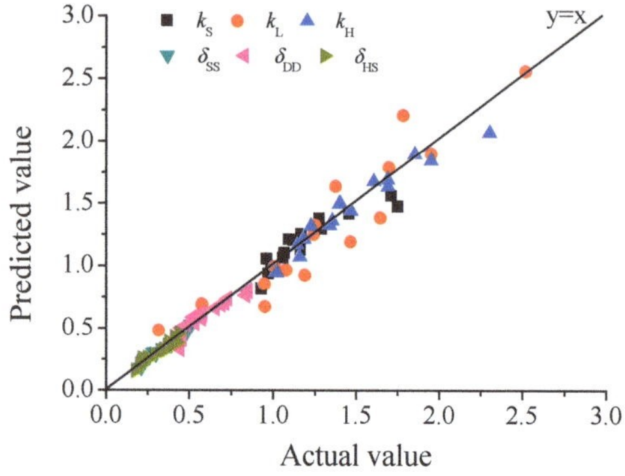

Figure 6. Comparison of actual values and model predicted values for k and δ.

2.3. *Effects of AA Pretreatment on the Enzymatic Hydrolysis of Pretreated Solid*

The CCD experimental results for effects of different parameters in AA pretreatment on enzymatic polysaccharide conversion (EPC) are shown in Table 1. Corresponding statistical analysis results are shown in Table 4. T and C_{SA} showed very significant effects ($p < 0.01$) on *EPC*, while C_{AA} and t showed significant effects ($p < 0.05$). The interactive effects of these variables showed no significant effects ($p > 0.05$), but the effects of quadric terms of T and C_{AA} were significant ($p > 0.05$), indicating that T and C_{AA} demonstrated significant non-linear effects on *EPC*. From the three-dimensional surface plots (Figure 7), it can be known that *EPC* increased with T, t, and C_{SA} continuously. However, the effect of C_{AA} showed a parabolic tendency with the maximum *EPC* obtained at around 75%. Further increase in C_{AA} oppositely decreased *EPC*. This was mainly because the severer acetylation of cellulosic solid took place at higher C_{AA}, leading to weakening of cellulose recognition by cellulase enzymes [18]. Therefore, based on the above optimization and kinetic results, AA concentration should be controlled at about 75% in order to obtain a compromised optimum *EPC*.

Table 4. ANOVA for the response surface quadratic model for enzymatic polysaccharide conversion.

Source	Sum of Squares	df	Mean Squares	*F* Value	*p*-Value $p > F$
Model	713.04	14	50.93	9.57	<0.0001
X_1	399.11	1	399.11	74.99	<0.0001
X_2	38.94	1	38.94	7.32	0.0163
X_3	38.53	1	38.53	7.24	0.0168
X_4	158.77	1	158.77	29.83	<0.0001
X_1X_2	1.84	1	1.84	0.35	0.5650
X_1X_3	0.030	1	0.030	0.0059	0.9414
X_1X_4	8.87	1	8.87	1.67	0.2164
X_2X_3	4.79	1	4.79	0.90	0.3580
X_2X_4	1.08	1	1.08	0.20	0.6593
X_3X_4	10.42	1	10.42	1.96	0.1821
X_1^2	26.45	1	26.45	4.97	0.0415
X_2^2	24.60	1	24.60	4.62	0.0483
X_3^2	3.33	1	3.33	0.62	0.4415
X_4^2	13.32	1	13.32	2.50	0.1344

Table 4. *Cont.*

Source	Sum of Squares	df	Mean Squares	*F* Value	*p*-Value $p > F$
Residual	79.83	15	5.32		
Lack of Fit	66.17	10	6.62	2.42	0.1705
Pure Error	13.66	5	2.73		
Cor Total	792.87	29			

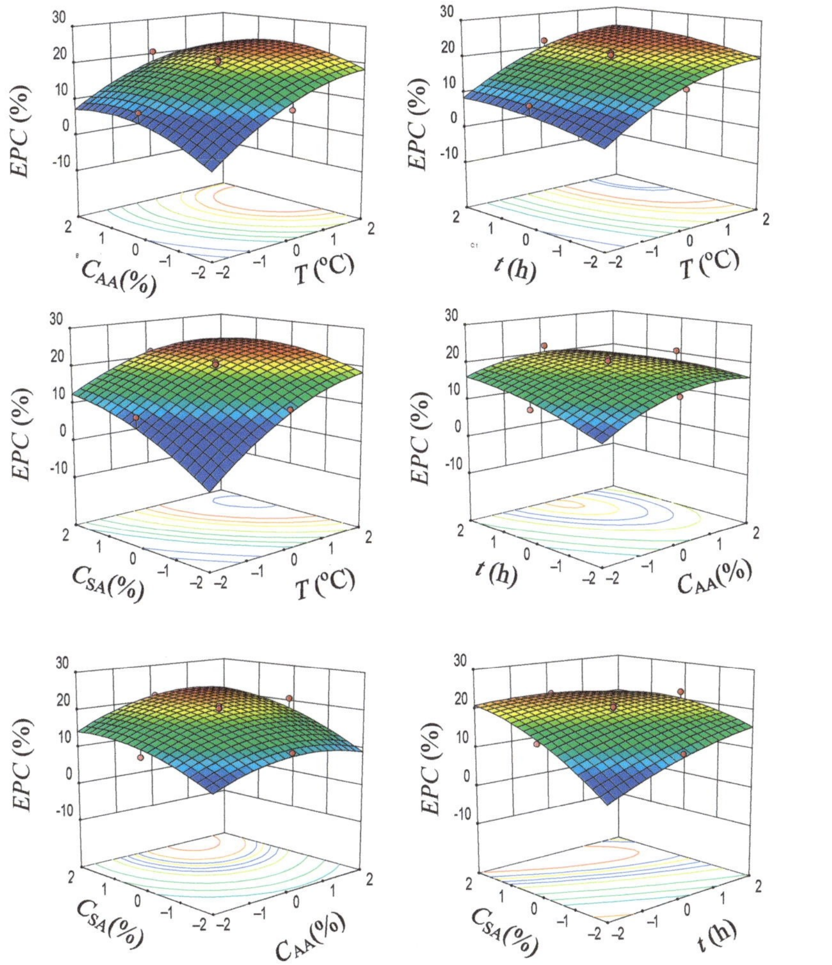

Figure 7. Three-dimensional surface plots of effects of temperature (T, °C), AA concentration (C_{AA}, wt%), pretreatment time (t, h), and sulfuric acid concentration (C_{SA}, wt%) on enzymatic polysaccharide conversion (*EPC*, %) of AA-pretreated substrates. The red color indicates a high level and the blue indicates a low level.

2.4. Effect of PAA Post-Treatment on Cellulose Hydrolysis

Table 1 showed that the obtained D_d was generally in the range of 20–70%, mainly depending on the AA concentration and reaction temperature, which increased the cellulose enzymatic digestibility to some extents. However, the *EPC* was still very low (less than 30%). This was mainly because the D_d was not high enough. AA-pretreatment was conducted under a relatively mild condition (<110 °C), and such a mild condition could not cause a significant modification of lignin structure and migration of lignin in the cell wall layers. Therefore, in order to obtain a greatly improved cellulose digestibility, a high D_d should be obtained. Moreover, the acetyl group introduced in AA-pretreatment also inhibited enzymatic hydrolysis. Therefore, it is necessary to remove the acetyl group by post-treatment. Previous works have demonstrated that alkalis, such as NaOH, $Ca(OH)_2$, or NH_4OH, are good reagents for deacetylation [28]. However, water washing is needed prior to alkaline deacetylation. In the present work, we employed peracetic acid (PAA) for post-treatment at PAA loading (based on initial dry bagasse weight) of 2.5–10%, which not only removed acetyl group, but also selectively removed lignin by oxidative degradation. The chemical compositions of PAA post-treated substrates are shown in Table 5. As expected, both further delignification and deacetylation were achieved by PAA post-treatment. The residual lignin content could be reduced by about 50%, while the degradation of hemicelluloses (xylan) was less than 20%, indicating that PAA was very selective towards delignification. For example, after sugarcane bagasse was pretreated by 70% AA with 0.3% SA at 110 °C for 2 h, followed by 10% PAA post-treatment, the lignin content could be reduced to 5.88%. However, to achieve a similar degree of delignification by single-stage PAA treatment, PAA loading should be higher than 40% [29].

Table 5. Chemical compositions of PAA post-treated substrates.

AA Pretreatment	PAA Loading (%) [a]	*SY* (%)	Holocellulose (%)	Cellulose (%)	Xylan (%)	Total Lignin (%)	*AGC* (%)
60% AA, 0.3% SA, 110 °C, 2 h	0	59.8	86.0	62.6	16.1	14.7	2.46
	2.5	57.6	88.9	67.5	15.9	12.1	1.87
	5.0	55.4	90.1	68.6	16.8	10.5	1.23
	7.5	53.1	90.9	70.2	17.0	8.42	0.78
	10	50.1	92.3	72.5	17.4	6.23	0.58
70% AA, 0.3% SA, 110 °C, 2 h	0	54.0	88.2	65.6	15.8	13.8	3.15
	2.5	52.1	89.5	68.9	15.7	10.2	2.59
	5.0	50.3	90.6	70.4	16.4	8.31	1.58
	7.5	49.7	92.2	74.3	15.9	6.18	1.22
	10	48.9	93.1	75.6	16.6	5.88	0.69
80% AA, 0.3% SA, 110 °C, 2 h	0	52.1	90.1	68.2	15.6	10.2	3.45
	2.5	50.5	92.4	70.6	16.6	8.27	3.00
	5.0	48.7	93.2	73.4	16.0	7.01	2.45
	7.5	47.2	94.0	76.5	15.4	6.23	1.66
	10	46.7	94.4	77.7	15.9	5.12	0.96

[a] based on initial dry bagasse weight.

The acetyl group of the AA-pretreated solid could be reduced from 2.4–3.4%, depending on AA pretreatment to less than 1% when 10% PAA loading (based on initial dry bagasse weight) was used. The enzymatic hydrolysis of PAA post-treated substrates (Figure 8) illustrated that PAA post-treatment could dramatically increase the *EPC* of the substrates. The highest *EPC* at 120 h was achieved with 70% AA pretreatment and 10% PAA post-treatment, reaching about 70%, compared with only 26% of the control (without PAA post-treatment). This was mainly because PAA post-treatment significantly removed the residual lignin, as well as the acetyl group, causing a high degree of delignification (>85%), with an associated liberation of cellulose fibers and an increase in the hydrophilicity of the substrates. Moreover, it has been known that cellulase enzymes could be non-productively adsorbed on the residual lignin by hydrophobic, hydrogen bonding, electrostatic interactions, and cation−π interactions [30]. Oxidative pretreatments, such as by PAA delignification, could well modify lignin structure to increase its hydrophilicity, thus reducing the non-productive adsorption of enzymes on lignin [20]. However, the mechanism on the improved digestibility has to be further investigated in terms of the cell wall microstructure changes and surface characteristics of the substrates. However, it should be noted that PAA post-treatment could be performed without water washing of the AA pretreated solid, and the decomposition of PAA is mainly water and oxygen, which is involve a green process without pollutants. Hence, compared with alkali-PAA pretreatment, the AA-PAA process would require less water and PAA consumption. It also should be noted that the cellulase used in the present work was not a specific enzyme formula for lignocellulose hydrolysis. Thus, the sub-enzyme components of the cellulase complex were not optimal for biomass degradation. Therefore, detailed comparison of the results of the present work with those reported by the literature is not possible. However, according to Zhao and Liu [19], when bagasse was pretreated, first, by 80–90% AA, followed by 1–4% NaOH (based on initial bagasse weight) deacetylation, the enzymatic glucan conversions with 20 FPU/solid of cellulase complex (Novozym Celluclast 1.5 L, specific cellulase for biomass degradation) and 40 CBU/g solid of supplemental β-glucosidase were in the range of 60–90%. The *EPC* obtained in the present work was in this range.

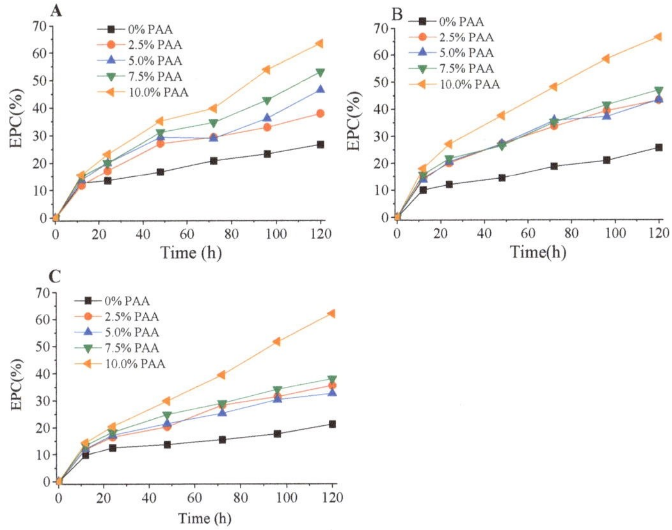

Figure 8. Enzymatic hydrolysis of PAA post-treated substrates. (**A**) 60% AA pretreatment; (**B**) 70% AA pretreatment; (**C**) 80% AA pretreatment.

3. Materials and Methods

3.1. Materials

The sugarcane bagasse used in the present work was collected in Guangxi Zhuang Autonomous Region in South China. It was ground and screened, and the fraction retained by a 20-mesh sieve was used in all pretreatment experiments. The main chemical compositions of the bagasse were determined to be 5.02% moisture, 71.26% holocellulose, 43.68% cellulose, 27.58% hemicellulose, 19.25% acid-insoluble lignin, and 1.90% acid-soluble lignin. The cellulase enzymes used for enzymatic hydrolysis of the pretreated cellulosic solid was Cellulase R-10 produced by Yakuh Honsha Co. Ltd. (Tokyo, Japan), with a filter paper activity of ~6000 FPU/g enzyme powder. The chemical agents used in the pretreatments, including acetic acid (AA), H_2O_2, and sulfuric acid, were purchased from Beijing Beihua Fine Chemicals Co., Ltd. (Beijing, China). Peracetic acid (PAA) was prepared by reaction of AA and 30 wt% H_2O_2, with a volume ratio of 2.5:1 at room temperature for 72 h. 3% (*w*/*w*) sulfuric acid was added as a catalyst, according to our previous kinetic modeling and optimization works [31–33]. The obtained PAA solution had a PAA concentration of about 2.2 mol/L. It should be noted that PAA is not stable and easy to decompose. Therefore, in order to ensure its purity and concentration, PAA solution was prepared before experiments and stored in fridge. Moreover, before preparation, the vessels used were carefully washed to avoid contamination by metal ions that can catalyze the decomposition of peracetic acid. The standard glucose, xylose, and arabinose were purchased from Sigma-Aldrich (Shanghai, China).

3.2. AA Pretreatment and PAA Post-Treatment

AA pretreatment was carried out in a 1000 mL three-neck glass flask heated by water bath or electric jacket under atmospheric pressure. An amount of 30 g of screened bagasse was packed into the flask followed by addition of 300 mL 60–90 wt.% AA solution with 0–0.5 wt.% sulfuric acid (in liquid phase). Electrical stirring with a Teflon paddle was used at 300 rpm to keep the system as homogeneous as possible. After AA pretreatment, the mixture was filtered using a Buchner funnel. The obtained solid was first washed with 300 mL 60–90 wt% AA solution and then filtered to remove as much liquid as possible. Typically, after filtration, the liquid content of the pretreated solid was 75–80%. When no PAA post-treatment was performed, the filtered solid was washed by running water until neutrality and filtered and oven-dried for further analysis of chemical compositions.

PAA post-treatment was carried out in a 1000 mL glass flask immersed in a water bath at 75 °C. The AA-washed and filtered pretreated solid was packed into the flask, and a certain volume of prepared PAA solution was directly added with PAA loading of 0–10% (based on initial dry bagasse weight before AA pretreatment). A Teflon® paddle was used for intermittently stirring to keep the system as uniform as possible. After PAA post-treatment, the solid was washed using running water until neutrality, and then it was filtered and oven-dried for further analysis of chemical compositions.

3.3. Enzymatic Hydrolysis of Pretrereated and Post-Treated Substrates

The AA-pretreated and PAA-post-treated substrates were digested by cellulase loading of 20 FPU/g solid at temperature 50 ± 0.5 °C, pH 4.8 (0.1 mol/L sodium acetate buffer), and 130 rpm in an air-bath shaker. Enzymatic digestibility, denoted as enzymatic polysaccharide conversion (*EPC*, %), was defined as the percentage of holocellulose converted to reducing sugar (glucose plus xylose) after incubation with cellulase enzyme.

3.4. Experimental Design

To optimize the AA pretreatment process, a response surface methodology (central composite design, CCD) was employed to study the effects of temperature (T, °C), AA concentration (C_{AA}, wt%), pretreatment time (t, h), and sulfuric acid concentration (C_{SA}, wt% or mol/L) on several response variables, including solid recovery yield (*SY*, %), degree of delignification (*DD*, %), solubilization of holocellulose (*HS*), acetyl group content (*AGC*, %), and enzymatic polysaccharide conversion (*EPC*, %). The levels of the variables are summarized in Table 6. A CCD with eight star points, as well as six replicates at the center points, leading to 30 runs, was employed for the optimization. The variables were coded according to the following equation:

$$X_i = \frac{x_i - x_0}{\Delta x},\ i = 1,\ 2 \ldots k \tag{24}$$

where; X_i is the dimensionless value of an variable; x_i is the real value of an variable; x_0 is the level value of X_i at the center point; and Δx is the step change. A quadratic polynomial equation (Equation (25)), including all interaction terms, was used to calculate the predicted response variable.

$$Y_i = \beta_0 + \sum_{i=1}^{4} \beta_i X_i + \sum_{i=1}^{4} \beta_{ii} X_i^2 + \sum_{i,j=1}^{4} \beta_{ij} X_i X_j \tag{25}$$

where; Y_i is the predicted response variable; X_i and X_j are the input variables; β_0 is the intercept term; β_i, β_{ii}, and β_{ij} are the regressed parameters for linear effects, squared effects (non-linear effects), and interactive effects, respectively. Design-Expert 9.0.6 software 8.0.7.1

(Stat-Ease, Inc., Minneapolis, MN, USA) was used to make the experimental design, regress the parameters, and make statistical analysis. The experimental design is shown in Table 6.

Table 6. Levels of variables used in the CCD experimental design.

Variables, Abbreviation and Units	Code	Levels				
		−2	−1	0	1	2
Temperature (T, °C)	X_1	70	80	90	100	110
AA concentration (C_{AA}, wt%)	X_2	55	65	75	85	95
Pretreatment time (t, h)	X_3	1.0	1.5	2.0	2.5	3.0
Sulfuric acid concentration (C_{SA}, wt%)	X_4	0.0	0.1	0.2	0.3	0.4

3.5. Analytic Methods

The chemical compositions of bagasse and pretreated substrates were analyzed in accordance with corresponding Chinese Standards, namely, moisture content, GB/T 2677.2–1993, ash, GB/T 2677.3–1993, hot water extractives, GB/T 2677.4–1993, 1% NaOH extractives, GB/T 2677.5–1993, benzene-ethanol extractives, GB/T 2677.6–1994, holocellulose, GB/T 2677.10–1995, Klason lignin, GB/T 2677.8–1994, and acid-soluble lignin, GB/T 747–2003. The cellulose content was measured by the nitric acid–ethanol method [34]. The determination of PAA concentration was in accordance with Chinese standard GB 19104–2008. The monosaccharides and ethanol were determined by Shimadzu (Tokyo, Japan) HPLC (LC-10AT) equipped with a SCL-10A system controller, a CTO-AS column oven, a RID-10A refractive index detector, and an Aminex HPX-87H column. The mobile phase was 0.005 M H_2SO_4 at a flow rate of 0.8 mL/min. The cellulase activity (filter paper activity) was determined according to Ghose [35], but, this was performed by using HPLC to determine the formed glucose concentration instead of using 3,5-dinitrosalicylic acid to measure the reducing sugar concentration.

4. Conclusions

A response surface methodology (central composite design, CCD) was employed to study the effects of several factors on pretreatment of sugarcane bagasse (SCB) by aqueous acetic acid (AA) with addition of sulfuric acid (SA) as a catalyst under mild condition (<110 °C). Several quadratic polynomial models were obtained, based on the CCD experimental results. As found in the experiments, temperature, AA concentration, time, and SA concentration showed significant effects on solid yield (*SY*), degree of delignification (*DD*), holocellulose solubilization (*HS*), acetyl group content (*AGC*), and enzymatic polysaccharide conversion (*EPC*). *SY*, *DE*, and *HS* were increased in relation to the levels of the factors. However, higher *AGC* was observed at high AA concentration because of cellulose acetylation. Kinetic modeling was further investigated for AA pretreatment using both Saeman's model and the potential degree of reaction (*PDR*) model. However, Saeman's model showed a great deviation from the experimental results, while the PDR model fit the experimental data very well, with determination coefficients of 0.95–0.99. Nevertheless, the enzymatic digestibility of the AA-pretreated substrates was still lower than 30%, mainly due to the relatively low degree of delignification and acetylation of cellulose. Post-treatment of the pretreated cellulosic solid by PAA with loading of 10% (based on initial dry sugarcane bagasse weight) well improved the cellulose digestibly by further selectively removing 50–60% of the residual linin and acetyl group. The enzymatic polysaccharide conversion increased from <30% for AA-pretreatment to about 70% for PAA post-treatment. Compared with alkaline deacetylation, PAA post-treatment avoided the water washing step after AA pretreatment.

Supplementary Materials: The following supporting information can be downloaded at: https://www.mdpi.com/article/10.3390/molecules28124689/s1, Figures S1–S4: 3D surface plots of effects of temperature (T, °C), AA concentration (C_{AA}, wt%), pretreatment time (t, h) and sulfuric acid concentration (C_{SA}, wt%) on solid recovery yield (SY, %), degree of delignification (DD, %), solubilization of holocellulose (*HS*, %), and acetyl group content (*AGC*, %) of pretreated substrates, respectively. Figures S5–S7: Plots of $\ln(1 - S_S)$, $\ln(1 - D_d)$, and $\ln(1 - H_S)$ with t at different temperatures, respectively. Tables S1–S4: ANOVA for Response Surface Quadratic model for *SY*, *DD*, *HS* and *AGC*, respectively.

Author Contributions: Conceptualization, Y.B. and X.Z.; methodology, Y.B. and M.T; software, Z.D.; validation, Y.B., M.T. and Z.D.; formal analysis, Y.B.; investigation, Y.B.; resources, Z.D.; data curation, Y.B.; writing—original draft preparation, Y.B.; writing—review and editing, X.Z.; visualization, M.T.; supervision, X.Z.; project administration, X.Z.; funding acquisition, Y.B. and X.Z. All authors have read and agreed to the published version of the manuscript.

Funding: This research was funded by the R&D Program of Beijing Municipal Education Commission (No. KM202310011008) and the National Natural Science Foundation of China (No. 22178197).

Institutional Review Board Statement: Not applicable.

Data Availability Statement: The data presented in this study are available on request from the corresponding author.

Conflicts of Interest: The authors declare no conflict of interest.

Sample Availability: Not applicable.

References

1. Reena, R.; Alphy, M.P.; Reshmy, R.; Thomas, D.; Madhavan, A.; Chaturvedi, P.; Pugazhendhi, A.; Awasthi, M.K.; Ruiz, H.; Kumar, V.; et al. Sustainable Valorization of Sugarcane Residues: Efficient Deconstruction Strategies for Fuels and Chemicals Production. *Bioresour. Technol.* **2022**, *361*, 127759. [CrossRef]
2. Algayyim, S.J.M.; Yusaf, T.; Hamza, N.H.; Wandel, A.P.; Fattah, I.M.R.; Laimon, M.; Rahman, S.M.A. Sugarcane Biomass as a Source of Biofuel for Internal Combustion Engines (Ethanol and Acetone-Butanol-Ethanol): A Review of Economic Challenges. *Energies* **2022**, *15*, 8644. [CrossRef]
3. Shabbirahmed, A.M.; Haldar, D.; Dey, P.; Patel, A.K.; Singhania, R.R.; Dong, C.-D.; Purkait, M.K. Sugarcane Bagasse into Value-Added Products: A Review. *Environ. Sci. Pollut. Res.* **2022**, *29*, 62785–62806. [CrossRef]
4. Alokika; Anu; Kumar, A.; Kumar, V.; Singh, B. Cellulosic and Hemicellulosic Fractions of Sugarcane Bagasse: Potential, Challenges and Future Perspective. *Int. J. Biol. Macromol.* **2021**, *169*, 564–582. [CrossRef] [PubMed]
5. Rezania, S.; Oryani, B.; Cho, J.; Talaiekhozani, A.; Sabbagh, F.; Hashemi, B.; Rupani, P.F.; Mohammadi, A.A. Different Pretreatment Technologies of Lignocellulosic Biomass for Bioethanol Production: An Overview. *Energy* **2020**, *199*, 117457. [CrossRef]
6. Marques, P.F.; Soares, A.K.L.; Lomonaco, D.; Alexandre e Silva, L.M.; Santaella, S.T.; de Freitas Rosa, M.; Carrhá Leitão, R. Steam Explosion Pretreatment Improves Acetic Acid Organosolv Delignification of Oil Palm Mesocarp Fibers and Sugarcane Bagasse. *Int. J. Biol. Macromol.* **2021**, *175*, 304–312. [CrossRef]
7. Wang, J.; Xu, Y.; Tang, B.; Zeng, M.; Liang, Z.; Jiang, C.; Lin, J.; Xiao, W.; Liu, Z. Enhanced Saccharification of Sugarcane Bagasse by the Optimization of Low Concentration of Naoh and Ammonia Pretreatment. *Ind. Crops Prod.* **2021**, *172*, 114016. [CrossRef]
8. Hashmi, M.; Sun, Q.; Tao, J.; Wells, T.; Shah, A.A.; Labbé, N.; Ragauskas, A.J. Comparison of Autohydrolysis and Ionic Liquid 1-Butyl-3methylimidazolium Acetate Pretreatment to Enhance Enzymatic Hydrolysis of Sugarcane Bagasse. *Bioresour. Technol.* **2017**, *224*, 714–720. [CrossRef]
9. Pin, T.C.; Nakasu, P.Y.S.; Mattedi, S.; Rabelo, S.C.; Costa, A.C. Screening of Protic Ionic Liquids for Sugarcane Bagasse Pretreatment. *Fuel* **2019**, *235*, 1506–1514. [CrossRef]
10. Zhang, H.; Zhang, J.; Xie, J.; Qin, Y. Effects of Naoh-Catalyzed Organosolv Pretreatment and Surfactant on the Sugar Production from Sugarcane Bagasse. *Bioresour. Technol.* **2020**, *312*, 123601. [CrossRef] [PubMed]
11. Zhang, H.; Fan, M.; Li, X.; Zhang, A.; Xie, J. Enhancing Enzymatic Hydrolysis of Sugarcane Bagasse by Ferric Chloride Catalyzed Organosolv Pretreatment and Tween 80. *Bioresour. Technol.* **2018**, *258*, 295–301. [CrossRef]
12. Zhao, X.; Cheng, K.; Liu, D. Organosolv Pretreatment of Lignocellulosic Biomass for Enzymatic Hydrolysis. *Appl. Microbiol. Biotechnol.* **2009**, *82*, 815–827. [CrossRef]
13. Savou, V.; Kumagai, S.; Saito, Y.; Kameda, T.; Yoshioka, T. Effects of Acetic Acid Pretreatment and Pyrolysis Temperatures on Product Recovery from Fijian Sugarcane Bagasse. *Waste Biomass Valorization* **2019**, *11*, 6347–6357. [CrossRef]
14. Pan, X.-J.; Sano, Y. Atmospheric Acetic Acid Pulping of Rice Straw Iv: Physico-Chemical Characterization of Acetic Acid Lignins from Rice Straw and Woods. Part 1. Physical Characteristics. *Holzforschung* **1999**, *53*, 511–518. [CrossRef]
15. Zhao, X.; Liu, D. Kinetic Modeling and Mechanisms of Acid-Catalyzed Delignification of Sugarcane Bagasse by Aqueous Acetic Acid. *Bioenergy Res.* **2013**, *6*, 436–447. [CrossRef]
16. Muurinen, E. Organosolv Pulping: A Review and Distillation Study Related to Peroxyacid Pulping. Ph.D. Thesis, University of Oulu, Oulu, Finland, 2000.
17. Zhang, K.; Pei, Z.; Wang, D. Organic Solvent Pretreatment of Lignocellulosic Biomass for Biofuels and Biochemicals: A Review. *Bioresour. Technol.* **2016**, *199*, 21–33. [CrossRef] [PubMed]

18. Pan, X.; Gilkes, N.; Saddler, J.N. Effect of Acetyl Groups on Enzymatic Hydrolysis of Cellulosic Substrates. *Holzforschung* **2006**, *60*, 398–401. [CrossRef]
19. Zhao, X.; Liu, D. Fractionating Pretreatment of Sugarcane Bagasse for Increasing the Enzymatic Digestibility of Cellulose. *Chin. J. Biotechnol.* **2011**, *27*, 384–392.
20. Zhou, Z.; Ouyang, D.; Liu, D.; Zhao, X. Oxidative Pretreatment of Lignocellulosic Biomass for Enzymatic Hydrolysis: Progress and Challenges. *Bioresour. Technol.* **2023**, *367*, 128208. [CrossRef] [PubMed]
21. Saeman, J.F. Kinetics of Wood Saccharification—Hydrolysis of Cellulose and Decomposition of Sugars in Dilute Acid at High Temperature. *Ind. Eng. Chem.* **1945**, *37*, 43–52. [CrossRef]
22. Zhao, X.; Zhou, Y.; Liu, D. Kinetic Model for Glycan Hydrolysis and Formation of Monosaccharides During Dilute Acid Hydrolysis of Sugarcane Bagasse. *Bioresour. Technol.* **2012**, *105*, 160–168. [CrossRef]
23. Zhao, X.; Morikawa, Y.; Qi, F.; Zeng, J.; Liu, D. A Novel Kinetic Model for Polysaccharide Dissolution During Atmospheric Acetic Acid Pretreatment of Sugarcane Bagasse. *Bioresour. Technol.* **2014**, *151*, 128–136. [CrossRef] [PubMed]
24. Dong, L.; Zhao, X.B.; Liu, D. Kinetic Modeling of Atmospheric Formic Acid Pretreatment of Wheat Straw with "Potential Degree of Reaction" Models. *Rsc. Adv.* **2015**, *5*, 20992–21000. [CrossRef]
25. Mcdonough, T.J. The Chemistry of Organosolv Delignification. *Tappi J.* **1994**, *76*, 186–193.
26. Vazquez, G.; Antorrena, G.; Gonzalez, J. Kinetics of Acid-Catalyzed Delignification of Eucalyptus-Globulus Wood by Acetic-Acid. *Wood Sci. Technol.* **1995**, *29*, 267–275. [CrossRef]
27. Vázquez, G.; Antorrena, G.; González, J.; Freire, S.; López, S. Acetosolv Pulping of Pine Wood. Kinetic Modelling of Lignin Solubilization and Condensation. *Bioresour. Technol.* **1997**, *59*, 121–127. [CrossRef]
28. Zhao, X.; Liu, D. Fractionating Pretreatment of Sugarcane Bagasse by Aqueous Formic Acid with Direct Recycle of Spent Liquor to Increase Cellulose Digestibility-the Formiline Process. *Bioresour. Technol.* **2012**, *117*, 25–32. [CrossRef] [PubMed]
29. Zhao, X.; Van der Heide, E.; Zhang, T.; Liu, D. Single-Stage Pulping of Sugarcane Bagasse with Peracetic Acid. *J. Wood Chem. Technol.* **2011**, *31*, 1–25. [CrossRef]
30. Ouyang, D.; Chen, H.; Liu, N.; Zhang, J.; Zhao, X. Insight into the Negative Effects of Lignin on Enzymatic Hydrolysis of Cellulose for Biofuel Production Via Selective Oxidative Delignification and Inhibitive Actions of Phenolic Model Compounds. *Renew. Energy* **2022**, *185*, 196–207. [CrossRef]
31. Zhao, X.; Zhang, T.; Zhou, Y.; Liu, D. Preparation of peracetic acid from hydrogen peroxide: Part I: Kinetics for peracetic acid synthesis and hydrolysis. *J. Mol. Catal. A Chem.* **2007**, *271*, 246–252. [CrossRef]
32. Zhao, X.; Cheng, K.; Hao, J.; Liu, D. Preparation of Peracetic Acid from Hydrogen Peroxide, Part II: Kinetics for Spontaneous Decomposition of Peracetic Acid in the Liquid Phase. *J. Mol. Catal. A Chem.* **2008**, *284*, 58–68. [CrossRef]
33. Zhao, X.; Zhang, T.; Zhou, Y.; Liu, D. Preparation of Peracetic Acid from Acetic Acid and Hydrogen Peroxide: Experimentation and Modeling. *Chin. J. Process Eng.* **2008**, *1*, 35–41.
34. Shi, S.; He, F. *Analysis and Detection in Pulping and Papermaking*; China Light Industry Press: Beijing, China, 2003.
35. Ghose, T.K. Measurement of Cellulase Activities. *Pure Appl. Chem.* **1987**, *59*, 257–268. [CrossRef]

molecules

Article

Development of Photothermal Catalyst from Biomass Ash (Bagasse) for Hydrogen Production via Dry Reforming of Methane (DRM): An Experimental Study

Ittichai Kanchanakul [1], Thongchai Rohitatisha Srinophakun [2,*], Sanchai Kuboon [3], Hiroaki Kaneko [4], Wasawat Kraithong [3], Masahiro Miyauchi [4] and Akira Yamaguchi [4]

1 Interdisciplinary of Sustainable Energy and Resources Engineering, Faculty of Engineering, Kasetsart University, Bangkok 10900, Thailand; ittichai.ka@ku.th
2 Department of Chemical Engineering, Kasetsart University, Bangkok 10900, Thailand
3 National Nanotechnology Center National Science and Technology Development Agency (NSTDA), Pathum Thani 12120, Thailand; sanchai@nanotec.or.th (S.K.); wasawat@nanotec.or.th (W.K.)
4 Department of Materials Science and Engineering, Tokyo Institute of Technology, Tokyo 152-8550, Japan; hkaneko@ceram.titech.ac.jp (H.K.); mmiyauchi@ceram.titech.ac.jp (M.M.); ayamaguchi@ceram.titech.ac.jp (A.Y.)
* Correspondence: fengtcs@ku.ac.th; Tel.: +66-2942-8555 (ext. 1214)

Abstract: Conventional hydrogen production, as an alternative energy resource, has relied on fossil fuels to produce hydrogen, releasing CO_2 into the atmosphere. Hydrogen production via the dry forming of methane (DRM) process is a lucrative solution to utilize greenhouse gases, such as carbon dioxide and methane, by using them as raw materials in the DRM process. However, there are a few DRM processing issues, with one being the need to operate at a high temperature to gain high conversion of hydrogen, which is energy intensive. In this study, bagasse ash, which contains a high percentage of silicon dioxide, was designed and modified for catalytic support. Modification of silicon dioxide from bagasse ash was utilized as a waste material, and the performance of bagasse ash-derived catalysts interacting with light irradiation and reducing the amount of energy used in the DRM process was explored. The results showed that the performance of 3%Ni/SiO_2 bagasse ash WI was higher than that of 3%Ni/SiO_2 commercial SiO_2 in terms of the hydrogen product yield, with hydrogen generation initiated in the reaction at 300 °C. Using the same synthesis method, the current results suggested that bagasse ash-derived catalysts had better performance than commercial SiO_2-derived catalysts when exposed to an Hg-Xe lamp. This indicated that silicon dioxide from bagasse ash as a catalyst support could help improve the hydrogen yield while lowering the temperature in the DRM reaction, resulting in less energy consumption in hydrogen production.

Keywords: hydrogen production; photothermal catalysis; dry reforming of methane; biomass waste; bagasse ash

Citation: Kanchanakul, I.; Srinophakun, T.R.; Kuboon, S.; Kaneko, H.; Kraithong, W.; Miyauchi, M.; Yamaguchi, A. Development of Photothermal Catalyst from Biomass Ash (Bagasse) for Hydrogen Production via Dry Reforming of Methane (DRM): An Experimental Study. *Molecules* **2023**, *28*, 4578. https://doi.org/10.3390/molecules28124578

Academic Editors: Javier Llanos and Mohamad Nasir Mohamad Ibrahim

Received: 30 April 2023
Revised: 30 May 2023
Accepted: 2 June 2023
Published: 6 June 2023

1. Introduction

Substitutional energy sources have drawn the public's attention in recent years due to the consequences of fossil fuel consumption. One of the alternative energy sources is hydrogen (H_2), an energy carrier that allows energy transport in a usable form from one place to another [1]. Clean hydrogen can be produced in several ways, such as through electrolysis. However, one of the lucrative approaches is the dry reforming of methane (DRM) process, which uses greenhouse gases as feedstock to produce synthesis gases (H_2 and CO). There are a few drawbacks to the DRM reaction. First, it needs more than 700 °C to reach optimal efficiency and conversion [2]. This is energy intense. Second, the coking formation occurs at high temperatures. Several transition metals and noble metals, including Rh, Ni, Ir, Ru, Pt, and Co, are known to be active as DRM catalysts. The active metals are usually dispersed as small (nanoscale) particles supported on porous ceramic supports, such as Al_2O_3, and SiO_2 [3], through catalytic synthesis methods. However, Ni is the most suitable active metal due to its comparatively lower cost to upscale production to the industrial scale. Despite the economical price of Ni, Ni catalysts suffer inevitably from rapid deactivation caused by coke deposition, active metal sintering, or both [4], and it is less stable compared to the other noble catalysts. The rate of carbon deposition was reported to decrease with rising reaction temperature [5]. However, the increment in temperature is not a reasonable way to stop carbon deposition due to energy use concerns. Several researchers reported an improvement of the temperature reduction in thermal catalysts for the DRM reaction, such as Rh, Ni, Pd, Co, Ir, and Ce supported on Al_2O_3 and SiO_2 [6–9]; nonetheless, these require a trade-off with the higher cost of noble metals. Recently, a unique way of designing catalysts using plasmonic nanoparticles (NPs) has appeared to be an attractive approach for the DRM reaction to reduce the operating temperature. The plasmonic/metal NPs interact with light incidents, such as sunlight or light sources with heat, by transferring photoexcited charge carriers from metal NPs to the reactants, leading to chemical transformations under less energy-intense conditions. In this case, it is ideally possible to target electronic excitation so that only DRM reactions are activated. This leads to sustainability goals by lowering the operating temperature that traditionally runs at high temperatures and by improving the selectivity of reactions that may undergo side reactions.

Sugar is one of the top export products from Thailand. However, despite all the profits from the sugar industry, sugar production generates massive waste materials, such as bagasse, press mud, and spent wash [10]. Biomass waste from sugar production, such as bagasse, could be used as fuel for thermal power plants or boiler stations to produce energy that can be fed back into sugar production. However, residuals ash from burning bagasse would be disposed of at landfills, which could bring about other environmental issues caused by the bagasse ash.

The current study focused on utilizing waste materials (bagasse ash) and modifying their properties as a catalyst support for photothermal catalyst in the DRM reaction compared with a commercial support catalyst (SiO_2). Furthermore, we used a conventional wet impregnation approach to obtain Ni/SiO_2 catalysts in a synthesis design strategy.

2. Results and Discussion

2.1. Fabrication of the Catalyst Support Preparation and Synthesis Catalyst

In this work, catalyst support was prepared by extracting SiO_2 from bagasse ash using an acidic extraction approach (3% HCl reagent grade), then modifying its properties by KOH activation at a ratio of 1:4 [11] to maximize the surface area of extracted SiO_2. For catalyst synthesis, Ni/SiO_2 was synthesized by conventional wet impregnation as shown in Figure 1. There are two mechanisms of wet impregnation. One relies on capillary action to draw the solution into the pores. The other is that the solution transport changes from a capillary action process to a diffusion process in the wet impregnation method [12].

2.2. Characterization

The synthesized catalyst surface area was analyzed using an Autosorb® iQ3 gas sorption analyzer (Anton Paar QuantaTec Inc.; Boynton Beach, FL, USA), in which adsorption-desorption isotherms take place with liquid nitrogen's help at −195 °C. The Brunauer-Emmett-Teller (BET) technique was used to calculate the catalyst surface area within the pressure range of 0.12 to 0.20. X-ray powder diffraction (XRD) patterns were recorded using a diffractor (XRD; Rigaku, Smartlab; Tokyo, Japan) axial diffractometer in the $2\theta = 10^{\circ}$ to 80° angular arrays with a step of $0.05^{\circ}s^{-1}$ and CuK-α1 radiation (the wavelength CuK-α1 = 1.5406 Å) of the diffractometer for the synthesized catalysts. Then, the synthesized catalysts were analyzed using X-ray fluorescence (Epsilon 1; Malvern Panalytical Ltd.; Malvern, UK) for the elemental analysis to confirm the chemical composition of the catalysts in percentage terms. The UV–vis DRS were measured using a JASCO V-670 spectrometer; JASCO Coorperation; Tokyo, Japan in the wavelength range from 200 to

800 nm with $BaSO_4$ as the reference. Additionally, scanning electron microscopy (SEM) images of the synthesized catalysts were measured using Schottky field emission scanning electron microscopy (FE-SEM; SU8030; Hitachi-High Tech Corp.; Tokyo, Japan). The SEM images captured the surface morphology of the catalysts, and the EDS mapping checked the dispersion of Ni particles and the composition of the catalysts. Transmission electron microscopy (TEM) images were taken on a Jeol-JEM-2100Plus; JEOL Ltd.; Tokyo, Japan operated at 200 kV. Specimens were prepared by suspending sample powders in ethanol; then, a drop of the suspension was deposited on copper grids.

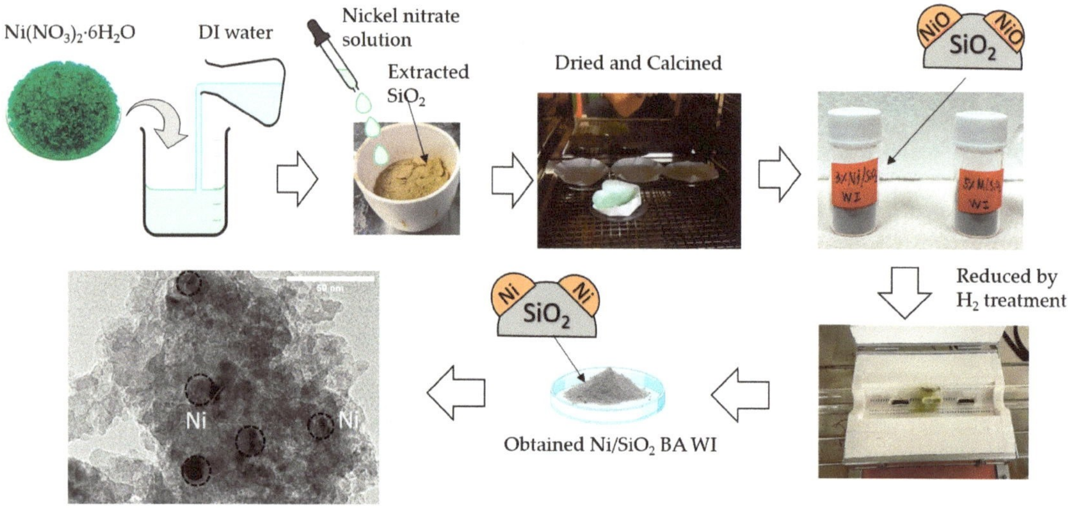

Figure 1. Schematic diagram of Ni/SiO_2 bagasse ash (BA) catalysts prepared using wet impregnation method (WI).

2.2.1. XRD Analysis

As shown in Figure 2a, extracted SiO_2 was extracted by acidic extraction using HCl and followed by KOH activation at various ratios 1:2–1:6. These samples were characterized by X-ray diffraction analysis. The amorphous SiO_2 can be detected at 24.3°. The ratio at 1:4 exhibited a high surface area and average pore size [11], which could be the optimal ratio of SiO_2: KOH for KOH activation. Additionally, the XRD pattern of $3Ni/SiO_2$ BA WI shows in Figure 2b, obtained by the Wet Impregnation approach. For a fresh $3Ni/SiO_2$ catalyst (light green line), the diffraction peaks appearing at 37.1°, 43.1°, and 62.5° can be attributed to the NiO phase (JCPDS 65-2901), while the broad peak at 24.3° can be identified to the SiO_2 phase (JCPDS 39-1425) [13]. The XRD pattern of the reduced $3Ni/SiO_2$ catalyst (blue line) also shows in Figure 2b which the diffraction peaks at 2θ = 44.3°, 51.4°, and 76.1°, which can be indicated by the crystal planes of (111), (200), and (220) of metallic nickel phase. After the catalytic activity test, the reduced $3Ni/SiO_2$ WI catalyst after the DRM process (red line) shows that the XRD pattern demonstrated identically to the reduced $3Ni/SiO_2$ WI catalyst (blue line). Therefore, we suspected coke formation after the DRM reaction. However, the reduced $3Ni/SiO_2$ WI catalyst after the DRM process (red line) exhibits XRD pattern without the appearance of the diffraction peaks of carbon.

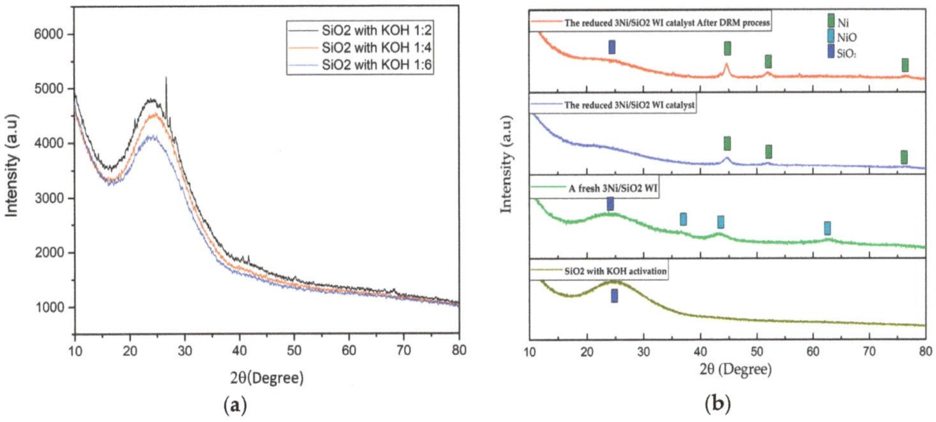

Figure 2. (**a**) The XRD pattern of extracted SiO_2 with KOH activation at various ratios of SiO_2 and KOH, (**b**) The XRD pattern from top to bottom of the reduced 3Ni/SiO_2 after the DRM process, the reduced 3Ni/SiO_2 WI fresh 3Ni/SiO_2 BA WI, a fresh reduced 3Ni/SiO_2 WI and SiO_2 after KOH activation.

2.2.2. BET Surface Analysis

The BET analysis results are shown in Table 1. The BET surface area of the commercial SiO_2 was 10.7 m^2/g, which changed slightly to 11.4 and 12.6 m^2/g in the 3 and 5 Ni/SiO_2 commercial WI samples, respectively. The BET surface area of the extracted SiO_2 with KOH activation at a ratio 1:4 was 185 m^2/g, which marginally declined to 163 and 157 m^2/g in the 3 and 5 Ni/SiO_2 BA WI, respectively (Table 1). From the Barrett–Joyner–Halenda (BJH) method, the average pore sizes of the commercial SiO_2, the extracted SiO_2 with KOH activation, 3 and 5 Ni/SiO_2 commercial WI, and 3 and 5 Ni/SiO_2 BA WI were 4.47, 20.2, 4.62, and 20.9 nm, respectively. The pore sizes of samples can play a crucial role in the diffusion of CH_4 and CO_2 molecules. In fact, pore size influences the diffusion of reactant molecules to the catalytically active sites within the catalyst material. If the pore size is too small, it can restrict the movement of reactants, leading to slower reaction rates.

Table 1. Physicochemical properties of extracted SiO_2 and Ni/SiO_2 WI catalyst.

Samples	BET [1] Surface Area (m^2/g)	Average Pore Size (nm)
Bare commercial SiO_2	10.7	4.47
Bare extracted SiO_2 from bagasse ash	42.3	36.0
Extracted SiO_2 BA, KOH activation 1:2	207	13.1
Extracted SiO_2 BA, KOH activation 1:4 *	185	20.2
Extracted SiO_2 BA, KOH activation 1:6	178	11.6
3Ni/SiO_2 commercial WI	11.4	4.62
5Ni/SiO_2 commercial WI	12.6	4.85
3Ni/SiO_2 bagasse ash WI	163	20.9
5Ni/SiO_2 bagasse ash WI	157	20.9

[1] The Brunauer-Emmett-Teller (BET) technique. * This ratio is used as catalyst support for catalysts synthesis.

2.2.3. SEM Analysis

As shown in Figure 3, the internal microstructures of the bare commercial SiO_2, extracted SiO_2, and the Ni/SiO_2 commercial WI were examined using SEM analysis, while EDS elemental mapping was used for the Ni/SiO_2 BA WI. The bare commercial SiO_2 is shown in Figure 3a, with a regular, smooth surface. Figure 3b shows an image of the extracted SiO_2 from BA with a complicated structure and a rough surface. Figure 3c,d represent synthesized Ni/SiO_2 commercial WI and Ni/SiO_2 BA WI, respectively. As seen, the Ni particles were successfully decorated on both the commercial SiO_2 and extracted SiO_2 BA surfaces. Furthermore, the extracted SiO_2 BA exhibited a rough structure with a high surface area, which would be beneficial for photothermal catalytic activity. The EDS element spectra of the Ni/SiO_2 BA composite are shown in Figure 3e, which confirmed not only the presence of Ni, Si, and O but also residual inorganic elements, such as Mg, Al, Fe, and K, on the synthesized catalyst. Figure 3f demonstrates the dispersion of Ni particles onto the catalyst support (SiO_2).

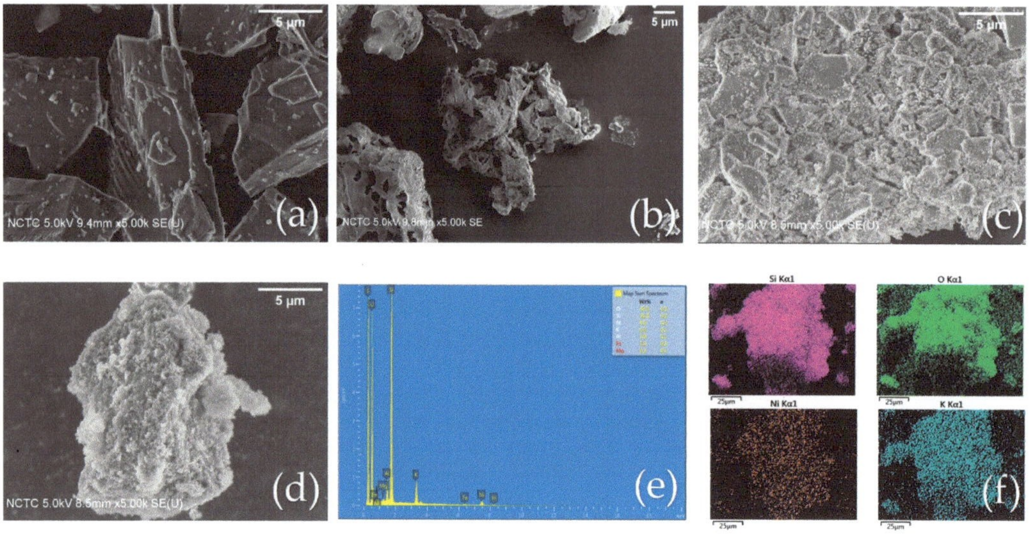

Figure 3. SEM images of (**a**) bare commercial SiO_2, (**b**) extracted SiO_2 from bagasse ash, (**c**) Ni/SiO_2 commercial WI, (**d**) Ni/SiO_2 BA WI. (**e**) EDS analysis of Ni/SiO_2 BA WI. (**f**) EDS mapping of Ni/SiO_2 BA WI.

2.2.4. XRF Analysis

Despite the EDS elemental mapping, XRF analysis helped to estimate the composition of the synthesized catalyst. The XRF analysis results indicated that Ni and SiO_2 were not only in the obtained samples, but there were also other residual elements, such as Al and K. In contrast, the XRF analysis of Ni/SiO_2 commercial WI only exhibited Ni and Si, as shown in Table 2. However, XRF analysis results did not include oxygen atoms in its calculation; thus, recalculating with the oxygen atoms will result in a 3% nickel by weight percentage.

Table 2. The elemental composition of 3Ni/SiO_2 BA WI and 3Ni/SiO_2 commercial WI.

Sample	Compounds			
	Si	Ni	Al	K
3Ni/SiO_2 BA WI	71.4%	11.7%	5.26%	11.7%
3Ni/SiO_2 commercial WI	88.0%	11.7%	-	-

2.2.5. Optical Properties

The optical properties of the synthesized photothermal catalysts were evaluated using ultraviolet-visible spectroscopy. The UV–Visible diffuse reflectance spectrum (UV–Vis DRS) of bare SiO_2 commercial, extracted SiO_2 BA, 3 and 5 Ni/SiO_2 commercial WI, and 3 and

5 Ni/SiO_2 BA WI are shown in Figure 4a,b. The UV–Vis DRS of the commercial SiO_2 had a non-absorption edge at all wavelengths (blue DRS line in Figure 4a). In contrast, after the wet impregnation method, 3 and 5 Ni/SiO_2 commercial WI exhibited an increment of absorption edge around 370 nm and strong absorption over a wide range of the UV and visible light regions (black and red DRS lines in Figure 4a). At the same time, the UV–Vis DRS of the extracted SiO_2 from BA had an absorption edge around 230 nm (blue line in Figure 4b). It is because not only extracted SiO_2 from bagasse ash contain pure SiO_2, but it also consists of inorganic residuals such as Al, K, Mg, and Fe, which could add light adsorption properties to our extracted SiO_2 sample. Furthermore, after the wet impregnation method, 3 and 5 Ni/SiO_2 BA WI demonstrated a similar trend to 3 and 5 Ni/SiO_2 commercial WI, with an increment of strong absorption over a wide range of the UV and visible light regions (black and red DRS lines in Figure 4b).

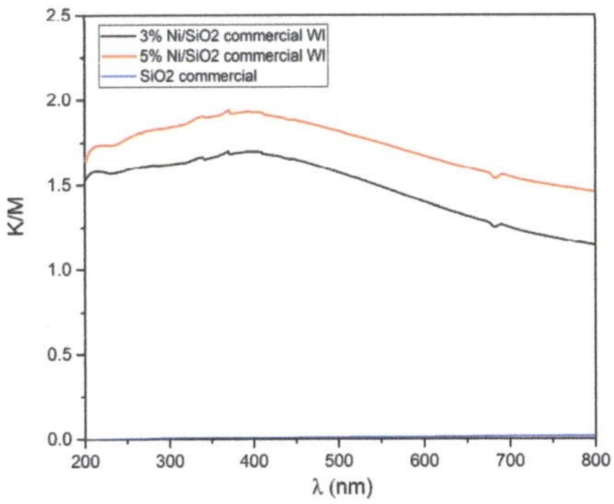

(a)

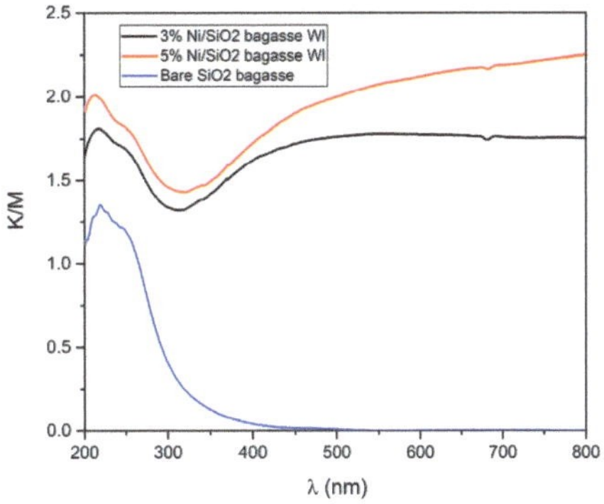

(b)

Figure 4. (**a**) UV–vis spectra of commercial SiO_2, 3 and 5 Ni/SiO_2 comm WI, (**b**) UV–vis spectra of extracted SiO_2 from bagasse ash, 3 and 5 Ni/SiO_2 BA WI.

2.3. Photothermal Catalytic Activity

2.3.1. Photothermal Catalytic Hydrogen Generation Results and Analysis

The photothermal catalytic activities of the synthesized catalysts were examined based on CH_4, CO_2 conversion, and H_2 yield under UV–visible light irradiation. We investigate the efficient catalytic activity between different catalyst supports (commercial SiO_2 and extracted SiO_2 from bagasse ash); their catalytic performance toward H_2 generation was carried out under UV-visible light from a Hg-Xe lamp. The light-adsorption ability of commercial SiO_2 demonstrates in Figure 4a blue line indicating that commercial SiO_2 cannot interact with UV-visible light. However, the light-adsorption ability is improved after the wet impregnation method, as shown in Figure 4a (red and black line). At the same time, the light-adsorption ability of the extracted SiO_2 BA was also enhanced by the wet impregnation method, which resulted in light-adsorption ability from 200 nm to 800 nm (Figure 4b red and black line) compared to the bare extracted SiO_2 (Figure 4b blue line) before the wet impregnation method, which only absorbed light in a range of 200–350 nm. Therefore, the catalysts could interact with UV-visible light after the wet impregnation method. Figure 5 shows the relative intensity of the Hg-Xe lamp, which was intense in the 200–450 nm range. The biomass-derived catalysts (Ni/SiO_2 BA WI) had higher activity than the commercial SiO_2 catalyst support one, as shown in Figure 6a,b because the surface area of Ni/SiO_2 BA WI was substantially higher than for the Ni/SiO_2 commercial WI, which could be attributed to higher activity.

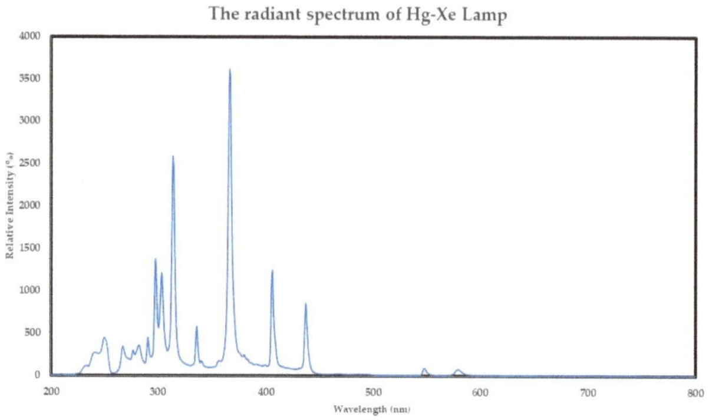

Figure 5. The radiant spectrum of Hg- Xe lamp.

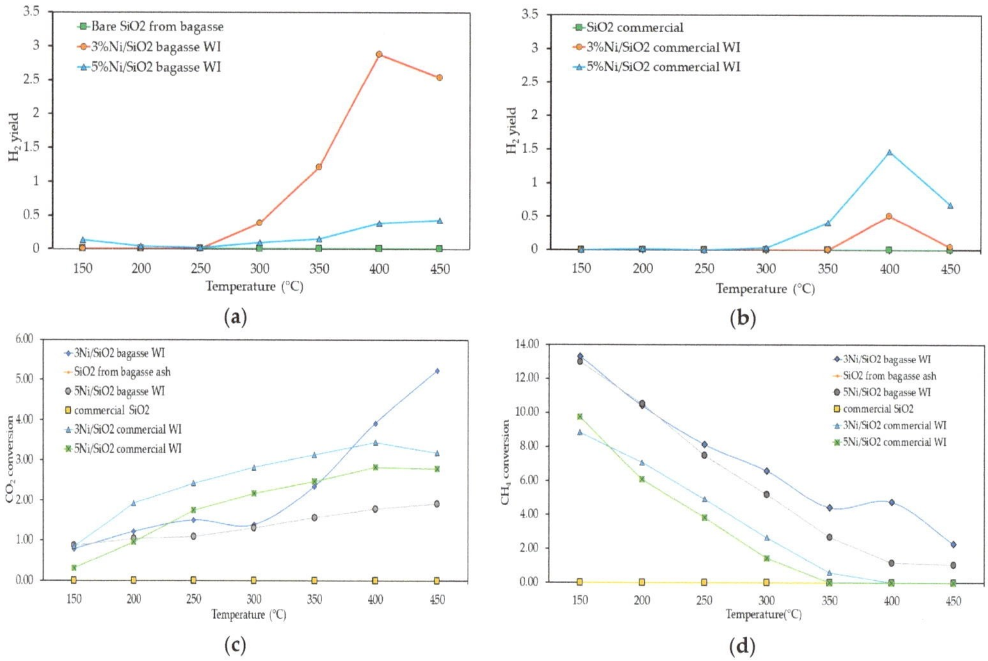

Figure 6. Photothermal catalytic activities of the synthesized catalysts at 4 h reaction time (**a**) H_2 yield of Ni/SiO_2 from bagasse ash (**b**) H_2 yield of Ni/SiO_2 commercial (**c**) CO_2 conversion of synthesized catalysts (**d**) CH_4 conversion of synthesized catalysts.

2.3.2. Band Alignment and Proposed Photothermal Catalytic Mechanism

Amorphous silica has a wide band gap energy of approximately 7.62–9.70 eV [14]; thus, the valence band electrons are relatively difficult to excite to the conduction band even when irradiated by UV light. However, the interband excitation in Ni particles was more favorable, with the photogenerated hot electrons overcoming the energy barrier. The previous study suggests that the hot carriers generated from the light-induced d-to-s interband excitation in Ni nanoparticles are proposed to mediate the transformation of photon energy to chemical energy [15,16].

To investigate the optimal Ni percentage amount on the synthesized catalysts with light irradiation, the photothermal catalytic performance of 3 and 5 Ni/SiO_2 BA WI and

3 and 5 Ni/SiO_2 commercial WI were tested in relation to the CH_4 and CO_2 conversions. Figure 6a,b revealed that 3Ni/SiO_2 BA WI had the highest catalytic activity. However, when the content of Ni increased to 5%, the CO_2 conversion and H_2 yield were reduced, perhaps due to the agglomeration of nickel particles on the catalyst support. In addition, the light adsorption property of Ni/SiO_2 BA WI could have increased the surface temperature on the catalyst support, leading to sintering, which subsequently resulted in catalyst deactivation and reduced 5Ni/SiO_2 BA WI performance.

In contrast, the CH_4 conversion of all synthesized catalysts showed a downward trend, indicating that the CH_4 reactant was consumed in the system, as shown in Figure 6d. An increase in the temperature resulted in reduced CH_4 conversion. Furthermore, H_2O appeared in the system.

The experimental results demonstrated that Ni particles on extracted SiO_2 from bagasse ash could interact with UV light to generate hydrogen at 300 °C because the UV–visible light provides photon energy to stimulate the Ni particles, leading to a plasmonic effect in the metal nanoparticles that creates electron-hole pairs (called hot carriers) and initiates the DRM reaction (Equation (1)). However, when the temperature increased, the H_2 yield dropped to 450 °C, and H_2O was detected in the system. Preferable reaction pathways, such as a reverse water gas shift reaction (Equation (2)) at low temperatures (200–350 °C), may have resulted in a side reaction and unwanted products, such as H_2O in this case.

$$CH_4 + CO_2 \leftrightarrow 2H_2 + 2CO_2 \ \Delta H^{\circ}{}_{298K} = +247 \text{ kJ mol}^{-1} \quad (1)$$

$$CO_2 + H_2 \leftrightarrow CO + H_2O \ \Delta H^{\circ}{}_{298K} = +41 \text{ kJ mol}^{-1} \quad (2)$$

3. Materials and Methods

3.1. Preparation of Bagasse Fly Ash

Bagasse fly ash was received from A Sugar Production Company (Thailand) after being burned as biomass fuel for the boiler. It contained a high moisture percentage; thus, it was necessary to dry the bagasse fly ash at 80 °C in a dry oven overnight, followed by drying at 105 °C in an oven for 2 h. After drying, the bagasse fly ash was fed into a crucible and burnt in a furnace at an initial temperature of 400 °C, a heating rate of 1 °C/min, and a holding time of 2 h to de-volatized any organic compounds. Then, the ash was heated to 900 °C to de-carbonize it and held for 2 h at the same heating rate to obtain crystalline silica.

3.2. Preparation of Silicon Dioxide Extraction

Bagasse fly ash was washed with 3% HCl reagent grade at a ratio of 12 mL 3% HCl per 1 g of bagasse fly ash as an extraction agent to reduce impurities other than SiO_2 in the bagasse fly ash. The washed ash was stirred using a magnetic stirrer at 240 rpm for 2 h on a hotplate at 200 °C. After mixing, the sample was cleaned using deionized water until the pH was neutral. Afterward, the sample was passed through a vacuum filter containing filter papers and dried at 80 °C in a dry oven overnight. Then, the sample was calcined at 400 °C at a heating rate of 1 °C/min and cooled in the oven.

3.3. Preparation of KOH Activation for Silicon Dioxide

Silica dioxide is chemically activated by potassium hydroxide (KOH) at SiO2-to-KOH ratios ranging from 1:1 to 1:8 to obtain SiO_2 particles with greater surface areas. The resulting mixed SiO_2 samples (~3 g) and KOH (1:1 to 1:8) were added to 100 mL of DI water and then stirred at 70 °C for 1 h. After mixing, the samples were dried in the oven at 80 °C overnight until all the liquid had evaporated. Subsequently, each sample was placed in a ceramic crucible and calcined under an air atmosphere at 800 °C (heating rate 5 °C/min) for 1 h. During the calcination process, porosity was induced in the silica structure by the combustion of K_2O derived from the KOH reaction. Next, the activated products were stirred with 2.5% HCl reagent grade to eliminate residual K_2SiO_3 in the samples. Afterward, the mixture was washed with DI water repeatedly to dissolve any

KCl until the pH was neutral, followed by vacuum filtration to separate the solids from the liquid. Finally, the solid product was dried at 80 °C overnight to obtain the activated SiO_2 with a high surface area.

3.4. Preparation of Catalysts

Wet Impregnation Method

Nickel(II) nitrate hexahydrate ($Ni(NO_3)_2 \cdot 6H_2O$) is dissolved with DI water to obtain a Nickle nitrate solution. Then, a Nickle nitrate solution is dropped on the activated SiO_2 powder and mixed with a spatula. Subsequently, the samples are dried at 100 °C overnight and calcined at 600 °C for 2 h to obtain a Ni/SiO_2-WI. Before use, a fresh catalyst was reduced with H_2 reduction treatment at 600 °C for 2 h.

3.5. Photothermal DRM Activity Test

The photothermal activities of the powder samples were measured under ambient pressure in a flow reactor with a quartz window [17], which enabled us to irradiate the powder samples with a 150 W Hg–Xe lamp (Hayashi-Repic, LA-410UV-5; Tokyo, Japan). Approximately 10 mg of catalyst powder was put into a reactor; in sequence, the gas mixture CH_4:CO_2:Ar = 1:1:98 in vol% was continuously supplied to the reactor at a flow rate of 10 $mLmin^{-1}$. The generated hydrogen was measured using a micro gas chromatograph (Agilent, 3000 Micro GC; Santa Clara, CA, USA).

4. Conclusions

The Ni particles on extracted SiO_2 from bagasse ash (Ni/SiO_2 BA WI) and commercial SiO_2, fabricated using acidic extraction and KOH activation, could drive the photothermal catalytic DRM reaction under UV light. Compared to Ni particles on commercial SiO_2 (Ni/SiO_2 commercial WI) with light irradiation, the Ni/SiO_2 BA WI could generate more H_2 yield. In addition, light irradiation lowered the initiation temperature of syngas generation to 300 °C. However, the maximum H_2 yield was only 3%. As a proposed assumption, the hot carriers generated from light-induced d to s interband excitation in the Ni particles played a vital role in our study, mediating the transformation of photon energy to chemical energy and driving the DRM reaction, despite the side reactions, such as the reverse water gas shift reaction. Therefore, we suggest that the use of extracted SiO_2 from bagasse ash as the catalyst support provides a perspective on substitute materials for a practical path to establishing a plasmonic or non-plasmonic hot carrier-based photothermal catalytic system. The concept presented in this study showed the possibility of the utilization of waste materials as a solution for lowering reaction temperature and utilizing UV–visible light to activate the dry reforming of methane, which leads to energy reduction in its process. We expect that this concept will contribute to the progress of green energy and photothermal catalysis in the field of heterogeneous catalysis.

Author Contributions: Conceptualization, M.M., S.K., T.R.S. and I.K.; Investigation, I.K. and H.K.; Methodology, I.K. and S.K.; Formal analysis, W.K., S.K., H.K. and I.K.; Supervision, M.M., A.Y. and T.R.S.; Funding acquisition, T.R.S.; Writing–original draft, I.K.; writing—review and editing, M.M. and T.R.S. All authors have read and agreed to the published version of the manuscript.

Funding: This research received funding through a TAIST-Tokyo Tech scholarship academic Year 2021 funding number: SCA-CO-2564-15429-TH, the National Science and Technology Development Agency (NSTDA), Thailand, and the Innovation Development fund 2022 funding number: 65/02/CHE/Innovation, the Faculty of Engineering, Kasetsart University, Bangkok, Thailand.

Institutional Review Board Statement: Not applicable.

Informed Consent Statement: Not applicable.

Data Availability Statement: The data presented in this study are available on request from the corresponding author. The data is not publicly available due to the program agreement between the author (student) and the university.

Conflicts of Interest: The authors declare no conflict of interest.

Sample Availability: Not applicable.

References

1. Bennett, S.; Remme, U. *The Future of Hydrogen*; Technology Report; IEA: Paris, France, 2019.
2. Arora, S.; Prasad, R. An overview on dry reforming of methane: Strategies to reduce carbonaceous deactivation of catalysts. *RSC Adv.* **2016**, *6*, 108668–108688. [CrossRef]
3. Kee, R.J.; Karakaya, C.; Zhu, H. Process intensification in the catalytic conversion of natural gas to fuels and chemicals. *Proc. Combust. Inst.* **2017**, *36*, 51–76. [CrossRef]
4. Huang, L.; Li, D.; Tian, D.; Jiang, L.; Li, Z.; Wang, H.; Li, K. Optimization of Ni-Based Catalysts for Dry Reforming of Methane via Alloy Design: A Review. *Energy Fuels* **2022**, *36*, 5102–5151. [CrossRef]
5. Sokolov, S.; Kondratenko, E.V.; Pohl, M.-M.; Barkschat, A.; Rodemerck, U. Stable low-temperature dry reforming of methane over mesoporous La_2O_3-ZrO_2 supported Ni catalyst. *Appl. Catal. B Environ.* **2012**, *113–114*, 19–30. [CrossRef]
6. Barama, S.; Dupeyrat-Batiot, C.; Capron, M.; Bordes-Richard, E.; Bakhti-Mohammedi, O. Catalytic properties of Rh, Ni, Pd and Ce supported on Al-pillared montmorillonites in dry reforming of methane. *Catal. Today* **2009**, *141*, 385–392. [CrossRef]
7. Ferreira-Aparicio, P.; Guerrero-Ruiz, A.; Rodríguez-Ramos, I. Comparative study at low and medium reaction temperatures of syngas production by methane reforming with carbon dioxide over silica and alumina supported catalysts. *Appl. Catal. A Gen.* **1998**, *170*, 177–187. [CrossRef]
8. Hou, Z.; Chen, P.; Fang, H.; Zheng, X.; Yashima, T. Production of synthesis gas via methane reforming with CO_2 on noble metals and small amount of noble-(Rh-) promoted Ni catalysts. *Int. J. Hydrogen Energy* **2006**, *31*, 555–561. [CrossRef]
9. Ranjekar, A.M.; Yadav, G.D. Dry reforming of methane for syngas production: A review and assessment of catalyst development and efficacy. *J. Indian Chem. Soc.* **2021**, *98*, 100002. [CrossRef]
10. Partha, N.; Sivasubramanian, V. Recovery of Chemicals from Pressmud—A Sugar Industry Waste. *J. Indian Chem. Eng.* **2006**, *48*, 160–163.
11. Kanaphan, Y.; Klamchuen, A.; Chaikawang, C.; Harnchana, V.; Srilomsak, S.; Nash, J.; Meethong, N. Interfacially Enhanced Stability and Electrochemical Properties of C/SiO_x Nanocomposite Lithium-Ion Battery Anodes. *Adv. Mater. Interfaces* **2022**, *9*, 2200303. [CrossRef]
12. Deraz, N.M. The Comparative Jurisprudence of Catalysts Preparation Methods: I. Precipitation and Impregnation Methods. *J. Ind. Environ. Chem.* **2018**, *2*, 19–21.
13. Ding, F.; Zhang, Y.; Yuan, G.; Wang, K.; Dragutan, I.; Dragutan, V.; Cui, Y.; Wu, J. Synthesis and Catalytic Performance of Ni/SiO_2 for Hydrogenation of 2-Methylfuran to 2-Methyltetrahydrofuran. *J. Nanomater.* **2015**, *2015*, 791529. [CrossRef]
14. Güler, E.; Uğur, G.; Uğur, Ş.; Güler, M. A theoretical study for the band gap energies of the most common silica polymorphs. *Chin. J. Phys.* **2020**, *65*, 472–480. [CrossRef]
15. Jiang, H.; Peng, X.; Yamaguchi, A.; Ueda, S.; Fujita, T.; Abe, H.; Miyauchi, M. Photocatalytic Partial Oxidation of Methane on Palladium-Loaded Strontium Tantalate. *Sol. RRL* **2019**, *3*, 1900076. [CrossRef]
16. Wu, S.; Wu, S.; Sun, Y. Light-Driven Dry Reforming of Methane on Metal Catalysts. *Sol. RRL* **2021**, *5*, 2000507. [CrossRef]
17. Takeda, K.; Yamaguchi, A.; Cho, Y.; Anjaneyulu, O.; Fujita, T.; Abe, H.; Miyauchi, M. Metal Carbide as A Light-Harvesting and Anticoking Catalysis Support for Dry Reforming of Methane. *Glob. Chall.* **2019**, *4*, 1900067. [CrossRef]

molecules

Article

Bioenergy Generation and Phenol Degradation through Microbial Fuel Cells Energized by Domestic Organic Waste

Asim Ali Yaqoob [1,*], Nabil Al-Zaqri [2], Muhammad Alamzeb [3], Fida Hussain [4], Sang-Eun Oh [5] and Khalid Umar [1,*]

1 School of Chemical Sciences, Universiti Sains Malaysia, Minden 11800, Penang, Malaysia
2 Department of Chemistry, College of Science, King Saud University, P.O. Box 2455, Riyadh 11451, Saudi Arabia; nalzaqri@ksu.edu.sa
3 Department of Chemistry, University of Kotli, Kotli 11100, Azad Jammu and Kashmir, Pakistan
4 Research Institute for Advanced Industrial Technology, College of Science and Technology, Korea University, Sejong 30019, Republic of Korea
5 Department of Biological Environment, Kangwon National University, Chuncheon-si 24341, Republic of Korea
* Correspondence: asimchem4@gmail.com (A.A.Y.); khalidumar4@gmail.com (K.U.)

Abstract: Microbial fuel cells (MFCs) seem to have emerged in recent years to degrade the organic pollutants from wastewater. The current research also focused on phenol biodegradation using MFCs. According to the US Environmental Protection Agency (EPA), phenol is a priority pollutant to remediate due to its potential adverse effects on human health. At the same time, the present study focused on the weakness of MFCs, which is the low generation of electrons due to the organic substrate. The present study used rotten rice as an organic substrate to empower the MFC's functional capacity to degrade the phenol while simultaneously generating bioenergy. In 19 days of operation, the phenol degradation efficiency was 70% at a current density of 17.10 mA/m^2 and a voltage of 199 mV. The electrochemical analysis showed that the internal resistance was 312.58 Ω and the maximum specific capacitance value was 0.00020 F/g on day 30, which demonstrated mature biofilm production and its stability throughout the operation. The biofilm study and bacterial identification process revealed that the presence of conductive pili species (*Bacillus* genus) are the most dominant on the anode electrode. However, the present study also explained well the oxidation mechanism of rotten rice with phenol degradation. The most critical challenges for future recommendations are also enclosed in a separate section for the research community with concluding remarks.

Keywords: microbial fuel cells; phenol; energy generation; domestic waste; rotten rice; wastewater

check for updates

Citation: Yaqoob, A.A.; Al-Zaqri, N.; Alamzeb, M.; Hussain, F.; Oh, S.-E.; Umar, K. Bioenergy Generation and Phenol Degradation through Microbial Fuel Cells Energized by Domestic Organic Waste. *Molecules* **2023**, *28*, 4349. https://doi.org/10.3390/molecules28114349

Academic Editor: Juan Carlos Colmenares

Received: 10 May 2023
Revised: 23 May 2023
Accepted: 24 May 2023
Published: 25 May 2023

1. Introduction

Due to the fast growth of industrialization, population, and urbanization, there is a growing worry over worldwide water pollution. Providing a clean and secure water supply for people all around the globe is one of today's most pressing environmental challenges [1]. Figure S1 demonstrates the different sources of water pollution, which is becoming a serious problem for environmental sustainability. Even at low concentrations, the presence of organic pollutants poses a threat to living organisms. Hazardous organic pollutants from groundwater and wastewater continue to require the extension of proficient and cost-effective methods [2]. Traditional water treatment technologies, such as air stripping, adsorption, oxidation, biotreatment, and chlorine action, have a number of drawbacks [3]. Industrial wastewater contains various organic pollutants such as hydroquinone, benzene, phenol, and others that must be remedied [4]. The organic pollutant that has been focused on in this investigation is phenol. The presence of phenol in water gives recipient bodies a carbolic stench and can be hazardous to aquatic flora and wildlife. Phenolic compounds are prevalent contaminants in aqueous streams due to their ubiquitous use, and they are designated priority pollutants since they are detrimental to organisms even in low quantities [5]. Due to its potential damage to human health, phenol is also a priority pollutant on the US Environmental Protection Agency's (EPA) list of contaminants to avoid, which requires phenol levels in wastewater and drinking water to be less than 1 mg/L and less than 0.002 mg/L, respectively [6].

Microbial fuel cells (MFCs) can convert low-strength effluent and lignocellulosic material into electricity. MFCs have the ability to produce an electric current by using a broad variety of soluble or dissolved complex organic wastes in addition to renewable biomass [7,8]. Microorganisms interact with electrodes in an MFC by exchanging electrons, which are either withdrawn or supplied via an electrical circuit. MFCs are the most common form of bio-electrochemical systems, which use the metabolic activity of microorganisms to convert biomass into energy [9–11]. MFCs are the most sustainable, effective, and safe method of producing bioenergy [12,13]. Electrons and protons are produced when an organic substrate is oxidized, which leads to the generation of electrons. In past research, a variety of organic wastes, including domicile wastewater, waste from municipalities, and sewage sludge, were used [14–16]. As a source of fuel in the anode chamber, such carbohydrates are often used. In recent times, there has been a lot of focus placed on the generation of sustainable energy and the biodegradation of pollutants via the use of MFCs from food waste. This is because of the massive amounts of wasted food [17]. Numerous studies have shown that it is possible to use food waste as the organic substrate in MFCs in order to generate power [18–21]. Rice is the main meal in Malaysia, and it is often eaten twice or sometimes three times a day. In 2018, the Solid Waste Management and Public Cleaning Corporation (SWCorp) released a report stating that the highest percentage of food that is wasted daily is rotten rice, which is generated by households. This percentage increases by more than 20% during festival seasons [22]. Malaysia produces enough rice to serve 12 million people three times a day; therefore, a lot of rice waste is collected every day [22,23]. Since a large portion of today's rotten rice is disposed of using traditional methods such as landfilling, composting, and incineration, which may contaminate groundwater, produce poisonous gases, and cause odour issues, this concern has led to serious ecological problems and health threats [24]. Conventional methods are both ineffective and impractical since they cannot fully or effectively use the energy sources from waste, such as rotting rice, that are desired. Therefore, from the standpoint of bioenergy recovery, using the rotten rice as a substrate in MFCs may be the ideal method for treating solid waste. As far as we are aware, MFCs have used food waste as a substrate of organic matter for the growth of bacterial populations; nevertheless, using phenol is a significant advancement. In this research, the focus is on how MFCs work on the degradation of phenol in the presence of rotten rice as anodic fuel. MFCs have been used to produce electricity from rotten rice as a substrate while at the same time accomplishing biological wastewater treatment (biodegradation of organic pollutant phenol). This organic pollutant also serves as a carbon source during degradation, indicating that energy production will be higher than expected.

2. Results and Discussion

2.1. Bioenergy Generation and Polarization Trend

The complete process of operating the MFCs with the aim of generating bioenergy took place over the course of one month. At the beginning of its operation, the MFCs' reactor was loaded with an external resistance of 1000 Ω. During the process of phenol degradation, an increase in the cell's potential (measured in mV) and current (measured in mA) were noted. As the experiment turned on, the voltage in the reactor increased as a result of the proliferation of bacteria that sprang from the rotten rice. It is possible that the addition of rotting rice may increase the demand for carbon and oxygen, which will be explored in a further study in connection with the performance of MFCs. The formation of biofilm on the surface of the anode has an effect on the amount of voltage that is produced. The formation of the biofilm, which was necessary in order to transport electrons to the anode, took many days. On day 19, a voltage that was significantly high at 199 mV was recorded. The findings suggested that the voltage produced by the MFCs technology may be explained by the fact that the microorganisms were feeding on rotten rice. During the first 19 days, the voltage steadily increased from 0 mV to 199 mV before entering a drop phase, as seen in Figure 1. This phase lasted for the remainder of the study. After day 19, there is a steady trend towards a lower voltage, which is eventually followed by a voltage that is constant. This indicates that all of the organic substrate has been completely oxidized, and that the bacterial species involved are unable to produce electrons [25]. The addition of phenol, which was intended to boost the voltage, did not have the anticipated effect, and instead caused the voltage to decrease. There was no longer a movement of electrons from the anode to the cathode, and the voltage remained stable, as shown in Figure 1. This finding suggests that the bacterial species have completed their life cycle.

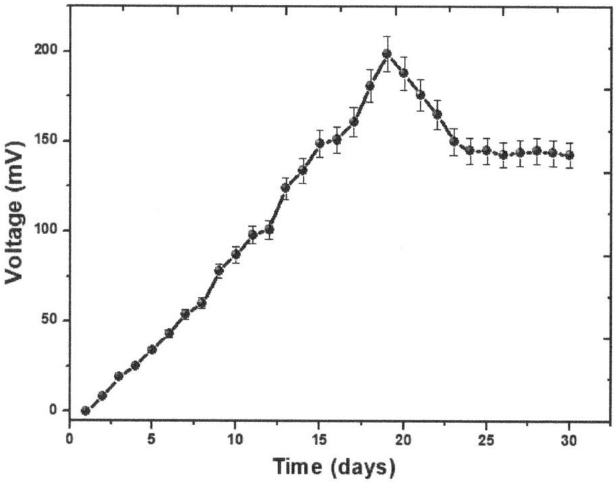

Figure 1. The voltage potential performance of phenol by MFCs.

This illustrates how stable voltage, current density, and power density are interconnected. According to an external source with variable external resistance, the recorded voltage is measured [26]. While the current or current density is measured in accordance with Ohm's law, the voltage is obtained by the variable resistor box [27]. As can be seen in Figure 2, the activation potential dropped from 200 mV to 13 mV while the resistance increased from 5 kΩ to 100 Ω. It is possible that this could be linked to the loss of heat energy that occurs when oxidation or reduction is initiated, or it might be due to the loss of electrons that occur from terminal proteins in bacterial cells to the anode surface [28]. The voltage dropped from 220 mV to 13 mV, while the rise in the external resistance brought it from 5 k Ω to 100 Ω. However, if the external resistance was reduced from 5 kΩ to 100 Ω, there was a significant increase in the rate at which the power density increased. This was calculated in the MFCs at an external resistance of 100 Ω, which served as the cell design point for the MFCs in the subsequent studies. The overall power density was found to be 0.85 mW/m^2, and the current density was found to be 17.10 mA/m^2. Although the achievable stability was not quick at reduced resistance, progress was being made towards power density. The voltage destabilization was quickly reduced, and even at low levels of external resistance, it remained fairly resilient. According to Abbas et al., [29], a strong electron discharge may be the cause of a persistent drop in potential and stability values at lower resistances. Reduced resistances allow electrons to move more rapidly across the circuit, which results in maximum current levels and power densities while having a lower degree of stability [30]. It is referred to as an ohmic over potential, and it was caused by the electrical resistance of the electrode as a result of the energy loss that occurred during the transport of protons via the cathode. Due to the enormous oxidative pressures that are present at the anode, the voltage will begin to drop as the current density continues to increase [31].

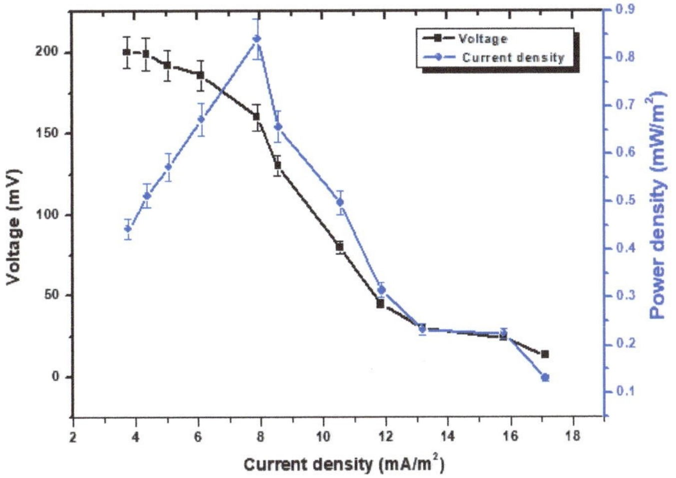

Figure 2. Polarization behavior with the relationship of voltage, power density vs. current density.

2.2. Conductivity Test

Figure 3 depicts the evolution of this phenomenon over time as the conductivity of the cell changes. The conductivity measurements were collected at a variety of times during the course of the 30-day experiment. From day 10 to the very final day, the value ranged from 1780 mS/cm to 3210 mS/cm, although day 20 had a value of 4300 mS/cm. The statistics are consistent with a large amount of power being produced on day 20, given the high conductivity. The largest voltage seems to have been produced on days 19 and 20 of the experiment based on the pattern of voltage generation. The relevance of the system remains true throughout an operation basically unaltered. Long-term exposure to environmental pressures such as pH shifts, organic substrate degradation, and temperature swings causes the system's efficiency to gradually decrease. A related study was also published recently by Rojas-Flores et al. [32] to explain the conductivity effect.

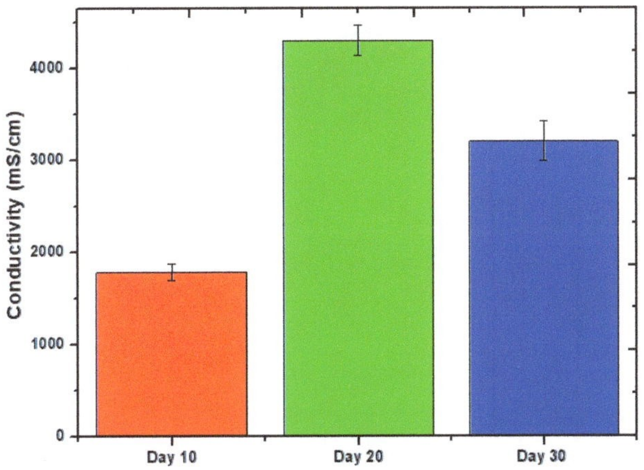

Figure 3. Conductivity of MFCs at different times.

2.3. Cyclic Voltammetry and Electrochemical Impedance Spectroscopy Study

It is an electrochemical approach that interprets the system's electrochemical behavior and is most typically used to characterize electron transfer interactions on the interface biofilm of the anode in MFCs [33,34]. In the anodic chamber of the MFCs, a forward peak was seen on day 30 at 0.8 V, showing a precise oxidizing property, whereas a reverse peak was seen at −0.5 V as shown in Figure 4. As a result, a solid redox loop in MFCs systems was attained to guarantee that active microbial communities can transfer electrons. On the 30th day, when the highest Faradaic current is present, the CV shows maximal reversible peaks exclusively owing to reduction electron-transfer. It has been suggested that the migration of an electron from a redox potential to the anode surface may be implicated in the process of controlled electron transfer [35]. Oxygen may be responsible for the highest levels of oxidation and reduction in MFCs [36]. The organic substrate in the MFCs was the main contributor to the oxidation rate. The exoelectrogenic density on the electrode grew with time, increasing the potential values. The amount of metabolites in the feedstock changes as exoelectrogenic levels grow, which affects the electrical solution's flexibility and conductivity [37]. It was investigated how various days—the first, tenth, twentieth, and thirtieth—affected the inoculum's electrocatalytic activity. In light of the fact that the 30th day curve has the highest peak of all the curves, its area may suggest that this day performs better than others in terms of stability and specific capacitance [38,39]. According to Table 1, this maximum peak denotes biofilm development over MFCs with the highest specific capacitance of 0.00020 F/g throughout all days of curves. The entire anode potential for MFCs grew as the anode's biofilm developed. The bacterial cell membrane's redox enzymes and the superficial charge storage capabilities in the bacterial cytoplasm may be responsible for this enhancement in anode capacity [40].

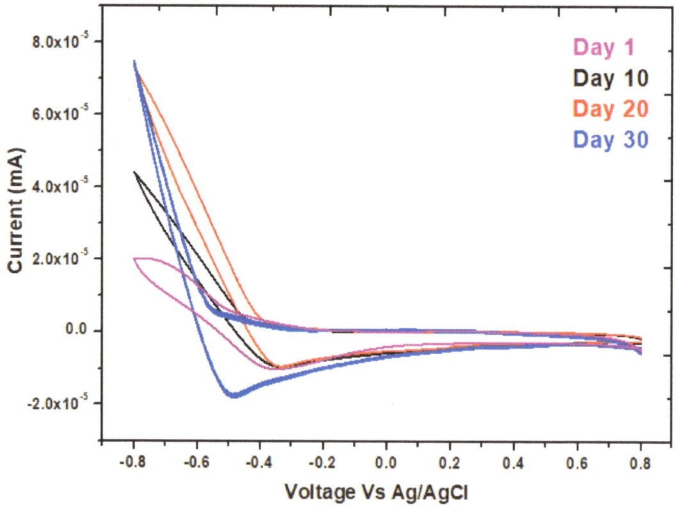

Figure 4. Cyclic voltammetry for the MFCs with phenol at day 1, 10, 20, and 30, respectively.

Table 1. The values of Cp at several specific times.

Measurement Days	Capacitance (F/g)
1st	0.00002
10th	0.00014
20th	0.00018
30th	0.00020

Figure 5 presents the Nyquist plot, which was derived using the EIS technique as an effective means of demonstrating the bio-electrochemical phenomena that took place in the MFCs. The Nyquist plot usually consists of either a straight line or a semicircle. In this plot, the straight line represents the diffusion-regulated reaction, and the diameter of the semicircle represents the resistance to the charge transfer through EIS [41]. This plot is displayed in Figure 5, as it was previously described by several researchers in the same pattern [42–44]. On the 19th day after the growth of the biofilm, the fuel cell was put through tests with the assistance of an EIS-potentiostat. When it applies to electrical accessibility, semi-circles or semi-bent lines imply tremendous mobility, but straight lines with a high Z'_{img} (Ohm) represent limited mobility [45]. On day 19, proof was supplied indicating the biofilm had entirely established itself and was now stable. This evidence was provided in the form of enhanced electronic mobility along with semi-bent curves. The MFCs internal resistance was 312.5 Ω. Having a low charge resistivity and a low internal resistance makes it feasible for electrons to move more readily. This is because charge resistivity is low [46]. After reaching its peak on day 19, the voltage output gradually started to decline in the days that followed. When there is internal resistance, there is a corresponding reduction in the quantity of electron mobility. The electrolytes and the efficiency of the organic substrate are additional factors that have an influence on the internal resistance [47].

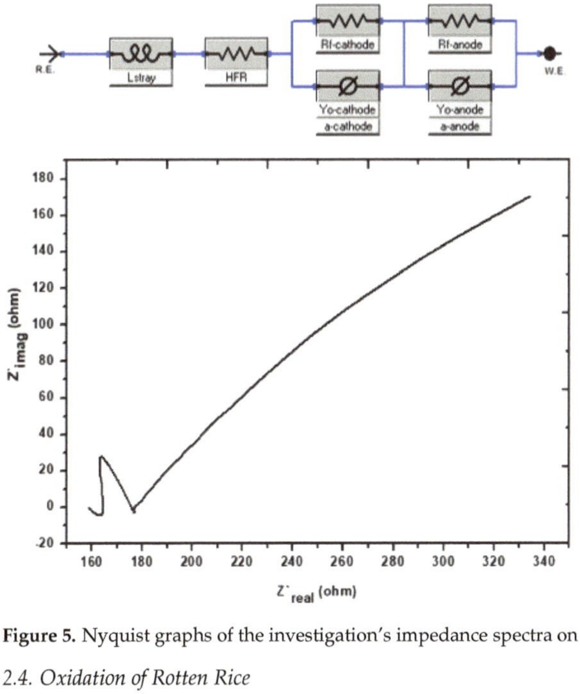

Figure 5. Nyquist graphs of the investigation's impedance spectra on day 30 of the experiment.

2.4. Oxidation of Rotten Rice

The MFCs technique is mostly dependent on the activities of the different bacterial species. In this work, researchers identified a number of bacterial species that are well-known exoelectrogens and organic pollutant-reducing species. The organic substrate in MFCs is oxidized by bacterial species, which results in the generation of bioenergy [48]. In this study, bacterial species oxidize the rotten rice waste, a polysaccharide sugar, which is first converted into simple sugar. The oxidation-reduction reaction can be written as:

$$\text{Anode reaction: Rotten rice} \rightarrow C_6H_{12}O_6 + 6H_2O \rightarrow 6CO_2 + 24H^+ + 24e^-$$

$$\text{Cathode reaction: } 24H^+ + 24e^- + 6O_2 \rightarrow 12H_2O$$

$$\text{Overall: } C_6H_{12}O_6 + 6O_2 \rightarrow 6CO_2 + 6H_2O + \text{Energy} + \text{Biomass}$$

The anode is responsible for producing electrons and protons, which are then delivered to the cathode. The proton is able to move directly from the anode to the cathode, but the electrons have to travel via an external circuit. In addition, there is a period of time during which electrons are transferred from the bacteria to the anode electrode before they are transported from the anode to the cathode. In the literature, the electron transfer mechanisms (bacteria to anode) are well explained, and they are also shown in Figure S3.

2.5. Biodegradation Performance of Phenol

The addition of rotten rice waste as an organic substrate improved the rate of biodegradation by enhancing the metabolism of bacteria. The present study achieved 70% degradation efficiency of phenol by this MFCs investigation. Two distinct microbes—biodegradative and electroactive microbes—participated in the biodegradation of the phenol. While electroactive bacteria contribute to the biodegradation of intermediates, the biodegradative bacteria target phenol and cause a full ring-cleavage process. Phenolic acid (intermediate) is produced via a complete carboxylation process mediated by the carboxylase phenol. However, subsequent phenol cleavage by electroactive bacteria results in the production of molecules of CO_2 and electrons, as shown in the electrochemical processes listed below. Figure 6 illustrates the evidence in support of the degradation of phenol and phenolic acid into the final product. The absorbance peaks of phenol at different time intervals are shown in Figure 7a. On the basis of UV curves, the calculated biodegradation efficiency is shown in Figure 7b. As can be seen on curve day 1, the absorbance peak of phenol at λ_{max} = 275 nm occurs prior to the operation of the reactor and is caused by the transition $\pi \rightarrow \pi^*$ of aromatic (C=C) [49]. Following the completion of the reactor cycle, the phenol peak of the curve (day 1) has moved to a maximum wavelength of 245 nm (day 10). The process of phenol's biodegradation into phenolic acid begins at this point [50]. Phenol is transformed into phenolic acid. As a consequence of the delocalization of electrons, the intermediate has a higher degree of conjugation than phenol [51]. The biodegradation of the intermediate phenolic acid was essentially degraded, as seen by the curve on days 20 and 30, which had a band at 245 nm. The biodegradation efficiency calculation also shows a gradual degradation trend. In light of these findings, the anodic and cathodic processes that take place in the MFCs are described by the electrochemical equation given below.

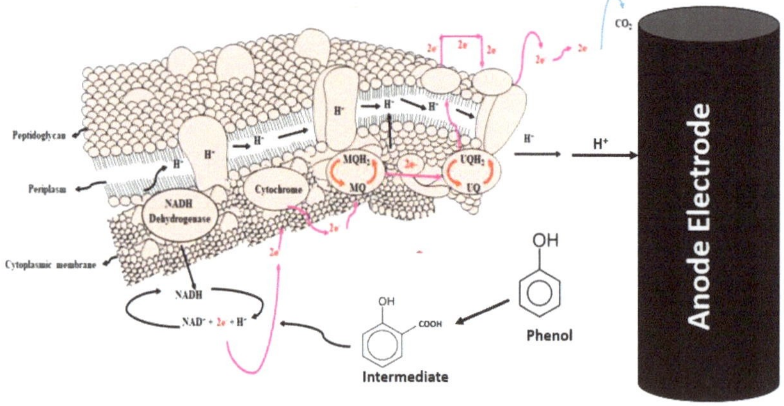

Figure 6. Graphic idea of phenol biodegradation through bacterial communities (modified according to present study from reference [52] with permission from Elsevier to reuse it).

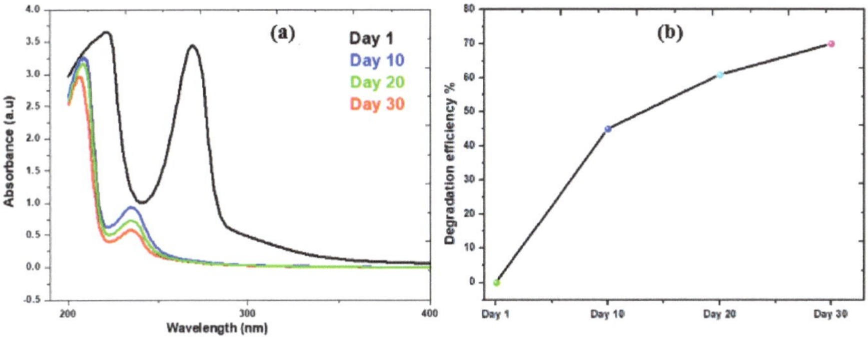

Figure 7. (**a**) UV-Spectra of Phenol Collected from MFCs; (**b**) Degradation percentage of phenol in this system.

Anode reaction:

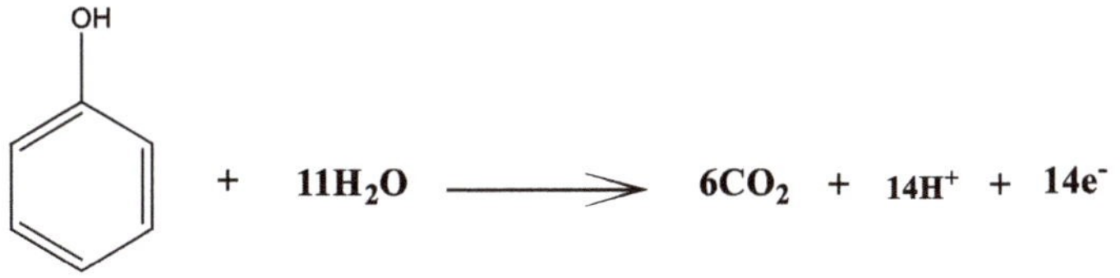

Cathode reaction:

$$3.5O_2 + 14H^+ + 14e^- \longrightarrow 7H_2O$$

Overall electrochemical reaction:

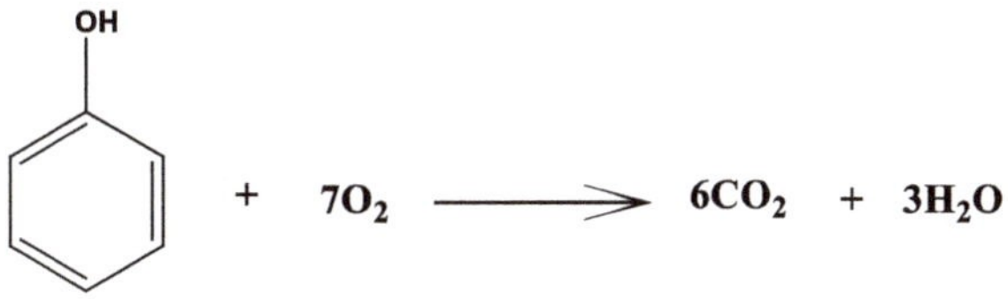

2.6. Biofilm Study

In addition, SEM analysis was carried out on the electrode (both the anode and the cathode) before and after the operation to analyze the bacterial morphology that was present on the surface of the electrode. A distinct distinction can be seen in the SEM of untreated graphite, treated anode, and treated cathode, as shown in Figure 8a–c. Figure 8a displays a planar morphological surface, due to the fact that it was a commercially available graphite electrode that had never been used. Meanwhile, the anode electrode had a thick filamentous structure on its surface, indicating the existence of bacterial species colonies that are most likely biodegradative and electroactive microorganisms. The cathode also had some clusters of bacteria; however, these clusters were not as dense or well developed in comparison to those on the anode electrode. On the surface of the anode electrode was a biofilm that was very thick and had developed to a mature state. This biofilm was scraped to identify the bacteria. The anode SEM had a morphology that resembled a tube or rod, and this form predominated in the pictures. It is evidence of the species of bacteria that are now known to possess conductive pili. Conductive pili are structures that are filamentous and take the form of rod-shaped wires; bacteria utilize these pili to transmit electrons from the cell to the anode electrode [53]. This is a significant and direct contact that takes place between the cell of the bacterium and the surface of the anode. There are a number of experiments that have been performed that have shown that the conductive pili are able to effectively transmit electrons from bacterial cells to the anode surface [54–56]. In addition, by referring to this research, it is possible to draw the conclusion that the dense development of biofilm is an indicator of the anode's high level of biocompatibility with regard to living organisms. Due to the high sugar content of the organic substrate that is given, the limited species of bacteria that are present are located in very dense conditions on the surface of the biofilm [57]. An EDX analysis of the biofilm was carried out in order to determine whether or not certain components were present on the anode's surface. Based on what we see in Figure 9, we are able to draw the conclusion that phenol does not have a negative impact on bacterial populations. Since the anodic biofilm solely contains carbon and oxygen at the end of operation. The results of the investigation demonstrated, on the whole, that the presence of the organic pollutant during the operation of the MFCs did not demonstrate any toxicity towards the bacterial population.

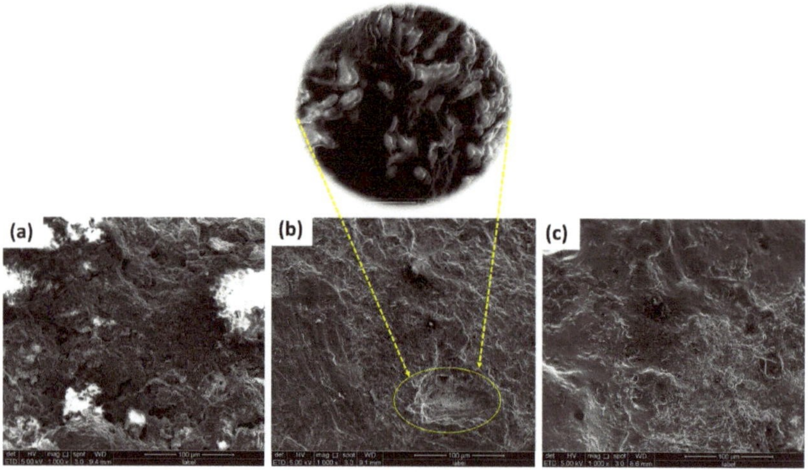

Figure 8. SEM images of (**a**) Graphite electrode before MFCs operation; (**b**) Graphite electrode as anode after MFCs operations; (**c**) Graphite electrode as cathode after MFCs operations.

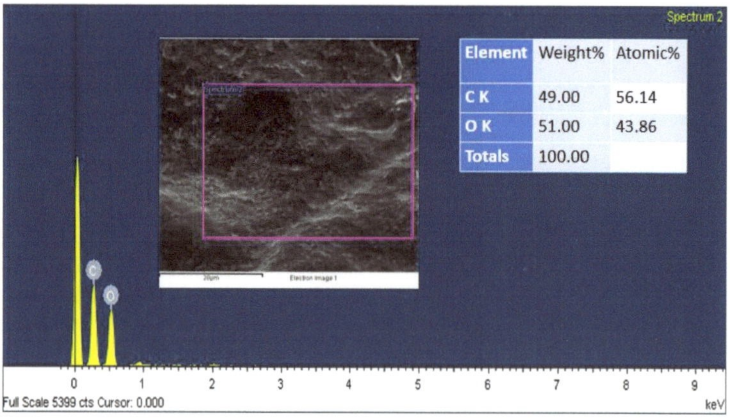

Figure 9. EDX spectra of anode electrode of the present study.

2.7. Bacterial Identification

In the framework of biological research, a process of bacterial isolation and identification was carried out in order to determine the species of bacteria. In an effort to determine the species of bacteria, the anodic biofilm was analyzed. The list of the most prevalent identified bacterial species is shown in Figure 10. There are different kinds of bacterial species that are classified as conductive pili-type species. It indicates that the rod-shaped appendages that may be observed in the SEM pictures are providing evidence that conductive pili-typed species exist. *Bacillus* is, in general, the most prevalent genus on the biofilm's surface. The detected bacteria are genus members that have previously been shown to be biodegradative and electroactive in a variety of earlier investigations and published works of literature. In addition, Nimje et al. [58] reported that the power density of the *Bacillus* strains when used as a biocatalyst was 0.000105 mW/m^2. However, a few investigations demonstrated that the found bacterial species (Figure 10) are well known as biodegradative and electroactive species, which we also found in our experiment [59,60].

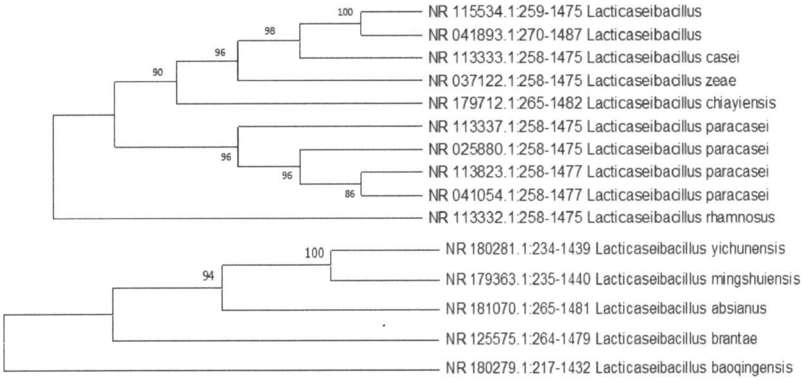

Figure 10. Phylogenetic study of bacterial species from anodes of the MFCs.

3. Experimentation

3.1. Chemical and Materials

The wastewater was collected from a pond located at Bayan Lepas, Penang, Malaysia (6.3195° N, 100.1132° E). The rotten rice waste was acquired from the Café at the Universiti Sains Malaysia (USM). Materials were acquired from a variety of sources, such as commercial graphite rods (FUDA 2B Lead; New York, NY, USA), distilled water, and phenol (Sigma-Aldrich, St. Louis, MO, USA).

3.2. Inoculation of MFCs

The pond effluent collected from a local pond (Bayan Lepas, Penang, Malaysia) was treated with 500 mg/L of phenol to prepare an inoculation of MFCs. In this study, as an organic substrate, 750 g of rotten rice waste was used. Synthetic wastewater was used to refer to the wastewater that included organic pollutants. Several pond effluent and synthetic wastewater physicochemical properties are listed in Table 2. We used a pH meter (EUTECH instrument-700, New York, NY, USA), a thermometer (GH, ZEAL LTD, London, UK), and an electrical conductivity meter (ESEL, Ambala, India) to measure the conductivity, pH, and temperature, respectively.

Table 2. Characteristics of pond effluent and synthetic wastewater.

Parameters	Pond Effluent	Synthetic Wastewater
Color	Cloudy	Cloudy
Electrical conductivity	15.40 μs/cm	24.10 μs/cm
pH	6.99	5.45
Temperature	24 °C	24 °C
Odor	Bad	Bad
Phenol	0 mg/L	500 mg/L

3.3. *MFCs Set Up and Operational Protocol*

In this experiment, a single-chamber microbial fuel cell (MFC) was developed to produce recycled water from wastewater and produce energy. For the degradation phenol, the MFC measured 16.5 cm and 13.5 cm in height and diameter, respectively. The total operational capacity of the MFC was 1000 mL; however, only 100 mL of phenol stock solution and 1000 mL of sewage were used to pollute 750 g of rotten rice. The MFC had a total of 1000 mL of wastewater. After that, commercial graphite rods measuring 11.5 cm × 1 cm (h × r) for the anode and the same length for the cathode were vertically put into the MFC in order to create the cathode and anode, respectively. There was a distance of 17 cm between the anode and the cathode. By adding a 1000 Ω external resistance, the electrodes were linked with copper wire. Experiments using rotten rice were carried out using MFCs for a period of one month at room temperature. Figure S2 depicts the MFC reactor used in this research.

3.4. *Electrochemical Measurements and Calculations*

The potential voltage between the anode and cathode was measured using a digital multimeter (UNI-T UT33A, China) once each day, and the resulting current was calculated using Ohm's law. Using Equations (1)–(4) below, we were able to determine the current density (CD), power density (PD), and internal resistance (r). The voltage unit is 'mV', CD has mA/m^2, and PD has mW/m^2:

$$\text{Voltage Output (V)} = \text{IR} \tag{1}$$

$$\text{Power Density} = \frac{V^2}{RA} \tag{2}$$

$$\text{Current Density} = \frac{I}{A} \tag{3}$$

$$\text{Internal Resistance} = \left[\frac{E - V}{V}\right]R \tag{4}$$

where "A" represents the anode electrode surface area; the "I" represents the current; the "R" represents the internal resistance; and "E" represents the electromotive force. Measurement of the electromotive force was carried out with the help of an open circuit voltage (OCV). The slope of the polarization curve was used in conjunction with a variable resistance box that had a range from 5000 to 100 Ω in order to determine the MFCs' internal resistance. Further, conductivity tests are also carried out every 10 days by using the conductometer.

3.4.1. Cyclic Voltammetry (CV)

The cyclic voltammetry (CV) technique was used to investigate the redox reactions occurring on the electrode surface. Using a scanning rate of 10 mV/s and a potential range of +0.8 V to 0.8 V, the electrode surface was measured at the 1st, 10th, 20th, and 30th day intervals, respectively. The platinum wire and the anode each took their place as the counter and working electrodes, while the Ag/AgCl mixture was used as the reference electrode. The reference electrode was found by using the potential of the electrodes. The particular capacitance, denoted by the notation Cp (unit is F/g), was defined as the sum of the data from the cathode and the anode when expressed as a fraction of a gramme. Cp was computed from CV data using the following Equation (5), where A represents the area of the CV curve, m represents the loaded sample amount in the CV instrument, k represents the CV scan rate in mV/s, and (V2 − V1) specifies the potential range of the CV:

$$C_P = \frac{A}{2mk\,(V_2 - V_1)} \quad (5)$$

3.4.2. Electrochemical Impedance Spectroscopy (EIS)

Electrochemical Impedance Spectroscopy (also known as EIS) was used to study the influence of the anode's resistance on voltage at many points in time during the experiment. On day 19, EIS research was carried out using MFCs in their operational mode, with the frequency range being from 100,000 Hz all the way down to 0.1 Hz. The amplitude of the alternating current (AC) was set to around 1 mV so as to prevent the adhesion of biofilm and to reduce the amount of disturbance caused to the steady-state system.

3.5. *Biological Characterization*

3.5.1. Biodegradation of Organic Pollutant

To assess biodegradation efficiency, phenol contents were measured under an ultraviolet–visible (UV-Vis) light source using a Shimadzu UV-Vis 2600 Spectrophotometer. Every 10 days, a sample of around 1 mL from the phenol MFCs was obtained, diluted to 50 ppm, and the organic pollutant level was calculated. Calculating the biodegradation efficiency % required using Equation (6), where T_o was denoted as transmittance of the standard, T as transmittance of water sample, and $[T_o]$ as concentration of the standard solution.

$$\text{Biodegradation efficiency \%} = \frac{T_o - T}{[T_o]} \times 100 \quad (6)$$

3.5.2. Bacterial Identification and Biofilm Studies

The purpose of this experiment was to ascertain the species of bacteria that are present in MFCs by isolating and identifying the bacteria. Obtaining pure bacterial cultures may be accomplished via the techniques of serial dilution and plating bacteria through the use of the streaking method. A material may be diluted into a solution using a method known as serial dilution. After the material is diluted, it is spread out on nutrient agar and left to rest for three days. The method of streak plate may be used to separate bacteria from a mixed population so that they can then be cultivated in a pure culture. The inoculum was spread throughout the agar surface in this manner, which resulted in the bacteria population becoming less dense. The polymerase chain reaction (PCR) technique was used in the process of creating bacterial genes. In order to amplify the genes, both a forward primer and a reverse primer were used. The cloning of the PCR-generated result was accomplished with the help of a cloning kit manufactured by Invitrogen and located in Carlsbad, CA, USA. After having their DNA sequenced, bacterial strains were uploaded to the GenBank database. Scanning electron microscopy (SEM) was used in order to investigate the development as well as the stability of the biofilm and the surface appearance. After going through the process of having the organic pollutant content on its surface reduced by synthetic wastewater, the anode in the SEM images shows signs of bacterial growth on its surface. At the conclusion of the procedure, an energy dispersive X-ray, abbreviated EDX, was taken from the biofilm-anode electrode in order to study the impact that the hazardous organic pollutant had on the biofilm.

4. Challenges and Future Perspectives

MFC technologies are environmentally viable ways to produce electricity and degrade organic contaminants from wastewater, according to modern research. MFC technologies have certain major drawbacks, including design, cost, and performance. Thus, they have never competed in renewable energy or wastewater treatment. MFCs can produce enough net energy to replace energy from organic material oxidation using waste and inorganic carbon under specific circumstances. MFC systems can biologically convert chemical energy into electrical form, enabling them to handle a broad variety of chemical substrates at varying concentrations. MFC technology helps researchers investigate electrochemical, biochemical, microbial, and material surface reactions under controlled circumstances [61]. They research how materials, chemical compounds, and feedstock substrates impact them. This approach helps us grasp MFCs' larger-scale difficulties. In order for MFC technology to be sustainable on a commercial scale, steps need to be taken to reduce the high operating costs and improve the technique's power generation. MFCs have a high degree of capital cost because they need costly electrode materials, such as catalyst, current collector, and separator materials. This contributes to the overall cost of the device [62]. Recently, attention has been paid to this issue by a number of scholars. The use of an electrode that is produced from biomass is one potential approach to resolving the cost concerns surrounding electrodes. In comparison to the literature, the findings that were obtained by a research team that concentrated on graphene-derived electrodes that were synthesized from waste biomass were much more favorable [63–65]. As a result, several types of biomasses and their most recent modifications might lead to an answer to this issue. There are several operating issues, such as pH, temperature control, potential range, etc., but several scholars optimized it at different ranges and mostly agreed to operate the MFCs at natural conditions, such as pH 7 at room temperature. They also believed that they could minimize the internal resistance by improving the MFCs design and reducing the distance between the anode and cathode [66]. In spite of these relatively minor concerns, the most important problems involve electron transportation and generation. The migration of electrons is closely connected to the development of electrodes, as mentioned above. The oxidation of the organic substrate determines whether or not it is favorable for bacterial species to oxidized substrate quickly, which is a prerequisite for electron production. One of the potential solutions is using material that has been generated from waste as an organic substrate [67–69]. Finally, integrating MFCs with other wastewater treatment methods could boost the efficacy of treatment and, as a result, significantly lower total power consumption [61].

5. Conclusions

The dual-purpose use of MFCs in environmentally friendly wastewater treatment is gaining in popularity. It leads to a more efficient degradation of organic compounds such as phenol and produces bioelectricity that can be channeled and ramped up to meet rising energy needs. In this work, we used MFCs driven by rotting rice to effectively generate a significant power density and efficiently biodegrade phenol. The optimization of variables such as internal resistance, polarization, specific capacitance, and voltage production is well discussed. As a consequence of the bacterial species adhering to the anode surface, a denser and more uniform biofilm was formed, according to SEM data. Bacterial species have developed the capacity to produce bioelectricity. The biofilm research and bacterial species identification methods showed that phenol is successfully degraded while producing energy with no negative side effects. According to all evidence, the present study followed the electron transfer mechanism (the mechanism is well explained in the previous literature) via conductive pili.

Supplementary Materials: The following supporting information can be downloaded at: https://www.mdpi.com/article/10.3390/molecules28114349/s1, Figure S1: Different sources of water pollution (Adapted from: https://www.filterwater.com/t-articles.water-pollution.aspx, accessed on

8 May 2023); Figure S2: MFCs Set-up for phenol degradation; Figure S3: Different mechanisms of the electron transfer from bacterial species to anode electrode in MFCs (Adapted from reference [70] with Elsevier permission). References [24,66,70–72] are cited in Supplementary Materials.

Author Contributions: K.U. and N.A.-Z.: Conceptualization visualization, investigation. A.A.Y. and K.U.: Methodology, writing—original draft preparation. S.-E.O. and F.H.: Writing—reviewing and editing, M.A. and A.A.Y.: Supervision, result investigation, validation. N.A.-Z.: Administrative duties, funding acquisition. All authors have read and agreed to the published version of the manuscript.

Funding: The authors extend their appreciation to the Deputyship for Research and Innovation, "Ministry of Education" in Saudi Arabia for funding this research (IFKSUOR3-209-1).

Institutional Review Board Statement: Not applicable.

Informed Consent Statement: Not applicable.

Data Availability Statement: All the data are included in the text.

Acknowledgments: The authors extend their appreciation to the Deputyship for Research and Innovation, "Ministry of Education" in Saudi Arabia for funding this research (IFKSUOR3-209-1).

Conflicts of Interest: The authors have stated that there are no competing interests that might influence their work.

Sample Availability: Not applicable.

References

1. Guerrero–Barajas, C.; Ahmad, A.; Ibrahim, M.N.M.; Alshammari, M.B. Advanced Technologies for Wastewater Treatment. In *Green Chemistry for Sustainable Water Purification*; Shahid-ul-Islam, Shalla, A.H., Shahadat, M., Eds.; Wiley: Hoboken, NJ, USA, 2023; pp. 179–202. ISBN 978-1-119-85229-2.
2. Tijani, J.O.; Fatoba, O.O.; Madzivire, G.; Petrik, L.F. A review of combined advanced oxidation technologies for the removal of organic pollutants from water. *Water Air Soil Pollut.* **2014**, *225*, 2102. [CrossRef]
3. Parveen, T.; Umar, K.; Mohamad Ibrahim, M.N. Role of nanomaterials in the treatment of wastewater: A review. *Water* **2020**, *12*, 495.
4. Idris, M.O.; Kim, H.-C. Exploring the effectiveness of microbial fuel cell for the degradation of organic pollutants coupled with bio-energy generation. *Sustain. Energy Technol. Assess.* **2022**, *52*, 102183. [CrossRef]
5. Kulkarni, S.J.; Kaware, J.P. Review on research for removal of phenol from wastewater. *Int. J. Sci. Res. Publ.* **2013**, *3*, 1–5.
6. Feng, Q.; Zhao, L.; Lin, J.-M. Molecularly imprinted polymer as micro-solid phase extraction combined with high performance liquid chromatography to determine phenolic compounds in environmental water samples. *Anal. Chim. Acta* **2009**, *650*, 70–76. [CrossRef]
7. Logan, B.E.; Hamelers, B.; Rozendal, R.; Schröder, U.; Keller, J.; Freguia, S.; Aelterman, P.; Verstraete, W.; Rabaey, K. Microbial fuel cells: Methodology and technology. *Environ. Sci. Technol.* **2006**, *40*, 5181–5192. [CrossRef] [PubMed]
8. Logan, B.E. Exoelectrogenic bacteria that power microbial fuel cells. *Nat. Rev. Microbiol.* **2009**, *7*, 375–381. [CrossRef]
9. Li, H.; Tian, Y.; Zuo, W.; Zhang, J.; Pan, X.; Li, L.; Su, X. Electricity generation from food wastes and characteristics of organic matters in microbial fuel cell. *Bioresour. Technol.* **2016**, *205*, 104–110. [CrossRef]
10. Logan, B.E.; Regan, J.M. Microbial fuel cells—Challenges and applications. *Environ. Sci. Technol.* **2006**, *40*, 5172–5180. [CrossRef]
11. Hossain, A.; Masud, N.; Roy, S.; Ali, M. Investigation of voltage storage capacity for the variation of electrode materials in microbial fuel cells with experimentation and mathematical modelling. *Int. J. Water Resourc. Environ. Eng.* **2022**, *14*, 97–109.
12. Idris, M.O.; Ahmad, A.; Daud, N.N.M. Removal of Toxic Metal Ions from Wastewater Through Microbial Fuel Cells. In *Microbial Fuel Cells for Environmental Remediation*; Springer: Berlin/Heidelberg, Germany, 2022; pp. 299–325.
13. Hossain, A.M.A.; Masud, N.; Yasin, M.S.; Ali, M. Analysis of the performance of microbial fuel cell as a potential energy storage device. *Proc. Int. Exch. Innov. Conf. Eng. Sci.* **2020**, *1*, 149–155.
14. Jia, J.; Tang, Y.; Liu, B.; Wu, D.; Ren, N.; Xing, D. Electricity generation from food wastes and microbial community structure in microbial fuel cells. *Bioresour. Technol.* **2013**, *144*, 94–99. [CrossRef] [PubMed]
15. Asefi, B.; Li, S.-L.; Moreno, H.A.; Sanchez-Torres, V.; Hu, A.; Li, J.; Yu, C.-P. Characterization of electricity production and microbial community of food waste-fed microbial fuel cells. *Process. Saf. Environ. Prot.* **2019**, *125*, 83–91. [CrossRef]
16. Yaqoob, A.A.; Mohamad Ibrahim, M.N.; Umar, K.; Bhawani, S.A.; Khan, A.; Asiri, A.M.; Khan, M.R.; Azam, M.; AlAmmari, A.M. Cellulose Derived Graphene/Polyaniline Nanocomposite Anode for Energy Generation and Bioremediation of Toxic Metals via Benthic Microbial Fuel Cells. *Polymers* **2021**, *13*, 135. [CrossRef] [PubMed]
17. Masud, N.; Hossain, A.-M.A.; Moresalein, M.J.; Ali, M. Performance Evaluation of Microbial Fuel Cell with Food Waste Solution as a Potential Energy Storage Medium. *Proc. Int. Exch. Innov. Conf. Eng. Sci.* **2021**, *2*, 96–102. [CrossRef]
18. Goud, R.K.; Babu, P.S.; Mohan, S.V. Canteen based composite food waste as potential anodic fuel for bioelectricity generation in single chambered microbial fuel cell (MFC): Bio-electrochemical evaluation under increasing substrate loading condition. *Int. J. Hydrog. Energy* **2011**, *36*, 6210–6218. [CrossRef]

19. Moharir, P.V.; Tembhurkar, A.R. Effect of recirculation on bioelectricity generation using microbial fuel cell with food waste leachate as substrate. *Int. J. Hydrog. Energy* **2018**, *43*, 10061–10069. [CrossRef]
20. Rikame, S.S.; Mungray, A.A.; Mungray, A.K. Electricity generation from acidogenic food waste leachate using dual chamber mediator less microbial fuel cell. *Int. Biodeterior. Biodegrad.* **2012**, *75*, 131–137. [CrossRef]
21. Asim, A.Y.; Mohamad, N.; Khalid, U.; Tabassum, P.; Akil, A.; Lokhat, D.; Siti, H. A glimpse into the microbial fuel cells for wastewater treatment with energy generation. *Desalination Water Treat.* **2021**, *214*, 379–389.
22. Azhari, N.W. *The Performance of Takakura Composting Using Food Waste from Makanan Ringan Mas Industry*; Universiti Tun Hussein Onn Malaysia: Parit Raja, Malaysia, 2019; pp. 1–150.
23. Goud, R.K.; Mohan, S.V. Pre-fermentation of waste as a strategy to enhance the performance of single chambered microbial fuel cell (MFC). *Int. J. Hydrog. Energy* **2011**, *36*, 13753–13762. [CrossRef]
24. Daud, N.N.M.; Ahmad, A.; Yaqoob, A.A.; Ibrahim, M.N.M. Application of rotten rice as a substrate for bacterial species to generate energy and the removal of toxic metals from wastewater through microbial fuel cells. *Environ. Sci. Pollut. Res.* **2021**, *28*, 62816–62827. [CrossRef] [PubMed]
25. Hassan, S.H.; Abd el Nasser, A.Z.; Kassim, R.M. Electricity generation from sugarcane molasses using microbial fuel cell technologies. *Energy* **2019**, *178*, 538–543. [CrossRef]
26. Simeon, M.I.; Asoiro, F.U.; Aliyu, M.; Raji, O.A.; Freitag, R. Polarization and power density trends of a soil-based microbial fuel cell treated with human urine. *Int. J. Energy Res.* **2020**, *44*, 5968–5976. [CrossRef]
27. Rabaey, K.; Verstraete, W. Microbial fuel cells: Novel biotechnology for energy generation. *Trends Biotechnol.* **2005**, *23*, 291–298. [CrossRef]
28. Prakash, O.; Mungray, A.; Kailasa, S.K.; Chongdar, S.; Mungray, A.K. Comparison of different electrode materials and modification for power enhancement in benthic microbial fuel cells (BMFCs). *Process. Saf. Environ. Prot.* **2018**, *117*, 11–21. [CrossRef]
29. Abbas, S.Z.; Rafatullah, M.; Ismail, N.; Nastro, R.A. Enhanced bioremediation of toxic metals and harvesting electricity through sediment microbial fuel cell. *Int. J. Energy Res.* **2017**, *41*, 2345–2355. [CrossRef]
30. Abbas, S.Z.; Rafatullah, M.; Ismail, N.; Shakoori, F.R. Electrochemistry and microbiology of microbial fuel cells treating marine sediments polluted with heavy metals. *RSC Adv.* **2018**, *8*, 18800–18813. [CrossRef]
31. Ahmad, A.; Alshammari, M.B. Basic principles and working mechanisms of microbial fuel cells. In *Microbial Fuel Cells: Emerging Trends in Electrochemical Applications*; IOP Publishing: Oxford, UK, 2022; pp. 1–30.
32. Rojas-Flores, S.; Benites, S.M.; La Cruz-Noriega, D.; Cabanillas-Chirinos, L.; Valdiviezo-Dominguez, F.; Quezada Álvarez, M.A.; Vega-Ybañez, V.; Angelats-Silva, L. Bioelectricity Production from Blueberry Waste. *Processes* **2021**, *9*, 1301. [CrossRef]
33. Fricke, K.; Harnisch, F.; Schröder, U. On the use of cyclic voltammetry for the study of anodic electron transfer in microbial fuel cells. *Energy Environ. Sci.* **2008**, *1*, 144–147. [CrossRef]
34. López Zavala, M.Á.; Gonzalez Pena, O.I.; Cabral Ruelas, H.; Delgado Mena, C.; Guizani, M. Use of cyclic voltammetry to describe the electrochemical behavior of a dual-chamber microbial fuel cell. *Energies* **2019**, *12*, 3532. [CrossRef]
35. Torres, C.I.; Marcus, A.K.; Lee, H.-S.; Parameswaran, P.; Krajmalnik-Brown, R.; Rittmann, B.E. A kinetic perspective on extracellular electron transfer by anode-respiring bacteria. *FEMS Microbiol. Rev.* **2010**, *34*, 3–17. [CrossRef] [PubMed]
36. Wen, H.; Zhu, H.; Xu, Y.; Yan, B.; Shutes, B.; Bañuelos, G.; Wang, X. Removal of sulfamethoxazole and tetracycline in constructed wetlands integrated with microbial fuel cells influenced by influent and operational conditions. *Environ. Pollut.* **2021**, *272*, 115988. [CrossRef] [PubMed]
37. Yaqoob, A.A.; Ibrahim, M.N.M.; Yaakop, A.S.; Ahmad, A. Application of microbial fuel cells energized by oil palm trunk sap (OPTS) to remove the toxic metal from synthetic wastewater with generation of electricity. *Appl. Nanosci.* **2021**, *11*, 1949–1961. [CrossRef]
38. Rodríguez-Couto, S.; Ahmad, A. Preparation, characterization, and application of modified carbonized lignin as an anode for sustainable microbial fuel cell. *Process. Saf. Environ. Prot.* **2021**, *155*, 49–60.
39. Inamdar, A.; Kim, Y.; Pawar, S.; Kim, J.; Im, H.; Kim, H. Chemically grown, porous, nickel oxide thin-film for electrochemical supercapacitors. *J. Power Sources* **2011**, *196*, 2393–2397. [CrossRef]
40. Dolatabadi, M.; Ahmadzadeh, S. A rapid and efficient removal approach for degradation of metformin in pharmaceutical wastewater using electro-Fenton process; optimization by response surface methodology. *Water Sci. Technol.* **2019**, *80*, 685–694. [CrossRef]
41. Yaakop, A.S.; Rafatullah, M. Utilization of biomass-derived electrodes: A journey toward the high performance of microbial fuel cells. *Appl. Water Sci.* **2022**, *12*, 99.
42. Serrà, A.; Bhawani, S.A.; Ibrahim, M.N.M.; Khan, A.; Alorfi, H.S.; Asiri, A.M.; Hussein, M.A.; Khan, I.; Umar, K. Utilizing biomass-based graphene oxide–polyaniline–ag electrodes in microbial fuel cells to boost energy generation and heavy metal removal. *Polymers* **2022**, *14*, 845.
43. Anwer, A.; Khan, M.; Khan, N.; Nizami, A.; Rehan, M.; Khan, M. Development of novel MnO2 coated carbon felt cathode for microbial electroreduction of CO2 to biofuels. *J. Environ. Manag.* **2019**, *249*, 109376. [CrossRef]
44. Bakar, M.A.B.A.; Kim, H.-C.; Ahmad, A.; Alshammari, M.B.; Yaakop, A.S. Oxidation of food waste as an organic substrate in a single chamber microbial fuel cell to remove the pollutant with energy generation. *Sustain. Energy Technol. Assess.* **2022**, *52*, 102–182.

45. He, Z.; Mansfeld, F. Exploring the use of electrochemical impedance spectroscopy (EIS) in microbial fuel cell studies. *Energy Environ. Sci.* **2009**, *2*, 215–219. [CrossRef]
46. Hung, Y.-H.; Liu, T.-Y.; Chen, H.-Y. Renewable coffee waste-derived porous carbons as anode materials for high-performance sustainable microbial fuel cells. *ACS Sustain. Chem. Eng.* **2019**, *7*, 16991–16999. [CrossRef]
47. Idris, M.O.; Noh, N.A.M. Sustainable microbial fuel cell functionalized with a bio-waste: A feasible route to formaldehyde bioremediation along with bioelectricity generation. *Chem. Eng. J.* **2023**, *455*, 140781. [CrossRef]
48. Nevin, K.P.; Kim, B.-C.; Glaven, R.H.; Johnson, J.P.; Woodard, T.L.; Methé, B.A.; DiDonato, R.J., Jr.; Covalla, S.F.; Franks, A.E.; Liu, A. Anode biofilm transcriptomics reveals outer surface components essential for high density current production in Geobacter sulfurreducens fuel cells. *PLoS ONE* **2009**, *4*, 96–128. [CrossRef]
49. Dearden, J.; Forbes, W. Light absorption studies: Part XIV. The ultraviolet absorption spectra of phenols. *Can. J. Chem.* **1959**, *37*, 1294–1304. [CrossRef]
50. Kowalski, R.; Kowalska, G. Phenolic acid contents in fruits of aubergine (*Solanum melongena* L.). *Pol. J. Food Nutr. Sci.* **2005**, *14*, 37–41.
51. Craft, B.D.; Kerrihard, A.L.; Amarowicz, R.; Pegg, R.B. Phenol-based antioxidants and the in vitro methods used for their assessment. *Compr. Rev. Food Sci. Food Saf.* **2012**, *11*, 148–173. [CrossRef]
52. Umar, M.F.; Rafatullah, M.; Abbas, S.Z.; Ibrahim, M.N.M.; Ismail, N. Bioelectricity production and xylene biodegradation through double chamber benthic microbial fuel cells fed with sugarcane waste as a substrate. *J. Hazard. Mater.* **2021**, *419*, 126469. [CrossRef]
53. Serrà, A.; Yaqoob, A.A.; Ibrahim, M.N.M.; Yaakop, A.S. Self-assembled oil palm biomass-derived modified graphene oxide anode: An efficient medium for energy transportation and bioremediating Cd (II) via microbial fuel cells. *Arab. J. Chem.* **2021**, *14*, 103–121.
54. Reimers, C.E.; Li, C.; Graw, M.F.; Schrader, P.S.; Wolf, M. The identification of cable bacteria attached to the anode of a benthic microbial fuel cell: Evidence of long distance extracellular electron transport to electrodes. *Front. Microbiol.* **2017**, *8*, 20–55. [CrossRef]
55. Lovley, D.R. Electrically conductive pili: Biological function and potential applications in electronics. *Curr. Opin. Electrochem.* **2017**, *4*, 190–198. [CrossRef]
56. Guang, L.; Koomson, D.A.; Jingyu, H.; Ewusi-Mensah, D.; Miwornunyuie, N. Performance of exoelectrogenic bacteria used in microbial desalination cell technology. *Int. J. Environ. Res. Public Health* **2020**, *17*, 1121. [CrossRef] [PubMed]
57. Aleid, G.M.; Alshammari, A.S.; Alomari, A.D.; Almukhlifi, H.A.; Ahmad, A.; Yaqoob, A.A. Dual role of sugarcane waste in benthic microbial fuel to produce energy with degradation of metals and chemical oxygen demand. *Processes* **2023**, *11*, 1060. [CrossRef]
58. Nimje, V.R.; Chen, C.-Y.; Chen, C.-C.; Jean, J.-S.; Reddy, A.S.; Fan, C.-W.; Pan, K.-Y.; Liu, H.-T.; Chen, J.-L. Stable and high energy generation by a strain of Bacillus subtilis in a microbial fuel cell. *J. Power Sources* **2009**, *190*, 258–263. [CrossRef]
59. Torlaema, T.A.M.; Ibrahim, M.N.M.; Ahmad, A.; Guerrero-Barajas, C.; Alshammari, M.B.; Oh, S.-E.; Hussain, F. Degradation of Hydroquinone Coupled with Energy Generation through Microbial Fuel Cells Energized by Organic Waste. *Processes* **2022**, *10*, 2099. [CrossRef]
60. Djukić-Vuković, A.; Meglič, S.H.; Flisar, K.; Mojović, L.; Miklavčič, D. Pulsed electric field treatment of Lacticaseibacillus rhamnosus and Lacticaseibacillus paracasei, bacteria with probiotic potential. *LWT* **2021**, *152*, 112304. [CrossRef]
61. Malik, S.; Kishore, S.; Dhasmana, A.; Kumari, P.; Mitra, T.; Chaudhary, V.; Kumari, R.; Bora, J.; Ranjan, A.; Minkina, T. A Perspective Review on Microbial Fuel Cells in Treatment and Product Recovery from Wastewater. *Water* **2023**, *15*, 316. [CrossRef]
62. He, L.; Du, P.; Chen, Y.; Lu, H.; Cheng, X.; Chang, B.; Wang, Z. Advances in microbial fuel cells for wastewater treatment. *Renew. Sustain. Energy Rev.* **2017**, *71*, 388–403. [CrossRef]
63. Yaqoob, A.A.; Ibrahim, M.N.M.; Umar, K. Biomass-derived composite anode electrode: Synthesis, characterizations, and application in microbial fuel cells (MFCs). *J. Environ. Chem. Eng.* **2021**, *9*, 106–111. [CrossRef]
64. Yaqoob, A.A.; Ibrahim, M.N.M.; Yaakop, A.S. Application of oil palm lignocellulosic derived material as an efficient anode to boost the toxic metal remediation trend and energy generation through microbial fuel cells. *J. Clean. Prod.* **2021**, *314*, 1280–1312. [CrossRef]
65. Idris, M.O.; Guerrero–Barajas, C.; Kim, H.-C. Scalability of biomass-derived graphene derivative materials as viable anode electrode for a commercialized microbial fuel cell: A systematic review. *Chin. J. Chem. Eng.* **2022**, *55*, 277–292. [CrossRef]
66. Ahmad, A.; Ibrahim, M.N.M.; Yaqoob, A.A.; Setapar, S.H.M. *Microbial Fuel Cells for Environmental Remediation*; Springer Nature: Berlin/Heidelberg, Germany, 2022.
67. Fadzli, F.; Ibrahim, M.; Yaakop, A. Benthic microbial fuel cells: A sustainable approach for metal remediation and electricity generation from sapodilla waste. *Int. J. Environ. Sci. Technol.* **2023**, *20*, 3927–3940.
68. Al-Zaqri, N.; Yaakop, A.S.; Umar, K. Potato waste as an effective source of electron generation and bioremediation of pollutant through benthic microbial fuel cell. *Sustain. Energy Technol. Assess.* **2022**, *53*, 102560.
69. Guerrero–Barajas, C.; Ibrahim, M.N.M.; Umar, K.; Yaakop, A.S. Local fruit wastes driven benthic microbial fuel cell: A sustainable approach to toxic metal removal and bioelectricity generation. *Environ. Sci. Pollut. Res.* **2022**, *29*, 32913–32928.
70. Fadzli, F.S.; Rashid, M.; Yaqoob, A.A.; Ibrahim, M.N.M. Electricity generation and heavy metal remediation by utilizing yam (Dioscorea alata) waste in benthic microbial fuel cells (BMFCs). *Biochem. Eng. J.* **2021**, *172*, 108067. [CrossRef]

71. Khatoon, A.; Mohd Setapar, S.H.; Parveen, T. Outlook on the role of microbial fuel cells in remediation of environmental pollutants with electricity generation. *Catalysts* **2020**, *10*, 819.
72. Ibrahim, M.N.M.; Yaqoob, A.A.; Ahmad, A. *Microbial Fuel Cells: Emerging Trends in Electrochemical Applications*; IOP Publishing: Oxford, UK, 2022.

molecules

MDPI

Article

Application of Alkali Lignin and Spruce Sawdust for the Effective Removal of Reactive Dyes from Model Wastewater

Kateřina Hájková [1,*], Michaela Filipi [2], Roman Fojtík [1] and Ali Dorieh [1]

1 Department of Wood Processing and Biomaterials, Faculty of Forestry and Wood Sciences, Czech University of Life Science Prague, Kamýcká 129, 165 00 Prague, Czech Republic; fojtikr@fld.czu.cz (R.F.); dorieh@fld.czu.cz (A.D.)

2 Institute of Chemistry and Technology of Macromolecular Materials, Faculty of Chemical Technology, University of Pardubice, Studentská 572, 532 10 Pardubice, Czech Republic; michaela.filipi@upce.cz

* Correspondence: hajkovakaterina@fld.czu.cz

Abstract: Today, the emphasis is on environmentally friendly materials. Alkali lignin and spruce sawdust are suitable natural alternatives for removing dyes from wastewater. The main reason for using alkaline lignin as a sorbent is the recovery of waste black liquor from the paper industry. This work deals with removing dyes from wastewater using spruce sawdust and lignin at two different temperatures. The decolorization yields were calculated as the final values. Increasing the temperature during adsorption leads to higher decolorization yields, which may be due to the fact that some substances react only at elevated temperatures. The results of this research are useful for the treatment of industrial wastewater in paper mills, and the waste black liquor (alkaline lignin) can be used as a biosorbent.

Keywords: alkali lignin; spruce sawdust; reactive dyes; decolorization

Citation: Hájková, K.; Filipi, M.; Fojtík, R.; Dorieh, A. Application of Alkali Lignin and Spruce Sawdust for the Effective Removal of Reactive Dyes from Model Wastewater. *Molecules* **2023**, *28*, 4114. https://doi.org/10.3390/molecules28104114

Academic Editors: Mohamad Nasir Mohamad Ibrahim and Dimitrios Bikiaris

Received: 5 April 2023
Revised: 10 May 2023
Accepted: 12 May 2023
Published: 16 May 2023

1. Introduction

There has been a growing worldwide interest in environmental issues related to emissions in recent years. This is because pollutants that affect water, soil, and air are constantly released into the environment, producing many residues. One of the most critical industries that is essentially considered a source of water pollution is the dye industry, whose products are used in many applications such as textiles, paper and pulp, wood composites, cosmetics, soap, colored glass, paints, ceramics, polymers, adhesives, art supplies, beverages, wax, and biomedicines [1–3].

The world's yearly production of dyes is estimated to be about 7×10^5 tons, and approximately 15% of this is released into the environment as liquid waste [4]. The producers and consumers of dyes are interested in the dye's stability and permanence and produce shades that do not easily disintegrate after use [5]. Due to the volume and complexity of its wastewater, the textile industry is among the most polluting treatments [6]. From the total worldwide consumption of dyes in the textile industry, it is approximated that 90% ends up in the fabrics [7]. Inefficiency in the dyeing of textiles leads to enormous amounts of dye that off-track directly to wastewater and harm flora and fauna [8–11]. Acid dyes are organic sulfonic acids; the commercially available forms are usually sodium salts, showing good water solubility [10]. Acid dyes are primarily effective on certain fiber types, such as wool, silk, and polypropylene fibers, and blends, such as cotton.

Reactive acid dyes, which contain a reactive functioning group such as chlorotriazine, are azo dyes or azo compounds associated with a metal complex and interact with cotton and wool to form a covalent bond. Reactive dyes hydrolyze easily, which presents another application problem. The hydrolyzed dye does not react with the substrate to form a covalent bond but stays in the dye bath, polluting the wastewater. Due to the high aromatic content of dye molecules and the stability of modern dyes, biological treatment is inefficient for their degradation. Many processes are available for the disposal of dyes using conventional treatment technologies, including biological and chemical oxidation, chemical coagulation, foam flotation, electrolysis, biodegradation, advanced oxidation, photocatalysis, and membrane adsorption processes [10,12–14].

Various adsorbents have been used to remove dyes from aqueous solutions [15–18]. The application of Aqai stalks (*Euterpe oleracea*) to remove textile dye from aqueous solutions was studied regarding the effect of pH, biosorbent doses and shaking time, and nitrogen adsorption–desorption curves [15]. The removal of reactive dye from wastewater using Brazilian pine (*Araucaria angustifolia*) shell was favorable at pH values around 2–7 [16]. Cupuassu (*Theobroma grandiflorum*) shell as a sorbent in the acidic pH region showed favorable biosorption of dyes [17]. Spirulina platensis microalgae were tested as a sorbent in industrial wastewater and were able to remove 94.4–99.0% of the dye from mixtures.

Zaidi et al. [19] investigated the use of *Artocarpus odoratissimus* leaves to remove malachite green dye from simulated wastewater. They demonstrated high adsorption efficiency with a maximum adsorption capacity of 254.93 mg g^{-1}. Compared to other adsorbents, the leaves of *Artocarpus odoratissimus* showed little effect on pH changes, supporting its potential for use in wastewater treatment.

The abundantly available *kangkong* root can be used to remove dyes; Lu et al. [20] used it to remove methyl violet. They were primarily concerned with the recovery and reuse of the dye and concluded that sodium-hydroxide-modified *kangkong* root could sustain dye removal after five consecutive cycles.

Lignin is a broad term for a large group of aromatic polymers formed by the combinatorial oxidative coupling of 4-hydroxy phenylpropanoids (Figure 1). The elemental composition and methoxyl content of lignin in spruce wood is 63.84% carbon, 6.04% hydrogen, 29.68% oxygen, and 15.75% methoxyl [19,20].

These polymers are mainly deposited in the walls of secondarily thickened cells, making them rigid and impermeable [21]. Lignin is one of the three main components of lignocellulosic biomass and is a renewable raw material with great potential [22]. Lignin is contained in the cell walls of almost all terrestrial plants. Lignin is the second most plentiful component in vegetation, surpassed only by cellulose. It is a polymer that does not consist of carbohydrate monomers but contains many aromatic compounds. Depending on the species, it contains up to three different phenylpropane monomers in different proportions. Coniferyl alcohol occurs in all species and is the dominant monomer in conifers.

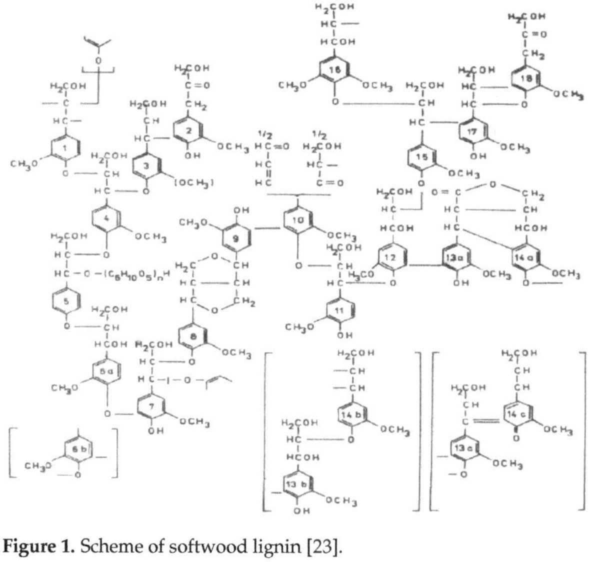

Figure 1. Scheme of softwood lignin [23].

Currently, the global production of technical lignin is approximately 100 million tons/year [24,25], with kraft-processed softwood lignin being the most produced technical lignin after lignosulfonate [26–28].

Materials such as activated carbon, rice husk, cotton seed shell, and oak wood waste are commonly used for dye removal. Activated carbon is a commercial adsorbent for eliminating pollution from wastewater. However, activated carbon's higher production cost and regeneration difficulty limit its widespread use [29]. Thus, developing cheaper, eco-friendly, and more efficient adsorbents is a subject of intensive research.

Sawdust is widely available in the timber and forest industries and has recently been studied as an adsorbent. Sawdust materials are biodegradable and have an affinity for water. The wood powder does not noticeably or appreciably decompose on prolonged contact with water [30]. Due to its lignocellulosic composition, sawdust can be a cheap adsorbent [31,32]. As most spruce monoculture is planted in Central Europe and is subsequently used for paper production, spruce sawdust supplied directly from the paper mill was used for the analysis.

This study was carried out to remove reactive dyes from model effluents. The research focused on two areas. The first was the sorption capacity of dye concentration at 25 and 90 °C. The second was the sorbent dosage at 25 and 90 °C. Since dye removal is essential for paper mills, material commonly found in paper mills, or in the case of black liquor (alkaline lignin), even waste material, was used.

2. Results and Discussion

This work focused on eliminating reactive dyes from model wastewater through chemisorption (Figure 2). Cheap lignocellulosic waste is a simple source of OH nucleophilic groups [33] that can be used for the chemical extraction of reactive dyes.

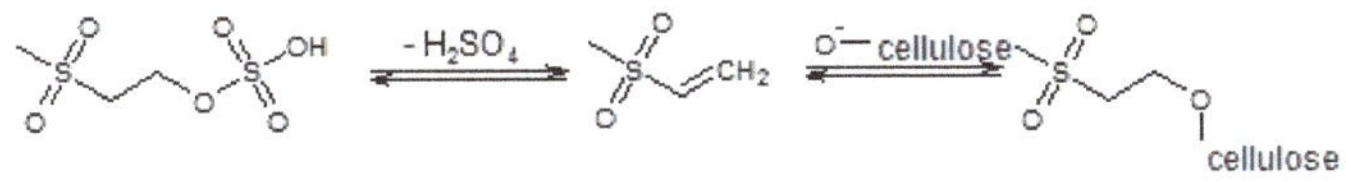

Figure 2. Elimination mechanism dyes [34].

The effect of different sorbents on dye chemisorption was investigated. The results for each sorbent are shown in Figures 3 and 4. Table 1 shows the decolorization equations for the individual dyes using spruce sawdust and alkaline lignin as sorbents.

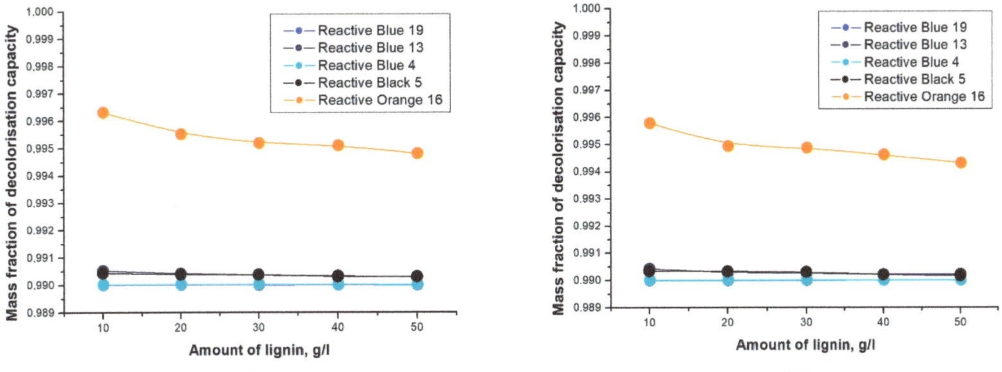

(**a**) (**b**)

Figure 3. Dependence of the decolorization capacity on the amount of alkaline lignin: (**a**) unheated sample (25 °C); (**b**) heated sample (90 °C).

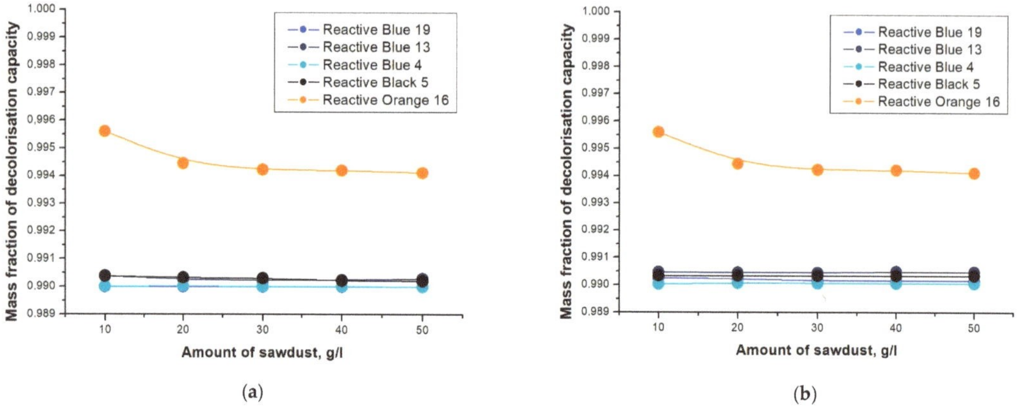

Figure 4. Dependence of the decolorization capacity on the amount of spruce sawdust: (**a**) unheated sample (25 °C); (**b**) heated sample (90 °C).

Table 1. Decolorization equation.

Sorbent	Dyes	Decolorization Equation
Alkaline lignin unheated sample	Reactive blue 13	X = 0.9803ρ + 0.0036
	Reactive blue 19	X = 0.9900ρ + 0.0002
	Reactive blue 4	X = 0.9801ρ + 0.0004
	Reactive black 5	X = 0.9804ρ + 0.0026
	Reactive orange 16	X = 0.9945ρ + 0.0203
Alkaline lignin heated sample	Reactive blue 13	X = 0.9902ρ + 0.0025
	Reactive blue 19	X = 0.9900ρ + 0.0001
	Reactive blue 4	X = 0.9900ρ + 0.0001
	Reactive black 5	X = 0.9901ρ + 0.0036
	Reactive orange 16	X = 0.9940ρ + 0.0202
Spruce sawdust unheated sample	Reactive blue 13	X = 0.9903ρ + 0.0000
	Reactive blue 19	X = 0.9900ρ + 0.0000
	Reactive blue 4	X = 0.9900ρ + 0.0000
	Reactive black 5	X = 0.9902ρ + 0.0001
	Reactive orange 16	X = 0.9938ρ + 0.0008
Spruce sawdust heated sample	Reactive blue 13	X = 0.9905ρ + 0.0000
	Reactive blue 19	X = 0.9901ρ + 0.0000
	Reactive blue 4	X = 0.9901ρ + 0.0000
	Reactive black 5	X = 0.9903ρ + 0.0000
	Reactive orange 16	X = 0.9938ρ + 0.0008

Figures 3 and 4 show that the highest removal of dye was for the reactive orange 16 dye. To remove this dye, alkaline lignin was more effective, but the result of spruce sawdust as a sorbent was also very high. The sorbent concentration for alkaline lignin was 10 $g \cdot L^{-1}$, and in the case of spruce sawdust, only 0.5 $g \cdot L^{-1}$ was sufficient. The values for the other dyes were very similar; see Table 1.

In Table 1, it can be seen from the decolorization equations that the dye reactive orange 16 was indeed the most captured. For reactive blue 13, higher decolorization occurred with spruce sawdust, similar to reactive blue 4 and reactive black 5. Reactive blue 19 had the same decolorization capacity when using alkaline lignin and spruce sawdust as sorbents. The decolorization results were very high for both sorbents used, so it is not possible to conclude whether it is preferable to use spruce sawdust or alkaline lignin.

The increasing chemisorption yield with biomaterial dose can be attributed to the increased surface area and the availability of more binding sites of the biosorbent. Studying the temperature dependence of the adsorption reactions provides insight into the changes during adsorption [35,36]. For this reason, the adsorption was higher for the heated sorbent. Adsorption is endothermic, with adsorption capacity increasing with increasing temperature. An increase in temperature can cause an increase in the endothermic adsorption, the mobility of dye molecules, and the number of active sites for adsorption [37]. For dyes and sorbents, specific structural changes may occur at elevated temperatures [38]. Water molecules displaced by the adsorbate gain more translational entropy than the adsorbate molecules lose, allowing randomness to dominate the system [39]. However, when attractive hydration forces prevail, an exothermic process associated with a decrease in entropy occurs. The increase in adsorption suggests that adsorption was already rapid at low adsorbent doses [40]. The reactive orange 16 dye was tested for comparison and achieved the best decolorization—Figure 5. Spruce sawdust and alkaline lignin were supplemented with another biomass-based material, namely flax, as adsorbents.

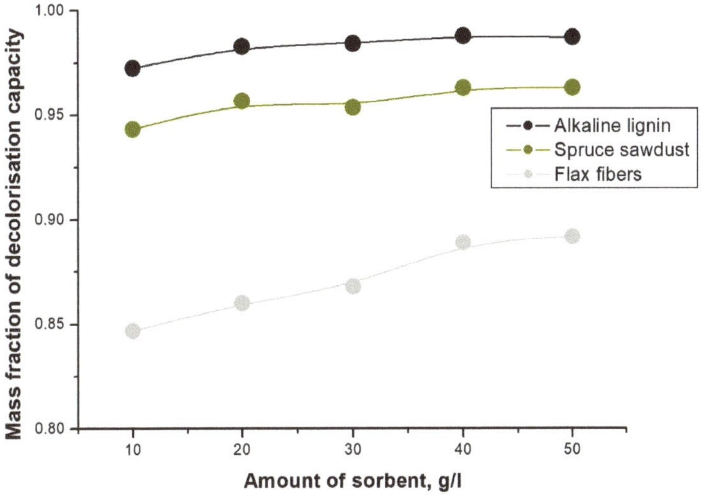

Figure 5. Dependence of decolorization capacity on sorbent loading at 70 °C.

Figure 5 shows that lignin was the most reactive sorbent for the most reactive dye, reactive orange 16, at a suspension temperature of about 70 °C. However, a very high decolorization result was also achieved for spruce sawdust. For comparison, flax fiber with the same weight as alkaline lignin and spruce sawdust was used, but the flax fiber did not show the same discoloration as spruce sawdust and alkaline lignin. This may be due to the chemical composition of the flax. Since spruce sawdust contains significantly more lignin than flax, this chemical component may be the appropriate sorbent.

3. Materials and Method

3.1. Materials

3.1.1. Spruce Sawdust

The spruce sawdust was milled using a pulp milling process from Mondi Štětí (Štětí, Czech Republic), a paper company that deals with kraft pulp cooked mainly from spruce monoculture. The dry matter content of the spruce sawdust was 95%. For the analysis, spruce sawdust samples were ground using a knife mill for about 20 s with an IKA MF 10 BASIC (Staufen, Germany) and then dried to a constant weight at 105 °C according to the Tappi T210 cm-03 standard [41].

3.1.2. Black Liquor

Black liquor from the flax biorefining process was obtained from Delfort Olšany, Czech Republic (dry matter content 42 wt.%). This paper production is based on the soda process, so the black liquor contains sodium hydroxide.

The black liquor's properties were analyzed and are shown in Table 2 to compare the variables that enter and exit the production. Therefore, physical quantities such as pH, density, viscosity, interfacial tension, and alkaline lignin concentration were analyzed.

Table 2. The properties of black liquor.

Properties of Black Liquor	Amount in Black Liquor
Total dry matter content, %	42.0
Ash content of total dry matter, %	27.3
Content of organic substances from total dry matter, %	14.7
Concentration of alkali lignin, $g \cdot dm^{-3}$	45.0
Sodium concentration, $g \cdot dm^{-3}$	39.0
Chemical oxygen consumption $\rho(O_2)$, $g \cdot dm^{-3}$	170.0
Density, $kg \cdot m^{-3}$	1062.0
Viscosity, mPa·s	1.4
Surface tension, $mN \cdot m^{-1}$	55.2
pH value	12.1

Alkaline lignin from black liquor was obtained by precipitation with sulfuric acid. Specifically, sulfuric acid was added to the black liquor in a slight excess, causing an exothermic reaction. Subsequently, the mixture was allowed to react, and after precipitation, the already solid alkali lignin was processed using a friction pan to powder material.

The alkaline lignin and sodium concentrations in black liquor were 45 $g \cdot dm^{-3}$ and 39 $g \cdot dm^{-3}$, respectively. The other properties of the black liquor were as follows: dry matter content 17.1%, of which ash constituted 65% and organic substances were 35%, and chemical oxygen consumption $\rho(O_2) = 170\ g \cdot dm^{-3}$. The absorption spectrum of alkaline lignin was measured on a GBC Cintra 10e spectrophotometer (GBC, Braeside, Australia). Based on the measurement of the absorption spectrum of alkali lignin, a wavelength of 280 nm was chosen—Figure 6.

The immediate solution of alkaline lignin was prepared by dissolving 100 mg of alkaline lignin in 1 L of aqueous NaOH solution with a concentration of 1 $mol \cdot L^{-1}$. The other solutions were prepared by dilution. They were found to absorb at a wavelength of 280 nm. These absorbances are shown in the graph in Figure 7. This dependence determined the concentration of alkali lignin in the sodium hydroxide solution. The reliance was derived based on calibrating the measured values with a spectrophotometer. The dependence of the absorbance A on the concentration of alkali lignin ρ ($g \cdot L^{-1}$) is shown by the equation:

$$A = 16.923\rho. \quad (1)$$

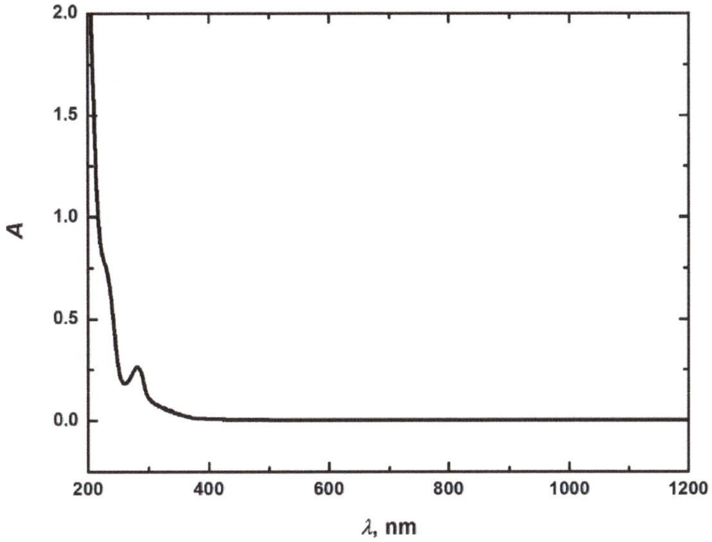

Figure 6. Absorption spectrum of alkaline lignin.

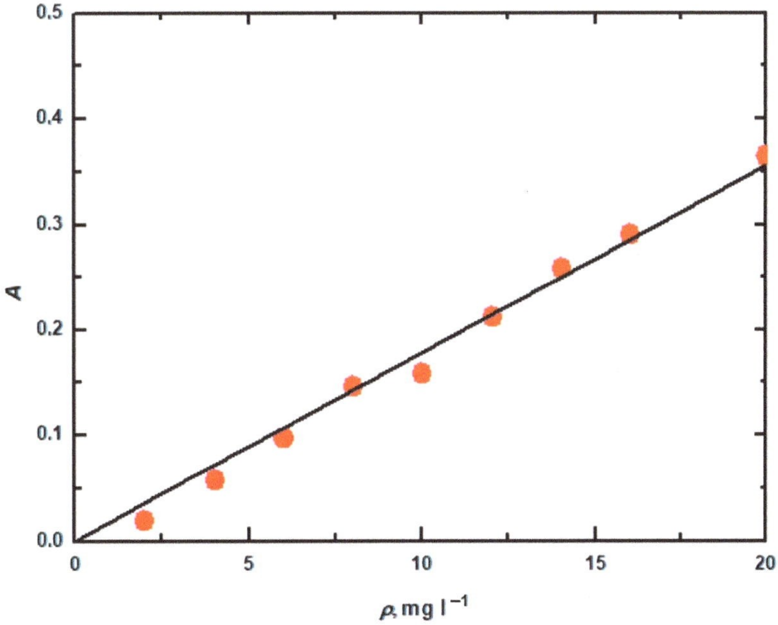

Figure 7. Spectrophotometer calibration dependence for alkali lignin at a wavelength of 280 nm.

3.1.3. Dyes

Merck reactive dyes were used for the experiments. Their specifications are listed in Table 3.

Table 3. Characterization of reactive dyes.

Reactive blue 13	Dye content 100 wt.%, A_{max} = 682 nm	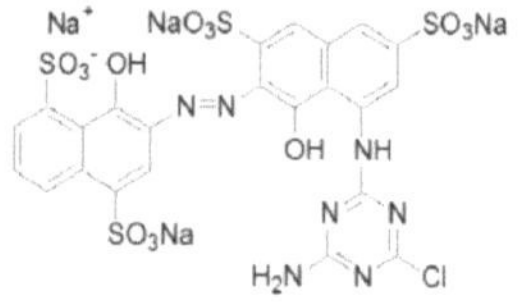
Reactive blue 19	Dye content 50 wt.%, A_{max} = 684 nm	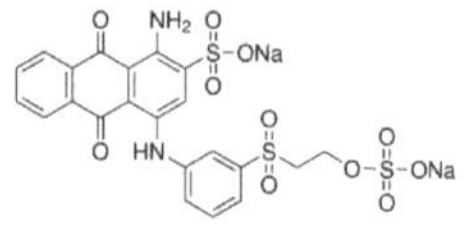
Reactive blue 4	Dye content 35 wt.%, A_{max} = 595 nm	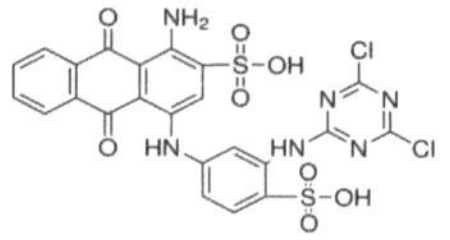
Reactive black 5	Dye content 50 wt.%, A_{max} = 597 nm	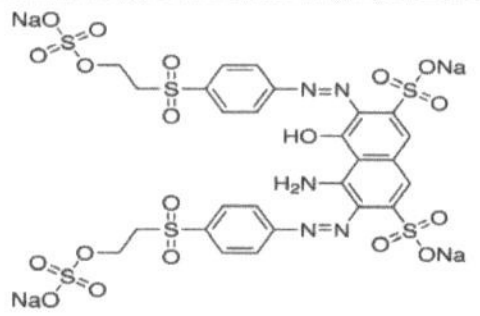
Reactive orange 16	Dye content 70 wt.%, A_{max} = 494 nm	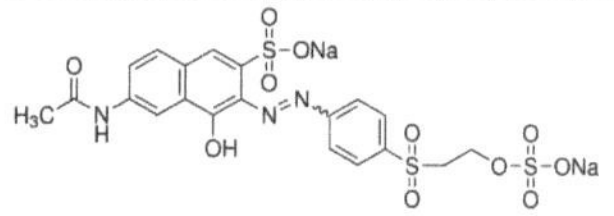

3.2. Experimental Methods

Aqueous solutions of the dyes were prepared by dissolving the respective dye in distilled water [42]. Spruce sawdust and alkaline lignin, precipitated from black liquor as already mentioned, were used as sorption materials. The dyes were mixed at different concentrations, and their absorbance was measured. The concentration–absorbance relationship of each dye was plotted on graphs and interspersed with calibration curves from which the ideal concentration for the experiment was evaluated, Table 4.

Table 4. Concentration of dyes.

Dyes	Concentration, mol·L^{-1}	Calibration Curve Equation
Reactive blue 13	0.01	ρ = 8317.7A + 0.184
Reactive blue 19	0.01	ρ = 11,686A + 0.498
Reactive blue 4	0.01	ρ = 20,886A + 0.511
Reactive black 5	0.01	ρ = 2712A + 0.019
Reactive orange 16	0.007	ρ = 2086.6A + 0.435

The dyes, at the concentrations shown in Table 4, were dissolved in a volume ratio of 1:24 in distilled water. The solutions thus prepared were mixed with alkaline lignin made from black liquor and then dissolved; the alkaline lignin was dosed in weights of 5, 10, 15, 20, and 25 g. Similar to the alkaline lignin, another batch of the solutions was mixed with 5, 10, 15, 20 and 25 g of spruce sawdust. The pH value was then adjusted to pH = 12 using sodium carbonate. The carbonate mixture was stirred vigorously for two hours at a laboratory temperature of 25 °C (unheated samples) and 90 °C (heated samples). After two hours of stirring, the pH of the suspension was acidified with 16% sulfuric acid.

The suspension volume was made up to 500 mL using distilled water, thus obtaining a sample for filtration. The filtered samples were analyzed using a spectrophotometer, and the dye concentration was calculated from the absorbance of the suspension [43–45]. Absorbance measurements were always carried out against a reference sample.

4. Conclusions

Wastewater treatment, especially in the paper industry's deinking area, is a topical issue; dyes must be removed, and biobased material should be used as the sorbent. Both alkaline lignin and spruce sawdust are materials that belong to the biomaterial category and have the advantage of being used in other operations in paper mills, so they are materials commonly used in paper mills.

The decolorization equation states that the two sorbents provide almost comparable decolorization even at a small sorbent dose in the effluent. Better values were obtained at elevated temperatures, but the results were favorable even at room temperatures.

The sorbents we tested were compared with flax fiber, which did not show the same decolorization capacity as alkaline lignin and spruce sawdust. This is probably due to the lower lignin content in annual plants.

Of the sorbents used, the most effective was alkaline lignin, obtained from black liquor, a waste from pulp cooking.

Therefore, alkaline lignin extracted from papermaking waste is a suitable alternative sorbent for industrial wastewater treatment.

Author Contributions: K.H.: conceptualization, supervision, investigation, writing—original draft, and writing—review and editing; M.F.: conceptualization, supervision, formal analysis, validation, investigation, visualization, and writing—original draft; R.F.: investigation and writing—review and editing; A.D.: writing—original draft and writing—review and editing. All authors have read and agreed to the published version of the manuscript.

Funding: This research was funded by "Advanced research supporting the forestry and wood-processing sector's adaptation to global change and the 4th industrial revolution", No. CZ.02.1.01/0.0/0.0/16_019/0000803 financed by OP.

Institutional Review Board Statement: Not applicable.

Informed Consent Statement: Not applicable.

Data Availability Statement: Data are available on request due to ethical restrictions. The data presented in this study are available on request from the corresponding author. The data are not publicly available due to (unfinished research).

Acknowledgments: The authors are grateful for the support of "Advanced research supporting the forestry and wood-processing sector's adaptation to global change and the 4th industrial revolution", No. CZ.02.1.01/0.0/0.0/16_019/0000803 financed by OP.

Conflicts of Interest: The authors declare no conflict of interest.

Sample Availability: Not applicable.

References

1. Dorieh, A.; Khan, A.; Selakjani, P.P.; Pizzi, A.; Hasankhah, A.; Meraj, M.; Pirouzram, O.; Abatari, M.N.; Movahed, S.G. Influence of wood leachate industrial waste as a novel catalyst for the synthesis of UF resins and MDF bonded with them. *Int. J. Adhes. Adhes.* **2021**, *111*, 102985. [CrossRef]
2. Robinson, T.; McMullan, G.; Marchant, R.; Nigam, P. Remediation of dyes in textile effluent: A critical review on current treatment technologies with a proposed alternative. *Bioresour. Technol.* **2001**, *77*, 247–255. [CrossRef]
3. Bassyouni, D.G.; Hamad, H.A.; El-Ashtoukhy, E.S.Z.; Amin, N.K.; Abd El-Latif, M.M. Comparative performance of anodic oxidation and electrocoagulation as clean processes for electrocatalytic degradation of diazo dye Acid Brown 14 in aqueous medium. *J. Hazard. Mater.* **2017**, *335*, 178–187. [CrossRef] [PubMed]

4. Yakubu, I.S.; Muhammad, A.R.; Lawan, U. Determination of heavy metals in tannery effluent. *Int. J. Adv. Acad. Res.* **2018**, *4*, 132–136.
5. Bassat, I.M.; Nigam, P.; Singh, D.; Marchant, R. Microbial decolorization of textile-dye-containing effluents: A review. *Bioresour. Technol.* **1996**, *58*, 217–227.
6. La Farre, M.; Pe'rez, S.; Kantiani, L.; Barcelo', D. Fate and toxicity of emerging pollutants, their metabolites and transformation products in the aquatic environment. *Trend Anal. Chem.* **2008**, *11*, 991–1007. [CrossRef]
7. Akceylan, E.; Bahadir, M.; Yılmaz, M. Removal efficiency of a calix[4]arene-based polymer for water-soluble carcinogenic direct azo dyes and aromatic amines. *J. Hazard. Mater.* **2009**, *162*, 960–966. [CrossRef]
8. Weber, J.; Stickney, V.C. Hydrolysis kinetics of reactive blue 19-vinyl sulfone. *Water Res.* **1993**, *27*, 63–67. [CrossRef]
9. Chung, T.; Fulk, G.E.; Andres, A.W. Mutagenicity testing of some commonly used dyes. *Appl. Environ. Microbiol.* **1981**, *42*, 641–648. [CrossRef]
10. Kavitha, D.; Namasivayam, C. Capacity of activated carbon in the removal of acid brilliant blue: Determination of equilibrium and kinetic model parameters. *Chem. Eng. J.* **2008**, *139*, 453–461. [CrossRef]
11. Dai, S.; Zhuang, Y.; Chen, Y.; Chen, L. Study on the relationship between structure of synthetic organic chemicals and their biodegradability. *Environ. Chem.* **1995**, *14*, 354–367.
12. Koprivanac, N.; Bozic, A.L.; Papic, S. Cleaner production processes in the synthesis of blue anthraquinone reactive dyes. *Dye. Pigment.* **2000**, *44*, 33–40. [CrossRef]
13. Vaidya, A.A.; Datye, K.V. Environmental pollution during chemical processing of synthetic fibres. *Colorage* **1982**, *14*, 3–10.
14. Soleymani Lashkenrai, A.; Najafi, M.; Peyravi, M.; Jahanshahi, M.; Mosavian, M.T.H.; Amiri, A.; Shahavi, M.H. Direct filtration procedure to attain antibacterial TFC membrane: A facile developing route of membrane surface properties and fouling resistance. *Chem. Eng. Res. Des.* **2009**, *149*, 158–168. [CrossRef]
15. Alencar, W.S.; Lima, E.C.; Royer, B.; dos Santos, B.D.; Calvete, T.; da Silva, E.A.; Alves, C.N. Application of aqai stalks as biosorbents for the removal of the dye Procion Blue MX-R from aqueous solution. *Sep. Sci. Technol.* **2012**, *47*, 513–526. [CrossRef]
16. Calvete, T.; Lima, E.C.; Cardoso, N.F.; Dias, S.L.P.; Pavan, F.A. Application of carbon adsorbents prepared from the Brazilian-pine fruit shell for removal of Procion Red MX 3B from aqueous solution—Kinetic, equilibrium, and thermodynamic studies. *Chem. Eng. J.* **2009**, *155*, 627–636. [CrossRef]
17. Cardoso, N.F.; Lima, E.C.; Pinto, I.S.; Amavisca, C.V.; Royer, B.; Pinto, R.B.; Alencar, W.S.; Pereira, S.F.P. Application of cupuassu shell as biosorbent for the removal of textile dyes from aqueous solution. *J. Environ. Manag.* **2011**, *92*, 1237–1247. [CrossRef] [PubMed]
18. Cardoso, N.F.; Lima, E.C.; Royer, B.; Bach, M.V.; Dotto, G.L.; Pinto, L.A.A.; Calvete, T. Comparison of Spirulina platensis microalgae and commercial activated carbon as adsorbents for the removal of Reactive Red 120 dye from aqueous effluents. *J. Hazard. Mater.* **2012**, *241–242*, 146–153. [CrossRef] [PubMed]
19. Zaidi, N.A.H.M.; Lim, L.B.L.; Usman, A.; Kooh, M.R.R. Efficient adsorption of malachite green dye using Artocarpus odoratissimus leaves with artificial neural network modelling. *Desalination Water Treat.* **2018**, *101*, 313–324. [CrossRef]
20. Lu, Y.C.; Kooh, M.R.R.; Lim, L.B.L.; Priyantha, N. Effective and Simple NaOH-Modification Method to Remove Methyl Violet Dye via Ipomoea aquatic Roots. *Adsorpt. Sci. Technol.* **2021**, *2021*, 593222. [CrossRef]
21. Boerjan, W.; Ralph, J.; Baucher, M. Lignin biosynthesis. *Annu. Rev. Plant Biol.* **2003**, *54*, 519–546. [CrossRef]
22. Ralph, J.; Lundquist, K.; Brunow, G.; Lu, F.; Kim, H.; Schatz, P.F.; Marita, J.M.; Hatfield, R.D.; Ralph, S.A.; Christensen, J.H. Lignins: Natural polymers from oxidative coupling of 4-hydroxyphenylpropanoids. *Phytochem. Rev.* **2004**, *3*, 29–60. [CrossRef]
23. Vanholme, R.; Demedts, B.; Morreel, K.; Ralph, J.; Boerjan, W. Lignin Biosynthesis and Structure. *Plant Physiol.* **2010**, *153*, 895–905. [CrossRef] [PubMed]
24. Pieter, J.; Fredrik, B.; Riwu Kaho, M. Valuasi ekonomi ekowisata terhadap pengembangan objek wisata kawasan pesisir pantai. *J. Ilmu Lingkung.* **2015**, *13*, 55–64. [CrossRef]
25. Červenka, E.; Král, Z.; Tomis, B. *Wood and Cellulose Chemistry*; University of Chemistry and Technology Pardubice: Pardubice, Czech Republic, 1980; p. 228. (In Czech)
26. Wu, X.; Jiang, J.; Wang, C.; Liu, J.; Pu, Y.; Ragauskas, A.; Li, S.; Yang, B. Lignin-derived electrochemical energy materials and system. *Biofuels Bioprod. Biorefin.* **2020**, *14*, 650–672. [CrossRef]
27. Kulas, D.G.; Thies, M.C.; Shonnard, D.R. Techno-economic analysis and life cycle assessment of waste lignin fractionantion and valorization using the ALPHA process. *ACS Sustain. Chem. Eng.* **2021**, *9*, 5388–5395. [CrossRef]
28. Kazzaz, A.E.; Fatehi, P. Technical lignin and its potential modification routes: A mini-review. *Ind. Crops Prod.* **2020**, *154*, 112732. [CrossRef]
29. Gondaliya, A.; Nejad, M. Lignin as a partial polyol replacement in polyurethane flexible foam. *Molecules* **2021**, *26*, 2302. [CrossRef]
30. Arefmanesh, M.; Vuong, T.V.; Nikafshar, S.; Wallmo, H.; Nejad, M.; Master, E.R. Enzymatic synthesis of kraft lignin-acrylate copolymers using an alkaline tolerant laccase. *Appl. Microbiol. Biotechnol.* **2022**, *106*, 2969–2979. [CrossRef]
31. Cifuentes, A.R.; Avila, K.; García, J.C.; Daza, C.E. The pyrolysis of rose stems to obtain activated carbons: A study on the adsorption of Ni(II). *Ind. Eng. Chem. Res.* **2013**, *52*, 16197–16205. [CrossRef]
32. Can, M. Investigation of the factors affecting acid blue 256 adsorption from aqueous solutions onto red pine sawdust: Equilibrium, kinetics, process design, and spectroscopic analysis. *Desalination Water Treat.* **2015**, *57*, 5636–5653. [CrossRef]

33. Wahab, M.A.; Jellali, S.; Jedidi, N. Ammonium biosorption onto sawdust: FTIR analysis, kinetics and adsorption isotherms modelling. *Bioresour. Technol.* **2010**, *101*, 5070–5075. [CrossRef]
34. Zou, W.; Bai, H.; Gao, S.; Li, K. Characterization of modified sawdust, kinetic and equilibrium study about methylene blue adsorption in batch mode. *Korean J. Chem. Eng.* **2013**, *30*, 111–122. [CrossRef]
35. Weidlich, T. Use of Organic-Based Reactants Halogen Derivatives for Obtaining Chemical Specialties Produced in the Czech Republic and New Options for Limiting Them Emissions. Proffesor's Thesis, University of Pardubice, Pardubice, Czech Republic, 2013. (In Czech)
36. Filipi, M.; Dušek, L. The use of spruce, birch, flax and rapeseed for the removal of vinyl sulfone dyes from wastewater. *Rostlinolékař* **2018**, *3*, 26–29. (In Czech)
37. Alkan, M.; Dogan, M. Adsorption kinetics of Victoria blue onto perlite. *Fresenius Environ. Bull. Adv. Food Sci.* **2003**, *125*, 418–428.
38. Salleh, M.A.M.; Mahmoud, D.K.; Karim, W.A.; Idris, A. Cationic and anionic dye adsorption by agricultural solid wastes: A comprehensive review. *Desalination* **2011**, *280*, 1–13. [CrossRef]
39. Senthilkumaar, S.; Kalaamani, P.; Subburaam, C.V. Liquid phase adsorption of crystal violet onto activated carbons derivated from male flowers of coconut tree. *J. Hazard. Mater.* **2006**, *136*, 800–808. [CrossRef]
40. Hema, M.; Arivoli, S. Comparative study on the adsorption kinetics and thermodynamics of dyes onto acid activated low-cost carbon. *Int. J. Phys. Sci.* **2007**, *2*, 10–17.
41. *Tappi Test Methods*; Tappi Press Atlanta: Atlanta, Georgia, 2004.
42. Bharathi, K.S.; Ramesh, S.T. Removal of dyes using agriculture waste as low-cost adsorbents: A review. *Appl. Water Sci.* **2013**, *3*, 773–790. [CrossRef]
43. Ofomaja, A.E.; Ho, Y.S. Equilibrium sorption of anionic dye from aqueous solution by palm kernel fibre as sorbent. *Dye. Pigment.* **2007**, *74*, 60–66. [CrossRef]
44. Prola, L.D.T.; Machado, F.M.; Bergmann, C.P.; de Souza, F.E.; Gally, C.R.; Lima, E.C.; Adebyao, M.A.; Dias, S.L.P.; Calvete, T. Adsorption of Direct Blue 53 dye from aqueous solutions by multi-walled carbon nanotubes and activated carbon. *J. Environ. Manag.* **2013**, *130*, 166–175. [CrossRef] [PubMed]
45. Filipi, M.; Milichovský, M. Adsorption Organic Cationic Dyes on Oxycelluloses and Linters. *J. Encapsulation Adsorpt. Sci.* **2014**, *4*, 43450. [CrossRef]

molecules

Article

Valoration of the Synthetic Antioxidant Tris-(Diterbutyl-Phenol)-Phosphite (Irgafos P-168) from Industrial Wastewater and Application in Polypropylene Matrices to Minimize Its Thermal Degradation

Joaquín Hernández-Fernández [1,2,3,*], Heidis Cano [4] and Ana Fonseca Reyes [5,*]

1 Chemistry Program, Department of Natural and Exact Sciences, San Pablo Campus, University of Cartagena, Cartagena 130015, Colombia
2 Chemical Engineering Program, School of Engineering, Universidad Tecnológica de Bolivar, Parque Industrial y Tecnológico Carlos Vélez Pombo, Km 1 Vía Turbaco, Turbaco 130001, Colombia
3 Department of Natural and Exact Science, Universidad de la Costa, Barranquilla 30300, Colombia
4 Department of Civil and Environmental, Universidad de la Costa, Barranquilla 080002, Colombia
5 Department of Mechanical Engineering, Universidad del Norte, Barranquilla 081007, Colombia
* Correspondence: jhernandezf@unicartagena.edu.co (J.H.-F.); fonsecama@uninorte.edu.co (A.F.R.)

Abstract: Industrial wastewater from petrochemical processes is an essential source of the synthetic phenolic phosphite antioxidant (Irgafos P-168), which negatively affects the environment. For the determination and analysis of Irgafos P-168, DSC, HPLC-MS, and FTIR methodologies were used. Solid phase extraction (SPE) proved to be the best technique for extracting Irgafos from wastewater. HPLC-MS and SPE determined the repeatability, reproducibility, and linearity of the method and the SPE of the standards and samples. The relative standard deviations, errors, and correlation coefficients for the repeatability and reproducibility of the calibration curves were less than 4.4% and 4.2% and greater than 0.99955, respectively. The analysis of variance (ANOVA), using the Fisher method with confidence in 95% of the data, did not reveal significant differences between the mentioned parameters. The removal of the antioxidant from the wastewater by SPE showed recovery percentages higher than 91.03%, and the chemical characterization of this antioxidant by FTIR spectroscopy, DSC, TGA, and MS showed it to be structurally the same as the Irgafos P-168 molecule. The recovered Irgafos was added to the polypropylene matrix, significantly improving its oxidation times. An OIT analysis, performed using DSC, showed that the recovered Irgafos-blended polypropylene (PP) demonstrated oxidative degradation at 8 min. With the addition of the Irgafos, the oxidation time was 13 min. This increases the polypropylene's useful life and minimizes the environmental impact of the wastewater.

Keywords: valoration; synthetic antioxidant; Irgafos P-168; industrial wastewater; polypropylene; thermal degradation

check for updates

Citation: Hernández-Fernández, J.; Cano, H.; Reyes, A.F. Valoration of the Synthetic Antioxidant Tris-(Diterbutyl-Phenol)-Phosphite (Irgafos P-168) from Industrial Wastewater and Application in Polypropylene Matrices to Minimize Its Thermal Degradation. *Molecules* **2023**, *28*, 3163. https://doi.org/10.3390/molecules28073163

Academic Editor: Mohamad Nasir Mohamad Ibrahim

Received: 8 March 2023
Revised: 27 March 2023
Accepted: 30 March 2023
Published: 2 April 2023

1. Introduction

Industrial processes are some of the productive forces that have significantly contributed to the development of today's society, developing products and providing services that have improved our quality of life [1–3]. However, one of the most significant problems revolving around industrial production is the pollution that it generates, since in every industrial process and each of its unit operations, specific residues are developed, most of which are toxic and highly polluting [2,4,5]. One of the products whose highest consumption has increased in the last century is plastic [6], thanks to its durability and great versatility [7]. With this increase in production come several industry challenges related to waste treatment [7]. The petrochemical industry is a producer of plastics or polymers, and the industry uses additives to increase or improve the properties of the plastic, such as increasing its useful life, lubricating the plastic, and increasing its resistance. Additionally, some plastics are non-slip, antistatic, or antioxidant [3,8,9]. All these additives, which are added in the manufacturing process, are not fully absorbed by the polymer and become process residues. Due to the dynamics of these processes, these residues are usually discharged into the wastewater, which may result in their discharge into different bodies of water [4,10,11]. These wastewaters usually have an unusually high chemical oxygen demand (DCO) and biological oxygen demand (BOD) because they contain some volatile organic compounds, phenols, minerals, polycyclic aromatics, and many others. Conventional treatment processes are not efficient or effective in these wastewaters [12,13], resulting in the presence of these chemical compounds in the aquatic environment. Although most of these additives can indeed be present in low concentrations, it is also correct that these low concentrations are already toxic enough to cause problems for both health and biodiversity [14,15]. The plastic industries that can dump this waste are characterized by their use of multiple additives that prevent the oxidation of their resins, which are synthesized by exposure to light or heat [16]. This oxidation negatively affects the polymers since their physical–mechanical capacities are significantly reduced. To avoid this, a family of additives called antioxidants (AO) [17,18] is used. PP is one of these polymers. PP has gained great importance in the industry since, thanks to its low density and high strength, it can be heated and cooled without losing any of its characteristics, such as its elevated mechanical and chemical stability [19–27]. Polypropylene is also a product of great economic interest since it is versatile in its application in packaging, textiles, household appliances, toys, and medical products, etc. [20,28–32]. For this reason, this product is in great demand (79 million tons in 2011) [33]. It may bring associated with certain environmental risks [34,35], since OAs are used in its production to avoid the oxidative degeneration of the polymer. OAs, such as Irgafos P-168, are used to improve the thermal stability and prolong the useful life of these polymers [36,37]. Irgafos 168 is manufactured by several companies worldwide, such as BASF, Toronto Research Chemicals, Spex, and Cymit Chemicals, among many others. Irgafos P-168 is a compound that acts as an oxidation inhibitor and helps to protect the polymer from the damaging effects of heat and light [37]. This protection, which is conferred to the polymers by OAs, makes them necessary in the polypropylene production processes, especially in the extrusion stages in which the polymer resins are dosed and mixed. However, some of these additives are not absorbed during the process: they are removed during the polypropylene washing and cleaning stages and are later discarded in wastewater and eventually in bodies of water [38]. Irgafos 168, which may be present in this discharge, is a compound with a chemical structure comprising an organophosphate (see Figure 1) attached to three rings of phenols that can be toxic to aquatic organisms and can affect water quality [39–42]. Irgafos P-168 can break down into potentially harmful secondary compounds [26,43–45]. Research on the health impact of Irgafos P-168 is not fully defined; however, the degradation products of Irgafos P-168, such as di-tertbutyl-phenol and tris(2,4-di-tert-butylphenyl) phosphate, are toxic. Some regulations have established their maximum concentrations between 5 and 40 mg/L. One study found concentrations of these Irgafos P-168 derivatives at 400 mg/L or more in bottles made with polymers that contained this additive [46–50]. In bodies of water, Irgafos P-168 can become hydrolyzed over time, and under various environmental conditions it can form di-terbutyl-phenol, which is highly toxic [51].

This di-terbutyl-phenol derived from Irgafos P-168 is cytotoxic, and it directly and significantly increases the expression of the P53 gene in more than one cell line. In rats, it caused hepatic and renal toxicity, increased organ weight, and histopathological changes [52]. For these rodents, the no-harm concentrations were between 5.01 and 20.03 mg kg^{-1} day^{-1}. Irgafos P-168 (tris(2,4-di-tert-butylphenyl) phosphite) has direct effects on human cells, such as preventing their growth, and can cause endocrine disruption [51,53]. The impact of these molecules on health and the environment has been determined thanks to the existence of analytical quantification techniques and previous sample treatments. Irgafos P-168 has been effectively extracted using extraction techniques such as solid phase microextraction (SPME), liquid–liquid extraction (LLE), supercritical fluid extraction (SFE), ultrasonic-assisted extraction (UAE) [54], solvent microextraction (MEPS), pressurized liquid fluid extraction (PLE) and dispersive solid phase extraction (dSPE) [55]. Some of the techniques used to quantify Irgafos P-168 in different processes and under other conditions have included mass spectrometry [56], pyrolysis-GC/MS [51], FTIR spectroscopy [57,58], DSC and TGA [59], NMR and UV–Vis [58], and HPLC [60–62].

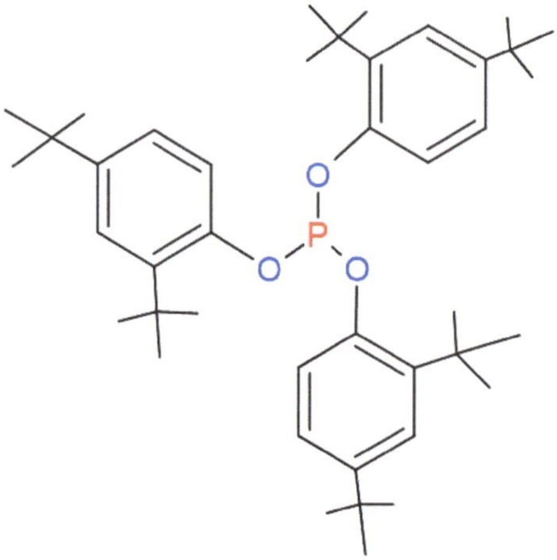

Figure 1. Chemical structure of Irgafos P-168.

In this investigation, we will evaluate the reliability of the Irgafos P-168 measurement method using HPLC. For this purpose, the repeatability, reproducibility, linearity, and recovery percentage are evaluated. The Irgafos molecule is quantified in industrial wastewater samples from polypropylene production plants. In this research, a methodology for extracting Irgafos from wastewater is proposed to minimize its environmental impact. The extracted Irgafos is purified and characterized using mass spectrometry, FTIR spectroscopy, TGA, and DSC. To guarantee the performance of this recovered Irgafos, it is added to a virgin PP matrix, and the resulting mixture is evaluated using FTIR spectroscopy, TGA, OIT and MFI.

2. Materials and Methods

2.1. Recovery of the Additive

2.1.1. Samples Collection

Samples were acquired at a polypropylene production facility that divides the production of the material into four steps (Figure 2): arrival, polymerization, additivation, and granulation, which are described in eight steps. In step 1, the material is received; in step 2, the raw material enters the reactor. It is mixed with a catalyst (Ziegler–Natta) that reduces the activation energy for the polymerization reaction. In step 3, darkening the extrusion, in order to lower the temperature and produce a more uniform grain, water is employed in the polymer extrusion process. After the polymer is produced, it is transferred to a purging column (step 4) for purification. The additives that improve and complement the properties of the PP are then added (step 5). In step 6, the extrusion and pelletization of the PP mixture containing the additives occur. The samples of interest in this inquiry primarily originated from the condensates of the extrusion and deodorization procedures (step 6). Every 12 min for 4 h, 200 mL of samples were taken, and the samples were kept at 4 °C in an amber glass bottle.

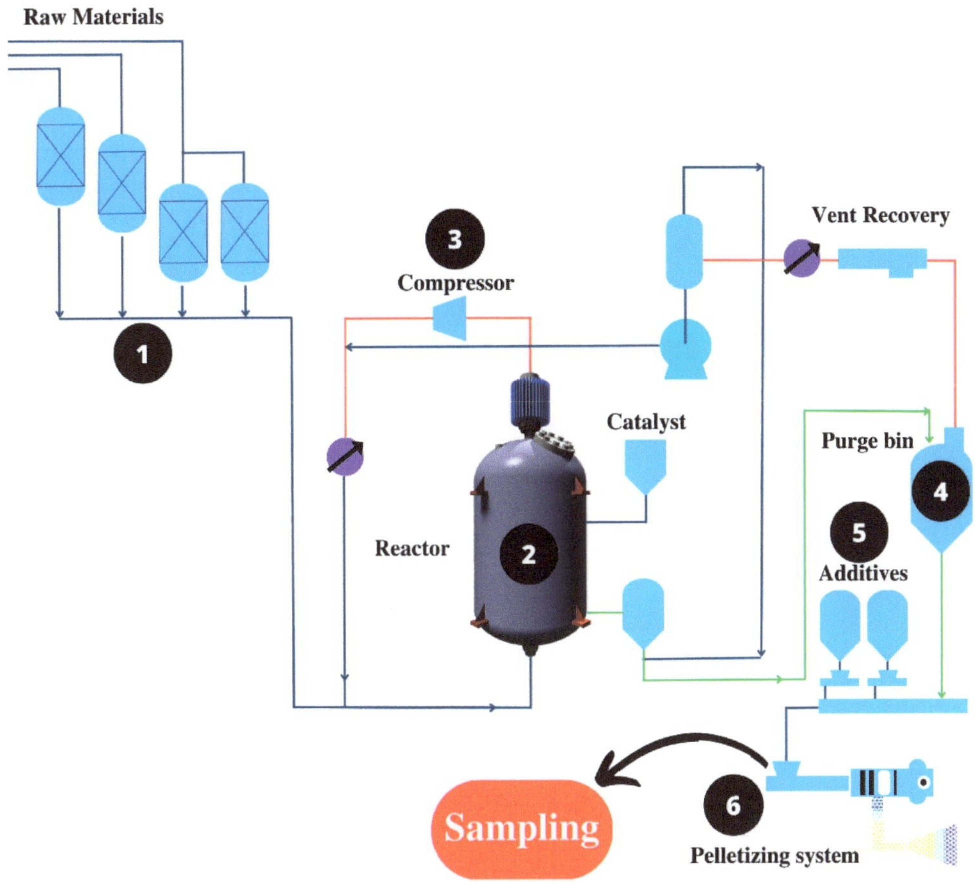

Figure 2. PP production flow diagram.

2.1.2. Extraction System for Sample and Standard

The collected sample was subjected to filtration through a PTFE filter before being extracted. Solid phase extraction (SPE) was performed with Strata-X tubes (33 µm). A total of 15 mL of the sample was used, and the operating flow rate was 1.1 mL min^{-1}. The SPE conditioning was carried out with methanol and distilled water with an 80:20 ratio of MeOH: H_2O. To carry out the elution of OA in SPE, 10 mL of acetonitrile (ACN) was used as a solvent. The eluted extract from the SPE was dissolved in ACN to 1 mL and separated by HPLC. When 500 mL of the extract was obtained, the pre-concentration process was repeated. The recovered solid was then dried (by recovering it with a stream of N_2) and subjected to FTIR spectroscopy, DSC, TGA, and HPLC-MS analysis. The obtained export product (Irgafos P-168) was stored in a controlled environment for later use in virgin PP resin. Figure 3 represents the general outline of the SPE process. In the SPE process, the residual water sample, which contained several residues of additives, such as Irgafos P-168 and even trace metals, was poured into the extraction cartridge. The extraction cartridge included a porous solid phase which, given its chemical structure, has a specific molecular region for non-polar substances. Although Irgafos has polarized regions in its chemical structure, the three Di-tert-butyl groups it has generate a steric hindrance, so the polar region does not have an affinity. This causes steric hindrance, so it has an affinity with the stationary phase of the extraction capsule.

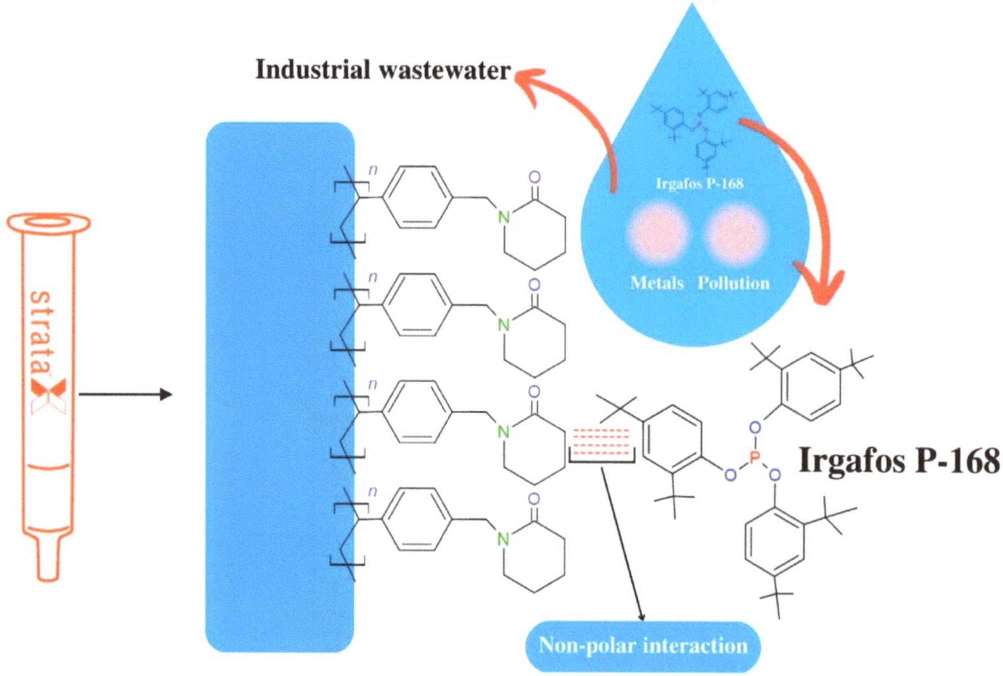

Figure 3. Mechanism of interaction and separation of the Irgafos P-168 in the stationary phase of the SPE.

2.1.3. Separation, Identification, and Characterization System by HPLC/DAD/MS/MS/MS

This analysis was conducted using a Micromass Quattro II triple quadrupole mass spectrometer and an Agilent 1200 HPCL. The system includes a Lichrosorb RP-18 column (4.6 m × 200 mm × 5 μ), syringes of 5 and 10 mL, a precision balance, a degasser (G1322A), a pump (G1311A), an automatic sampler (G1313A), a column carrier (G1316A), and these components. The chromatographic modifications were made using the Irgafos P-168 solution, which was produced in ACN, as a starting point. The solvents ACN and H_2O were combined in the following ratios to form the mobile phase: 83 and 17% (1 min, 15 mL/min); 93 and 7% (2 min, 2 mL/min); 95 and 5% (3.5 min, 3.5 mL/min); and 100 and 0% (8 min, 3.5 mL/min). The temperature in the column was 50 °C. Irgafos 168 was identified using MS and MS/MS fragments.

2.2. Reincorporation of Wastewater Additive in the PP Production Process

A virgin PP resin with no additives was used. It was separately combined with the recovered PP-Irgafos 168 samples and pure PP-Irgafos 168 samples to produce recovered PP-Irgafos 168 samples. The PP powder was premixed with 0.1 weight percent of recovered Irgafos P-168 and neat Irgafos 168. An ordinary Prodex Henschel 115JSS mixer was used for this process, running at 800 revolutions per minute for 7 min. Melt extrusion was used for mixing with a Welex-200 24.1 extruder. (see Figure 4) The extruder has five heating zones that guaranteed the correct homogenization of the PP and the additive mixtures. The working temperatures of the extruder were 190, 195, 200, 210, and 220 °C.

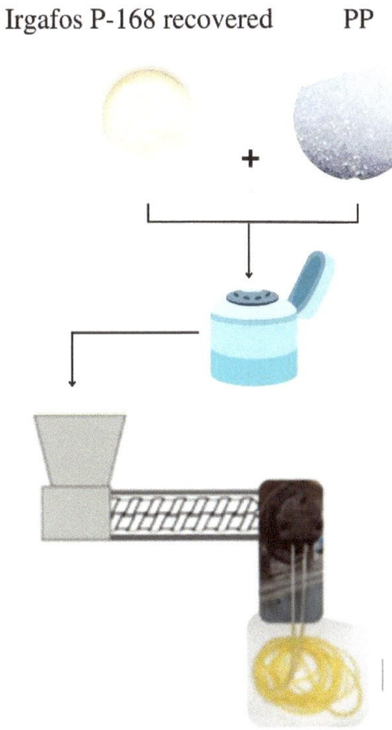

Figure 4. Mixing and extrusion of recovered Irgafos P-168 and pure Irgafos P-168 with virgin PP.

The PP solid produced by the extruder die was granulated, and these samples were spiked with 1000 mg kg^{-1} of recovered Irgafos P-168 and 1000 mg kg^{-1} of pure Irgafos P-168. Irgafos P-168 was diluted in ACN to a 500 mg/L to create a standard, and 5 mL of this solution was then dissolved in 100 mL to create a 25 mg/L solution.

2.3. *Sample Evaluation*

2.3.1. Heat Flux Characterization System (DSC)

To ascertain the samples' oxidation induction time (OIT), a DSC Q2000 V24.11 Build 124 device was used for calorimetric analysis. Results were obtained using a 6.1 mg sample under nitrogen and oxygen ambient conditions. The experiment was conducted to investigate the material's oxidation and volatility effects. The sample was first heated isothermally for 5 min at 60 °C and then for 20 min at 200 °C in a nitrogen atmosphere moving at 50 mL/min. The atmosphere was then altered to 50 mL/min of airflow and 30 min of oxidation at 200 °C. It was feasible to determine the OIT value, which corresponds to the instant at which the change in pitch happens, due to this change in atmosphere's revelation of a difference in the slope of the exothermic heat.

2.3.2. Spectrometric Characterization System

For the spectrometric characterization, Nicolet 6700 reference infrared spectroscopy equipment was used, with readings between 400 and 600 cm^{-1} and a resolution of 2 cm^{-1} (reflection mode).

2.3.3. Thermogravimetric Characterization System

Using a Perkin Elmer TGA7 thermobalance and a N_2 environment flowing at a rate of 50 mL/min, a TGA was conducted. The temperature at which there was a 5% mass loss served as the benchmark for the initial degradation temperature, and the DTG curve served as the benchmark for the maximum degradation temperature.

2.3.4. Melt Flow Index (MFI)

A Tinius Olsen MP1200 plastometer was used to determine the MFI. The working temperature inside the equipment cylinder was 230 °C, and a 2.16 kg piston was used to move the molten material.

3. Analysis and Discussion

3.1. Standards Calibration Curve for Irgafos P-168

We proceeded to establish the repeatability and reproducibility of the quantification method in which the concentration of the Irgafos P-168 recovered using the HPLC-MS method would be measured. For this, a calibration curve was generated with known amounts of concentrations between 0 and 5000 ppm. To carry out this calibration and guarantee its reliability, seven different samples were analyzed for five days by the same analyst. The tests were also carried out with five various analysts, each analyzing seven samples in different concentrations, as shown in Table 1. The maximum relative standard deviations for repeatability and reproducibility were 3% and 4.4%, respectively, and their respective maximum errors were 4.2% and 2.6%.

Table 1. Irgafos P-168 repeatability and reproducibility calibration curve data compilation.

Values for Irgafos P-168 Quantitation Model by HPLC-MS							
Sequence Intraday (Same Day)							
Standard	**1**	**2**	**3**	**4**	**5**	**6**	**7**
Theoretical	0	500	1000	1500	2000	3000	5000
Analyst 1	0	485	990	1521	1991	2945	4945
Analyst 1	0	477	1010	1488	1868	2929	4932
Analyst 1	0	488	977	1493	1900	2845	4900
Analyst 1	0	490	965	1495	1944	2737	4993
Analyst 1	0	475	981	1466	1968	2910	4831
Average	0	483.0	984.6	1492.6	1934.2	2873.2	4920.2
Deviation	0	6.7	16.8	19.6	50.0	85.1	60.0
RSD	0	1.4	1.7	1.3	2.6	3.0	1.2
Error	0	3.4	1.5	0.5	3.3	4.2	1.6
Sequence Interday (different day)							
Standard	**1**	**2**	**3**	**4**	**5**	**6**	**7**
Theoretical	0	500	1000	1500	2000	3000	5000
Analyst 1	0	443	934	1435	1976	2990	4956
Analyst 2	0	456	967	1521	1982	2985	4880
Analyst 3	0	482	973	1491	1895	2856	5014
Analyst 4	0	495	1011	1399	1989	3011	4876
Analyst 5	0	475	986	1458	2013	2867	4779
Average	0	470.2	974.2	1460.8	1971	2941.8	4901
Deviation	0	20.7	28.1	47.5	44.7	74.1	89.1
RSD	0	4.4	2.9	3.3	2.3	2.5	1.8
Error	0	6	2.6	2.6	1.5	1.9	2

The acceptance value for validating the calibration of the deviation obtained with respect to the expected values was less than 15% [62]; in addition, an ANOVA was applied using the Fisher grouping method, with which it was possible to establish that the difference in the measurements was not significant. The reliability of the process was established with a confidence of 95%, and a significance value of $\alpha = 0.05$ was established. Figure 2 shows the confidence intervals, comparing pairs of data columns for the same analyst (Figure 5a) and for different analysts (Figure 5b). These are the repeatability and reproducibility. The semi-continuous line perpendicular to all intervals indicates the position of zero in a break. When this semi-continuous line intersects with any of the intervals, it indicates that zero belongs to the gap, showing that there are no appreciable differences between the data pairs being examined. Therefore, it is concluded that there are no significant differences. Consequently, the method is repeatable and reproducible.

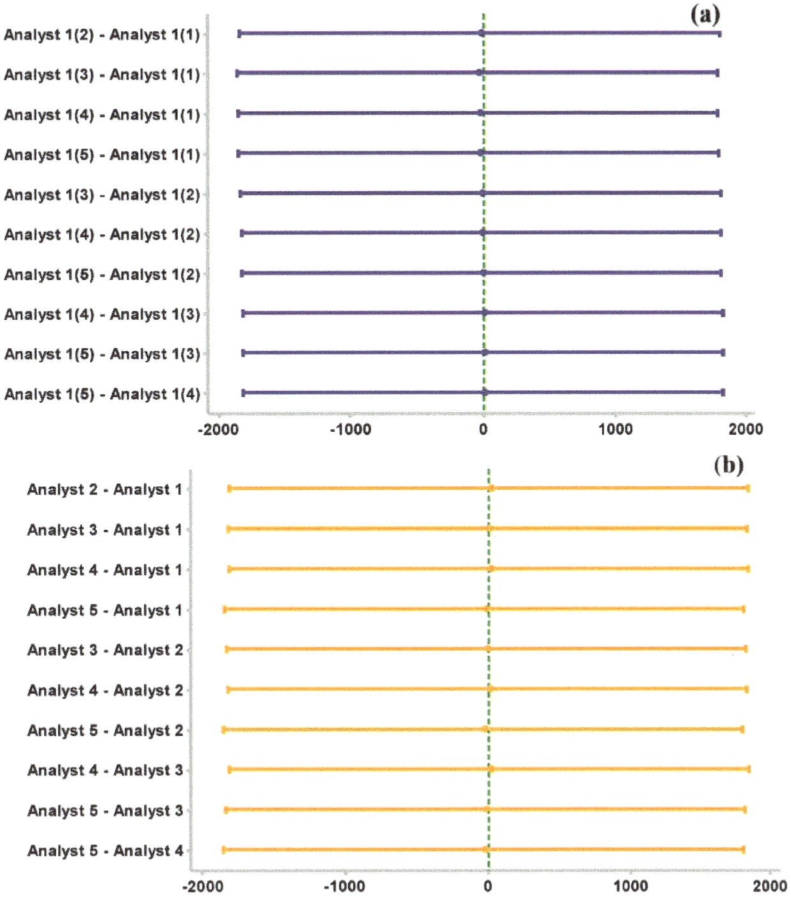

Figure 5. (**a**) Difference of means compared one to one for repeatability of the calibration curve. (**b**) Difference of means compared one to one for calibration reproducibility.

3.2. SPE of Standards with Acetonitrile

The values for the variables studied in the extraction are presented in Table 2. Here (Table 2), the maximum errors for repeatability and reproducibility were 4.8 % and 5.3 %, respectively, the maximum relative standard deviations of repeatability and reproducibility were 3.2 % and 2.5%, and the complete extractions were 96% and 98%.

Figure 6 shows the confidence intervals, comparing pairs of data columns for the same analyst (Figure 6a) and for different analysts (Figure 6b), i.e., the repeatability and reproducibility of the SPE. The significance of the difference for the intraday and interday data of the SPE was also analyzed using the Fisher method, with an $p = 0.05$ value and 95% confidence. Under the same analysis made in Section 3.1, Figure 6a,b, which show the repeatability and reproducibility of the standards extraction, respectively, show that there are no appreciable variances in the data.

Table 2. Irgafos P-168 repeatability and reproducibility extraction data compilation.

Values for Irgafos P-168 Extraction Method by SPE							
Sequence Intraday (Same Day)							
STDA	**1**	**2**	**3**	**4**	**5**	**6**	**7**
Theoretical	0	500	1000	1500	2000	3000	5000
Analyst 1	0	465	933	1396	1943	2950	4911
Analyst 1	0	476	969	1477	1922	2854	4795
Analyst 1	0	471	972	1482	1896	2889	4811
Analyst 1	0	485	985	1396	1958	2756	4824
Analyst 1	0	483	932	1484	1876	2949	4779
Average	0	476	958.2	1447	1919	2879.6	4824
Deviation	0	8.3	24.2	46.6	33.5	80.3	51.5
RSD	0	1.7	2.5	3.2	1.7	2.8	1.1
Error	0	4.8	4.2	3.5	4.1	4	3.5
% Recovery	0	95	96	96	96	96	96
Sequence Intraday (different day)							
STDA	**1**	**2**	**3**	**4**	**5**	**6**	**7**
Theoretical	0	500	1000	1500	2000	3000	5000
Analyst 1	0	481	927	1533	1911	2944	4890
Analyst 2	0	477	933	1478	1934	2920	4869
Analyst 3	0	469	957	1461	1873	2894	4835
Analyst 4	0	481	981	1432	1894	2904	4943
Analyst 5	0	493	935	1491	1933	2911	4905
Average	0	480.2	946.6	1479	1909	2914.6	4888.4
Deviation	0	8.7	22.3	37.4	26.1	19	40.3
RSD	0	1.8	2.4	2.5	1.4	0.7	0.8
Error	0	4	5.3	1.4	4.6	2.8	2.2
% Recovery	0	96	95	99	95	97	98

3.3. Linearity Analysis for the Calibration Curve and SPE of Standards

Using Origin, the average interday and intraday data were plotted against theoretical concentrations for the calibration curve and SPE. Each underwent a linear regression to obtain the parameters of R^2 and the correlation coefficient (r). Figure 7a,b show the linear regressions for the calibration curve, in which intraday and interday values of 0.99966 and 0.99997 were obtained for R^2 and 0.99983 and 0.99999 were obtained for the correlation coefficient. Figure 7c,d provide the same information as the previous paragraphs for the SPE unemployment. Here, intraday and interday values of 0.99999 and 0.99985 for R2 and 0.99999 and 0.99993 for the correlation coefficient were obtained.

3.4. Analysis of Concentration and Recovery of Irgafos 168 in Samples

To analyze the recovered Irgafos P-168, 40 samples were taken over 40 consecutive days. These samples were subjected to analysis by five different analysts on the same day, and the data are listed in Table 3. From the data, we can see that the RSDs were less than 15%; more specifically, the maximum was 4.7%, and the concentration of the recovered samples was in all cases was greater than 91.03%. The highest concentration, which was taken on day 26, was 99.38%, with a value average of 2548 ppm.

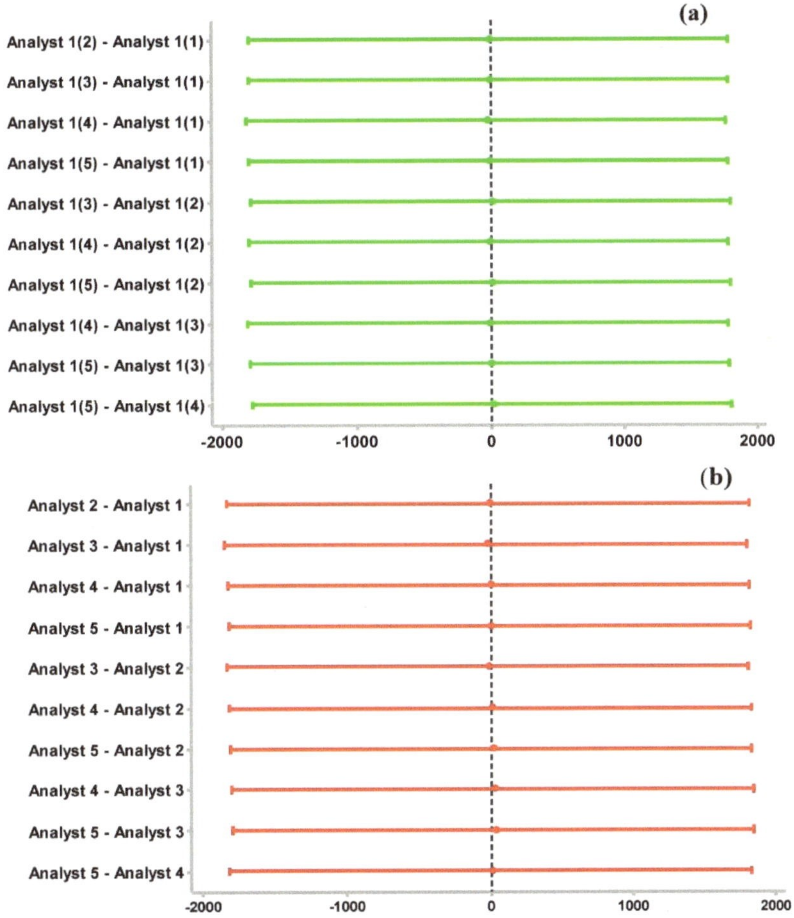

Figure 6. (**a**) Difference of means compared one to one for SPE repeatability. (**b**) Difference of means compared one to one for reproducibility of SPE.

Table 3. Quantification of Irgafos 168 in wastewater samples.

Values for the Extraction Method on Samples of Irgafos P-168 by SPE											
Day	Sample	Analyst 1	Analyst 2	Analyst 3	Analyst 4	Analyst 5	Average	Deviation	RSD	Recovery	% Recovery
1	1	1500	1512	1524	1538	1542	1523.2	17.6	1.2	1502	98.61
2	2	2100	2104	2075	2096	2104	2095.8	12.1	0.6	2000	95.43
3	3	1978	1875	1865	1910	1902	1906	44.3	2.3	1853	97.22
4	4	975	963	942	912	925	943.4	26	2.8	901	95.51
5	5	745	702	649	688	700	696.8	34.4	4.9	685	98.31
6	6	462	463	475	435	450	457	15.1	3.3	416	91.03
7	7	412	405	418	435	433	420.6	13.1	3.1	402	95.58
8	8	514	524	546	575	502	532.2	28.9	5.4	520	97.71
9	9	912	975	902	934	942	933	28.5	3.1	912	97.75
10	10	777	800	715	764	738	758.8	33.2	4.4	733	96.6
11	11	1000	975	956	1012	943	977.2	29	3	925	94.66

Table 3. *Cont.*

Values for the Extraction Method on Samples of Irgafos P-168 by SPE											
Day	Sample	Analyst 1	Analyst 2	Analyst 3	Analyst 4	Analyst 5	Average	Deviation	RSD	Recovery	% Recovery
12	12	1426	1354	1378	1300	1322	1356	49.2	3.6	1286	94.84
13	13	1542	1523	1567	1500	1586	1543.6	34.2	2.2	1508	97.69
14	14	947	977	968	942	975	961.8	16.2	1.7	952	98.98
15	15	871	822	809	839	826	833.4	23.6	2.8	826	99.11
16	16	766	743	709	785	768	754.2	29.4	3.9	733	97.19
17	17	700	705	766	745	735	730.2	27.7	3.8	716	98.06
18	18	974	955	961	908	911	941.8	30.3	3.2	904	95.99
19	19	850	824	809	834	870	837.4	23.6	2.8	813	97.09
20	20	352	342	365	385	366	362	16.2	4.5	342	94.48
21	21	711	702	742	733	746	726.8	19.4	2.7	709	97.55
22	22	911	915	908	977	934	929	28.7	3.1	913	98.28
23	23	1201	1245	1286	1300	1311	1268.6	45.3	3.6	1246	98.22
24	24	1542	1575	1563	1646	1602	1585.6	40.1	2.5	1546	97.5
25	25	2142	2092	2105	2134	2158	2126.2	27.1	1.3	2106	99.05
26	26	2535	2564	2571	2509	2641	2564	49.6	1.9	2548	99.38
27	27	3102	3152	3136	3184	3172	3149.2	32.2	1	3086	97.99
28	28	4150	4135	4106	4172	4108	4134.2	28.1	0.7	4087	98.86
29	29	3012	2975	2942	2955	3001	2977	29.6	1	2869	96.37
30	30	2807	2645	2711	2908	2938	2801.8	125.2	4.5	2716	96.94
31	31	2148	2299	2108	2276	2354	2237	104.4	4.7	2165	96.78
32	32	1545	1506	1562	1575	1536	1544.8	26.4	1.7	1502	97.23
33	33	2015	2106	2185	2153	2137	2119.2	64.8	3.1	2076	97.96
34	34	3102	3245	3371	3262	3312	3258.4	100.3	3.1	3154	96.8
35	35	2571	2534	2516	2509	2511	2528.2	25.9	1	2459	97.26
36	36	3014	3105	3275	3209	3315	3183.6	123.7	3.9	3077	96.65
37	37	1091	1142	1162	1085	1108	1117.6	33.3	3	1046	93.59
38	38	872	877	908	913	873	888.6	20.2	2.3	875	98.47
39	39	1075	1011	1134	1126	1088	1086.8	49.1	4.5	1042	95.88
40	40	1542	1500	1612	1573	1546	1554.6	41.4	2.7	1514	97.39

In order to assess the model's error, a linear regression was conducted. The results showed a proportionate association between the concentration determined by the analysts and the concentration of the Irgafos P-168 retrieved. As shown graphically in Figure 8, the model is predictive since it has an R^2 of 0.99955 and a correlation coefficient of 0.99978.

3.5. FTIR Analysis of the Recovered Irgafos

Samples of Irgafos recovered from the wastewater were taken for analysis by FTIR spectroscopy. The results indicated that the recovered Irgafos P-168 presented significant similarities with the spectrum of the compound in its pure state, as shown in Figure 9, suggesting a high purity of the recovered Irgafos P-168. Some differences are noted in the spectra due to noise from the equipment and the low concentrations of the samples used in this equipment.

In analyzing the spectrogram graph corresponding to the pure and recovered Irgafos P-168, an absorption band is found between 3000 and 2800 cm^{-1} that is related to the alcohol and ester groups. Precisely at 2868 cm^{-1}, a band of overlapping absorption corresponds to CH. At 2961 cm^{-1}, there is another band corresponding to CH_3. In the Irgafos P-168, the aromatic C=C bond is tenuously situated at 1602 cm^{-1} and 1491 cm^{-1}. At 1398 cm^{-1}, there is a corresponding deformed CH and CH_3 absorption band. Finally, at 1212 cm^{-1}, there is a band of the phosphite group (C-O-P. All these particularities are typical of Irgafos P-168 [63].

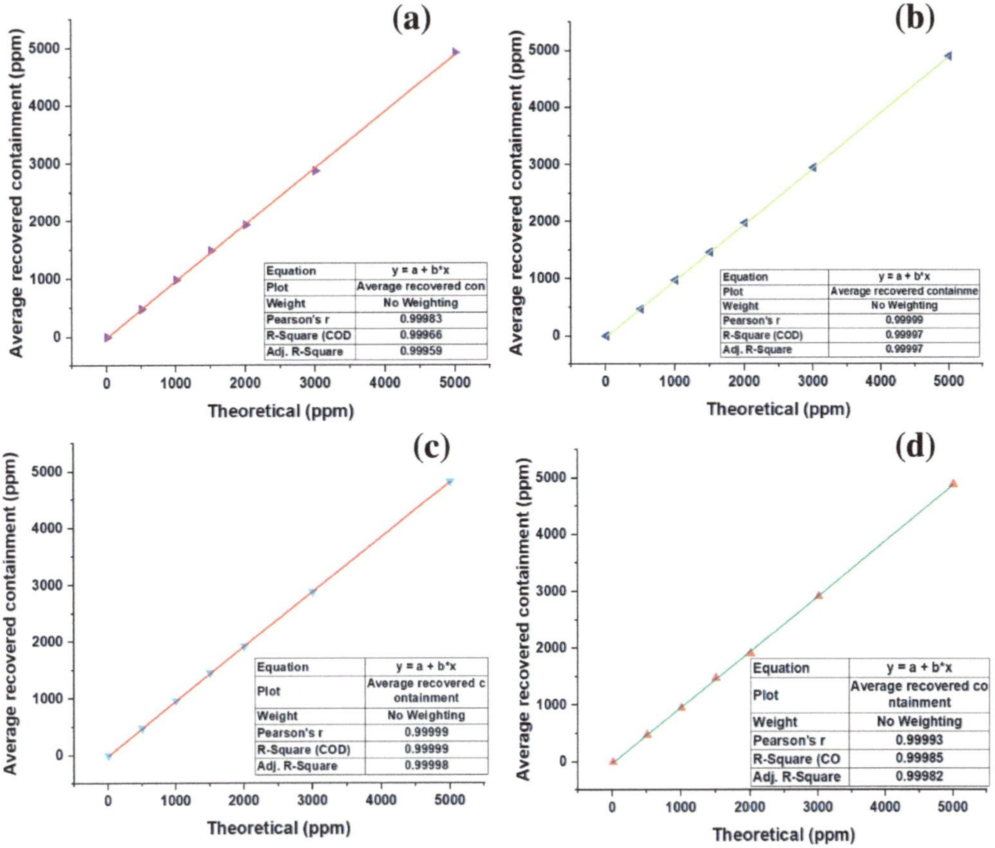

Figure 7. Linearity: (**a**) repeatability in calibration curve; (**b**) reproducibility in calibration curve; (**c**) repeatability for SPE; (**d**) reproducibility for SPE.

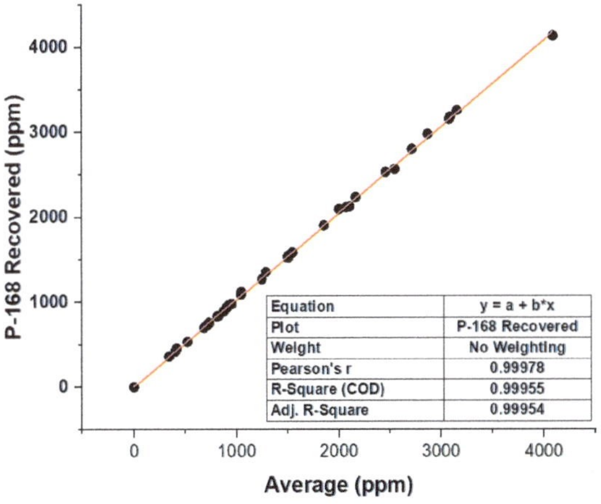

Figure 8. Average SPE concentrations of standards versus average concentration recovered from the samples.

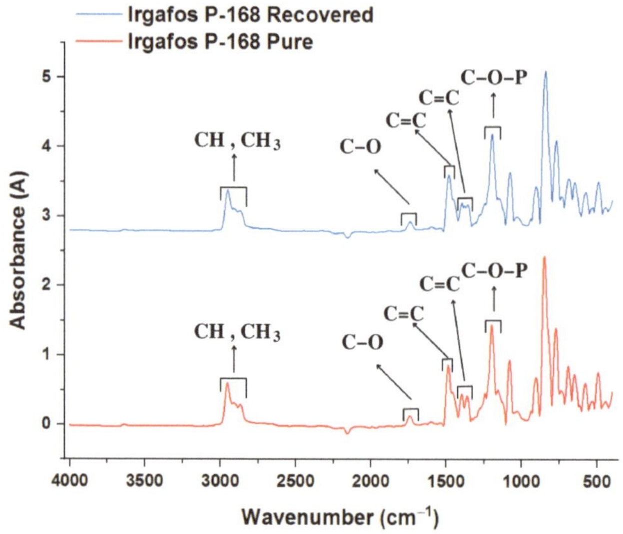

Figure 9. IR spectrogram for pure Irgafos P-168 and IR spectrogram for recovered Irgafos P-168.

3.6. TGA Analysis of the Recovered Irgafos

The physicochemical properties of the recovered material were analyzed by measuring its thermal stability. To measure its thermal decomposition point, scanning thermography tests were carried out on the Irgafos P-168 that was recovered after use in the PP process. Temperature changes were applied to the recovered and pure Irgafos samples at a rate of roughly 0.3 °C/s. It was observed that the thermal resistance of both samples was quite similar since in both cases, the mass did not vary substantially until it reached 280 °C. After this point, it began to decay; the results obtained suggest that the recovery process of the Irgafos P-168 did not significantly affect its thermal stability as its thermal decomposition point remained similar to that of pure Irgafos P-168, as can be seen in Figure 10.

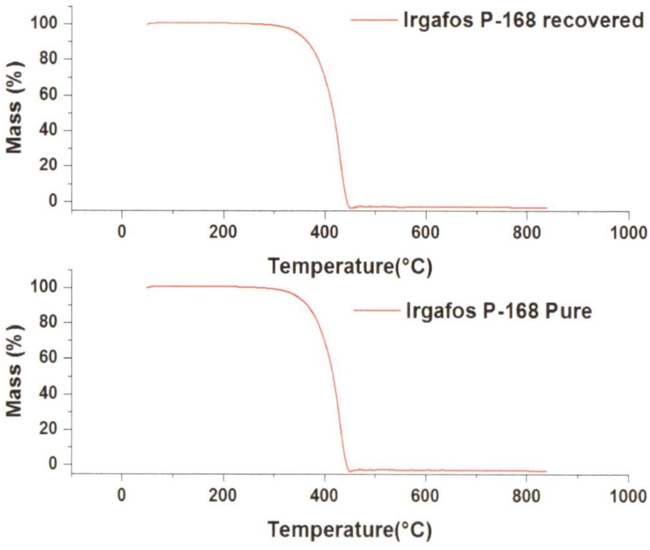

Figure 10. Pure Irgafos P-168 TGA and recovered Irgafos P-168 TGA.

To achieve a clearer vision of the similarity of the data, the derivative of the previous data was graphed (Figure 11) to show the similarity of the data. In these graphs, it can be seen that both the recovered Irgafos P-168 and the pure Irgafos P-168 demonstrate a change in the inflection point of the thermal curve at the same value, approximately 425 °C.

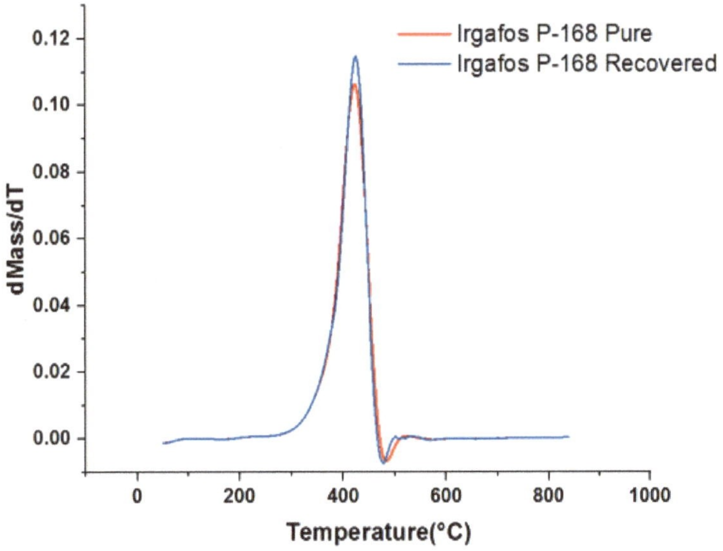

Figure 11. Derivative TGA of pure Irgafos and derivative TGA of recovered Irgafos.

Table 4 shows the mass of the ions and molecules obtained from the fragmentation of Irgafos P-168 by the detection technique of mass spectrometry coupled with TGA. Figure 12 shows the chemical and molecular structure of some of the ions generated from the fragmentation of the Irgafos P-168A molecular ion with a mass of 206; M+ can be seen in the 2,4-DTBP mass spectrum. The molecular ion at m/z 191 changes into an intense fragment ion after losing the methyl group. The phosphite mass spectrum shows the presence of the M+ molecular ion, and the loss of the 2,4-ditest group from the molecular ion is assumed to be butyl phenoxy, which produces the peak with a maximum intensity at m/z 441. It was then discovered that the sharp peak at m/z 647 in the phosphite spectrum is caused by a fragment ion created when the M+ molecular ion lost a methyl group. They contain the compound 2,4-di-tert-butylphenol in the phosphite spectrum at m/z 206. The fragment ions associated with the tropylium ion at m/z 91 and the tert-butyl fragment at m/z 57 are present in all ranges.

Table 4. Fragment mass and approximate relative abundance of pure and recovered Irgafos.

Qualification Ions of the Irgafos Pure (*m*/*z*)	Qualification Ions of the Irgafos Recovered (*m*/*z*)	Approximate Relative Abundance
57	57	49.0%
91	91	4.8%
147	147	29.8%
191	191	11.8%
237	237	4.7%
308	308	7.8%
329	329	8.0%
385	385	12.1%
441	441	100%
646	646	12.28%

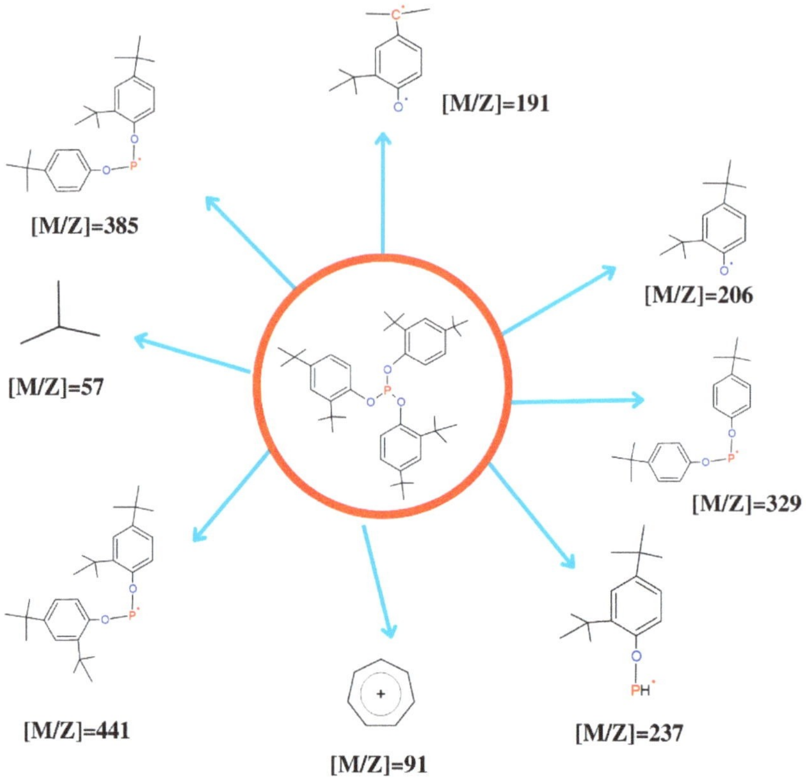

Figure 12. Structure of molecular ions from the fragmentation of Irgafos P-168.

3.7. DSC Analysis of the Recovered Irgafos

The purity of a chemical compound such as Irgafos P-168 can be analyzed using various scientific techniques. In this case, DSC was used to assess the purity of the Irgafos recovered from wastewater. In the test carried out, a temperature change was applied at a rate of 19.7 °C/min. The data obtained indicated that the melting point of the sample was close to 187 °C, and it reached a maximum point of heat absorption of approximately 40 mW. When comparing these results with those obtained by subjecting the compound in its pure state to the same conditions, it was possible to conclude that the purity of the recovered Irgafos is still relatively high, given the similarity of the data (see Figure 13).

3.8. FTIR Analysis of PP Plus Pure Irgafos P-168 and PP Plus Recovered Irgafos P-168

To generate the most accurate analysis possible, the reuse of the recovered Irgafos P-168 in the production of polypropylene was carried out to compare the characteristics of the final material with both the pure and the recovered additive.

Both materials were subjected to FTIR spectroscopy studies (see Figura 14) with resultant graphs. We can see that the absorbance regions are the same for both materials, meaning that the performance of P-168 is high since the characteristics of the obtained PP are the same as those obtained with the pure additive. The observed spectra (Figure 14) display two bands between 1305 and 1055 cm^{-1} that correspond to the symmetric and asymmetric stretches of the ester group of Irgafos P-168 and a peak at roughly 1737 cm^{-1} that is suggestive of the ester group (C=O) present in the structure of the Irgafos P-168. Between 2953 and 2975 cm^{-1}, one can see the typical CH_3 group band. Irgafos P-168's chemical composition also reveals these classes. Another indication of the existence of aromatic groups in the spectrum is the relative strength of the absorption, which is typical of the spectra of aromatic compounds and occurs in the range of 1450–1500 cm^{-1}.

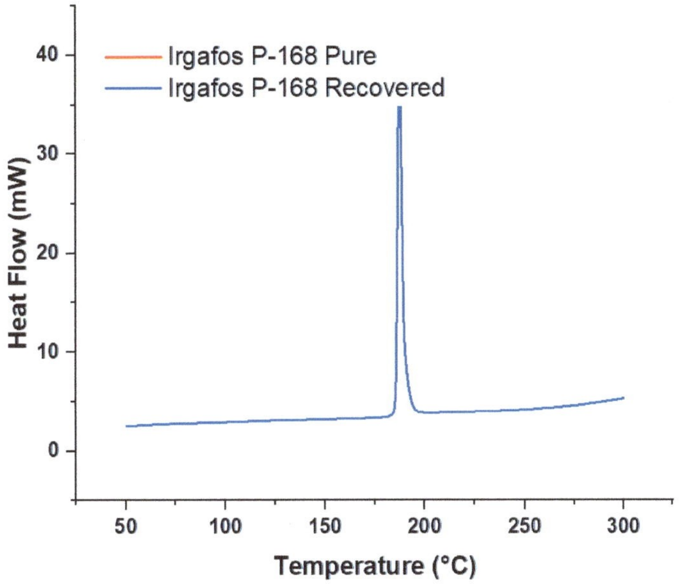

Figure 13. DSC Irgafos 168 pure and DSC Irgafos 168 recovered.

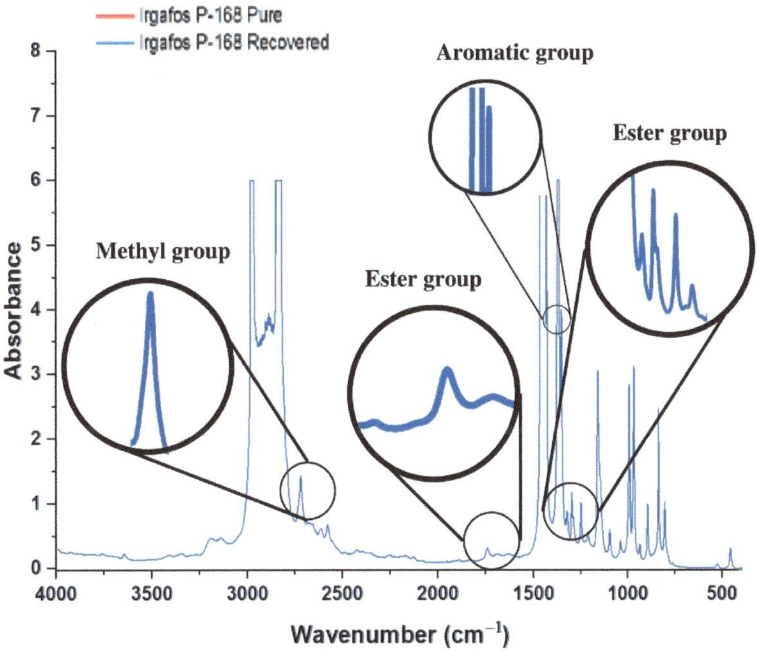

Figure 14. PP + P-168 pure and PP + P-168 recovered.

3.9. Melt Flow Index (MFI) and Molecular Weight Distribution

The thermal properties of polypropylene are some of the most critical qualities for its processing and the evaluation of multiple changes that can affect its quality during application. Using the MFI (see Figure 15), we can estimate the melt processing of the PP; the higher the MFI values, the higher the melt flowability, and the lower the MFI values, the lower the melt flowability of the PP. The MFI measurements of each sample were performed in triplicate. The MFI results are shown in Figure 15, and it can be seen that the MFI of the PP decreased with the dosage of the recovered P-168 and pure P-168. Concentrations of 0.1% of pure or recovered P-168 significantly improve the MFI. Without additives, PP has an MFI of 6.5. This MFI decreased to 4.1 and 4.3 with the addition of pure P-168 and recovered P-168, respectively.

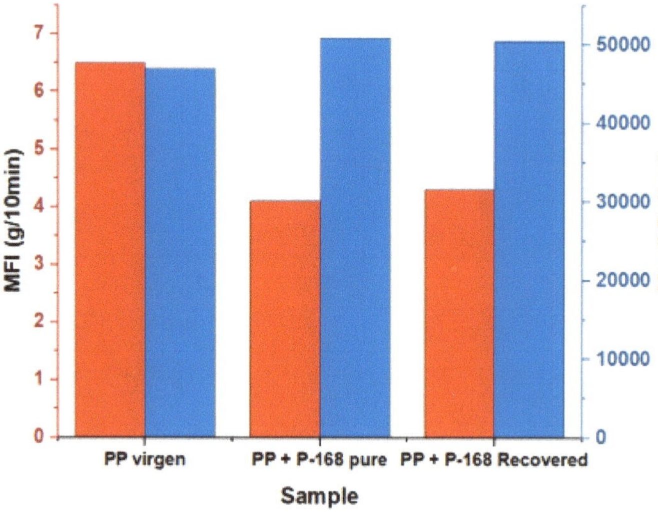

Figure 15. MFI PP + P-168 pure and PP + P-168 recovered.

This molecular weight was determined experimentally and theoretically using the Bramner equation as a reference (Equation (1)):

$$\overline{M_v} = -8480.6 \times Ln \text{ MFI} + 62,836 \tag{1}$$

The Bramner equation shows the relationship between the MFI and the average molecular weight (Mw). In Figure 15, an inverse relationship between the MFI and Mw is observed. Samples with an MFI of 6.5, 4.1, and 4.3 g/10 min presented Mw values of 46,961, 50, 870, and 50,466 kDaltons. It was observed that the addition of Irgafos P-168 stabilized the PP, preventing its complete degradation, decreasing its fluidity, and increasing its molecular weight distribution.

3.10. OIT

To determine the oxidation time, the changes in the slope of the curve generated by the DSC concerning the expected time and the heat flux were taken into account. As shown in Figure 16, an endothermic peak was observed, but with time and the oxygen atmosphere, the slope of the curves changed, revealing a new exothermic behavior consistent with oxidation. It can be seen that the non-stabilized PP had a significantly shorter oxidation time than the PP with additives. The OIT value for the unstabilized PP was 0.9 min, taking the change in slope at 17.3 min as reference, and the OIT for pure PP + P-168 and recovered PP + P-168 were 8 and 7.8 min, respectively. This demonstrates how the presence of P-168 minimizes the oxidation processes of PP and improves its thermal stability due to its coupling in the polymeric matrix.

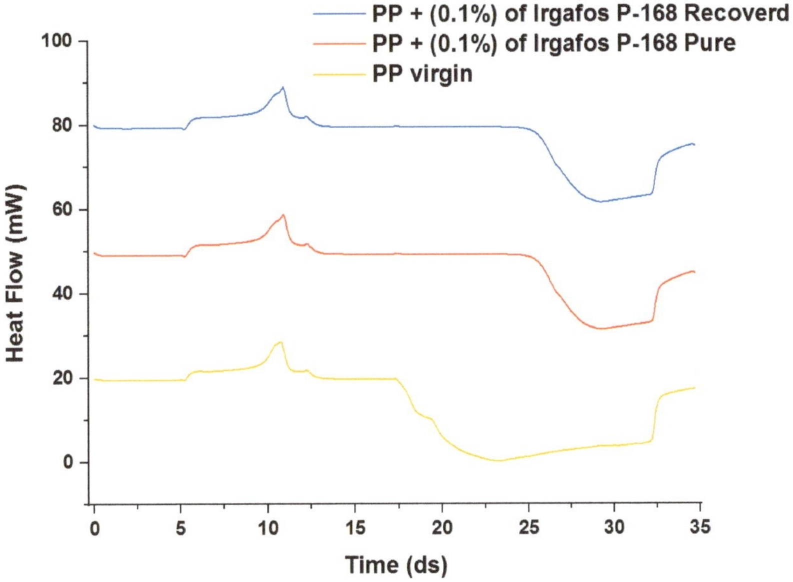

Figure 16. OIT PP, PP + P-168 pure and PP + P-168 recovered.

4. Conclusions

According to the findings of this study, it is possible to recover Irgafos P-168 with a high degree of purity in amounts greater than 91.03%, and its integration into the PP matrix enhances the material's thermal and thermo-oxidative stability. Without thermal stabilizers, PP cannot be used since it would entirely deteriorate during the extrusion process. As a result, the recovery of this Irgafos P-168 at such high purity levels exemplifies a critical methodology for use in the industrial sector, promotes the use of sustainable raw materials, and makes a significant contribution to the circular economy of the PP polymer industry.

Author Contributions: Conceptualization, H.C. and J.H.-F.; methodology, A.F.R.; validation, J.H.-F., A.F.R. and H.C.; formal analysis, A.F.R.; investigation, H.C. and J.H.-F.; writing—original draft preparation, J.H.-F.; writing—review and editing, H.C. and A.F.R.; supervision, J.H.-F.; project administration, A.F.R. All authors have read and agreed to the published version of the manuscript.

Funding: This research received no external funding.

Institutional Review Board Statement: Not applicable.

Informed Consent Statement: Not applicable.

Data Availability Statement: Not applicable.

Conflicts of Interest: The authors declare no conflict of interest.

References

1. Hazarika, M.; Dixit, U.S.; Davim, J.P. History of Production and Industrial Engineering through Contributions of Stalwarts. *Manuf. Eng. Educ.* **2019**, 1–29. [CrossRef]
2. Cao, L.; Lin, C.; Gao, Y.; Sun, C.; Xu, L.; Zheng, L.; Zhang, Z. Health risk assessment of trace elements exposure through the soil-plant (maize)-human contamination pathway near a petrochemical industry complex, Northeast China. *Environ. Pollut.* **2020**, *263*, 114414. [CrossRef] [PubMed]

3. Gebbink, W.A.; van Leeuwen, S.P. Environmental contamination and human exposure to PFASs near a fluorochemical production plant: Review of historic and current PFOA and GenX contamination in the Netherlands. *Environ. Int.* **2020**, *137*, 105583. [CrossRef] [PubMed]
4. Nie, M.; Nie, H.; He, M.; Lin, Y.; Wang, L.; Jin, P.; Zhang, S. Immobilization of biofilms of Pseudomonas aeruginosa NY3 and their application in the removal of hydrocarbons from highly concentrated oil-containing wastewater on the laboratory scale. *J. Environ. Manag.* **2016**, *173*, 34–40. [CrossRef]
5. Kumar, L.; Chugh, M.; Kumar, S.; Kumar, K.; Sharma, J.; Bharadvaja, N. Remediation of petrorefinery wastewater contaminants: A review on physicochemical and bioremediation strategies. *Process Saf. Environ. Prot.* **2022**, *159*, 362–375. [CrossRef]
6. van Oosterhout, L.; Dijkstra, H.; Borst, D.; Duijndam, S.; Rehdanz, K.; van Beukering, P. Triggering sustainable plastics consumption behavior: Identifying consumer profiles across Europe and designing strategies to engage them. *Sustain. Prod. Consum.* **2023**, *36*, 148–160. [CrossRef]
7. Pivato, A.F.; Miranda, G.M.; Prichula, J.; Lima, J.E.; Ligabue, R.A.; Seixas, A.; Trentin, D.S. Hydrocarbon-based plastics: Progress and perspectives on consumption and biodegradation by insect larvae. *Chemosphere* **2022**, *293*, 133600. [CrossRef]
8. Sridharan, S.; Kumar, M.; Saha, M.; Kirkham, M.; Singh, L.; Bolan, N.S. The polymers and their additives in particulate plastics: What makes them hazardous to the fauna? *Sci. Total Environ.* **2022**, *824*, 153828. [CrossRef]
9. Marcato, B.; Guerra, S.; Vianello, M.; Scalia, S. Migration of antioxidant additives from various polyolefinic plastics into oleaginous vehicles. *Int. J. Pharm.* **2003**, *257*, 217–225. [CrossRef]
10. Liao, B.; Ji, G.; Cheng, L. Profling of microbial communities in a bioreactor for treating hydrocarbon-sulfde-containing wastewater. *J. Environ. Sci.* **2008**, *20*, 897–899. [CrossRef]
11. Goswami, L.; Manikandan, N.A.; Dolman, B.; Pakshirajan, K.; Pugazhenthi, G. Biological treatment of wastewater containing a mixture of polycyclic aromatic hydrocarbons using the oleaginous bacterium Rhodococcus opacus. *J. Clean. Prod.* **2018**, *196*, 1282–1291. [CrossRef]
12. Fernández, J.H.; Cano, H.; Guerra, Y.; Polo, E.P.; Ríos-Rojas, J.F.; Vivas-Reyes, R.; Oviedo, J. Identification and Quantification of Microplastics in Effluents of Wastewater Treatment Plant by Differential Scanning Calorimetry (DSC). *Sustainability* **2022**, *14*, 4920. [CrossRef]
13. Chaudhry, A.; Bashir, F.; Adil, S.F.; Saif, S.; Shaik, M.R.; Hatshan, M.R.; Shaik, B. Ascorbic acid-mediated Fe/Cu nanoparticles and their application for removal of COD and phenols from industrial wastewater. *J. King Saud Univ.-Sci.* **2022**, *34*, 101927. [CrossRef]
14. Lwanga, E.H.; van Roshum, I.; Munhoz, D.R.; Meng, K.; Rezaei, M.; Goossens, D.; Bijsterbosch, J.; Alexandre, N.; Oosterwijk, J.; Krol, M.; et al. Microplastic appraisal of soil, water, ditch sediment and airborne dust: The case of agricultural systems. *Environ. Pollut.* **2023**, *316*, 120513. [CrossRef] [PubMed]
15. Franco, A.; Arellano, J.; Albendín, G.; Rodríguez-Barroso, R.; Quiroga, J.; Coello, M. Microplastic pollution in wastewater treatment plants in the city of Cádiz: Abundance, removal efficiency and presence in receiving water body. *Sci. Total Environ.* **2021**, *776*, 145795. [CrossRef]
16. Hernández-Fernandez, J.; Rodríguez, E. Determination of phenolic antioxidants additives in industrial wastewater from polypropylene production using solid phase extraction with high-performance liquid chromatography. *J. Chromatogr. A* **2019**, *1607*, 460442. [CrossRef]
17. Allen, N.S.; Edge, M.; Hussain, S. Perspectives on yellowing in the degradation of polymer materials: Inter-relationship of structure, mechanisms and modes of stabilisation. *Polym. Degrad. Stab.* **2022**, *201*, 109977. [CrossRef]
18. Cifuentes-Cabezas, M.; Mendoza-Roca, J.A.; Vincent-Vela, M.C.; Álvarez-Blanco, S. Management of reject streams from hybrid membrane processes applied to phenolic compounds removal from olive mill wastewater by adsorption/desorption and biological processes. *J. Water Process Eng.* **2022**, *50*, 103208. [CrossRef]
19. Alsabri, A.; Tahir, F.; Al-Ghamdi, S.G. Environmental impacts of polypropylene (PP) production and prospects of its recycling in the GCC region. *Mater. Today Proc.* **2021**, *56*, 2245–2251. [CrossRef]
20. Irshidat, M.R.; Al-Nuaimi, N.; Rabie, M. Hybrid effect of carbon nanotubes and polypropylene microfibers on fire resistance, thermal characteristics and microstructure of cementitious composites. *Constr. Build. Mater.* **2021**, *266*, 121154. [CrossRef]
21. Nascimento, E.M.D.; Eiras, D.; Pessan, L.A. Effect of thermal treatment on impact resistance and mechanical properties of polypropylene/calcium carbonate nanocomposites. *Compos. Part B Eng.* **2016**, *91*, 228–234. [CrossRef]
22. Pavon, C.; Aldas, M.; López-Martínez, J.; Hernández-Fernández, J.; Arrieta, M. Films Based on Thermoplastic Starch Blended with Pine Resin Derivatives for Food Packaging. *Foods* **2021**, *10*, 1171. [CrossRef]
23. Joaquin, H.-F.; Juan, L.-M. Autocatalytic influence of different levels of arsine on the thermal stability and pyrolysis of polypropylene. *J. Anal. Appl. Pyrolysis* **2022**, *161*, 105385. [CrossRef]
24. Hernández-Fernández, J. Quantification of oxygenates, sulphides, thiols and permanent gases in propylene. A multiple linear regression model to predict the loss of efficiency in polypropylene production on an industrial scale. *J. Chromatogr. A* **2020**, *1628*, 461478. [CrossRef]
25. Hernández-Fernández, J.; Lopez-Martinez, J.; Barceló, D. Quantification and elimination of substituted synthetic phenols and volatile organic compounds in the wastewater treatment plant during the production of industrial scale polypropylene. *Chemosphere* **2021**, *263*, 128027. [CrossRef] [PubMed]

26. Hernández-Fernandez, J.; Lopez-Martinez, J.; Puello-Polo, E. Recovery of (Z)-13-Docosenamide from Industrial Wastewater and Its Application in the Production of Virgin Polypropylene to Improve the Coefficient of Friction in Film Type Applications. *Sustainability* **2023**, *15*, 1247. [CrossRef]
27. Joaquin, H.-F.; Juan, L. Quantification of poisons for Ziegler Natta catalysts and effects on the production of polypropylene by gas chromatographic with simultaneous detection: Pulsed discharge helium ionization, mass spectrometry and flame ionization. *J. Chromatogr. A* **2020**, *1614*, 460736. [CrossRef] [PubMed]
28. Hernández-Fernández, J.; Cano-Cuadro, H.; Puello-Polo, E. Emission of Bisphenol A and Four New Analogs from Industrial Wastewater Treatment Plants in the Production Processes of Polypropylene and Polyethylene Terephthalate in South America. *Sustainability* **2022**, *14*, 10919. [CrossRef]
29. Hernández-Fernández, J.; Cano, H.; Aldas, M. Impact of Traces of Hydrogen Sulfide on the Efficiency of Ziegler–Natta Catalyst on the Final Properties of Polypropylene. *Polymers* **2022**, *14*, 3910. [CrossRef]
30. Hernández-Fernández, J.; Guerra, Y.; Espinosa, E. Development and Application of a Principal Component Analysis Model to Quantify the Green Ethylene Content in Virgin Impact Copolymer Resins During Their Synthesis on an Industrial Scale. *J. Polym. Environ.* **2022**, *30*, 4800–4808. [CrossRef]
31. Hernández-Fernández, J.; Guerra, Y.; Puello-Polo, E.; Marquez, E. Effects of Different Concentrations of Arsine on the Synthesis and Final Properties of Polypropylene. *Polymers* **2022**, *14*, 3123. [CrossRef] [PubMed]
32. Hernández-Fernández, J.; Vivas-Reyes, R.; Toloza, C.A.T. Experimental Study of the Impact of Trace Amounts of Acetylene and Methylacetylene on the Synthesis, Mechanical and Thermal Properties of Polypropylene. *Int. J. Mol. Sci.* **2022**, *23*, 12148. [CrossRef] [PubMed]
33. Lavrenov, A.V.; Saifulina, L.F.; Buluchevskii, E.A.; Bogdanets, E.N. Propylene production technology: Today and tomorrow. *Catal. Ind.* **2015**, *7*, 175–187. [CrossRef]
34. Tähkämö, L.; Ojanperä, A.; Kemppi, J.; Deviatkin, I. Life cycle assessment of renewable liquid hydrocarbons, propylene, and polypropylene derived from bio-based waste and residues: Evaluation of climate change impacts and abiotic resource depletion potential. *J. Clean. Prod.* **2022**, *379*, 134645. [CrossRef]
35. Hernández-Fernández, J.; Castro-Suarez, J.R.; Toloza, C.A.T. Iron Oxide Powder as Responsible for the Generation of Industrial Polypropylene Waste and as a Co-Catalyst for the Pyrolysis of Non-Additive Resins. *Int. J. Mol. Sci.* **2022**, *23*, 11708. [CrossRef]
36. Petrovics, N.; Kirchkeszner, C.; Tábi, T.; Magyar, N.; Székely, I.K.; Szabó, B.S.; Nyiri, Z.; Eke, Z. Effect of temperature and plasticizer content of polypropylene and polylactic acid on migration kinetics into isooctane and 95 v/v% ethanol as alternative fatty food simulants. *Food Packag. Shelf Life* **2022**, *33*, 100916. [CrossRef]
37. Hermabessiere, L.; Receveur, J.; Himber, C.; Mazurais, D.; Huvet, A.; Lagarde, F.; Lambert, C.; Paul-Pont, I.; Dehaut, A.; Jezequel, R.; et al. An Irgafos® 168 story: When the ubiquity of an additive prevents studying its leaching from plastics. *Sci. Total Environ.* **2020**, *749*, 141651. [CrossRef]
38. Vera, P.; Canellas, E.; Su, Q.-Z.; Mercado, D.; Nerín, C. Migration of volatile substances from recycled high density polyethylene to milk products. *Food Packag. Shelf Life* **2023**, *35*, 101020. [CrossRef]
39. Kung, H.-C.; Hsieh, Y.-K.; Huang, B.-W.; Cheruiyot, N.K.; Chang-Chien, G.-P. An Overview: Organophosphate Flame Retardants in the Atmosphere. *Aerosol Air Qual. Res.* **2022**, *22*, 220148. [CrossRef]
40. Onoja, S.; Nel, H.A.; Abdallah, M.A.-E.; Harrad, S. Microplastics in freshwater sediments: Analytical methods, temporal trends, and risk of associated organophosphate esters as exemplar plastics additives. *Environ. Res.* **2022**, *203*, 111830. [CrossRef]
41. Xiao, L.; Zheng, Z.; Irgum, K.; Andersson, P.L. Studies of Emission Processes of Polymer Additives into Water Using Quartz Crystal Microbalance—A Case Study on Organophosphate Esters. *Environ. Sci. Technol.* **2020**, *54*, 4876–4885. [CrossRef]
42. Li, A.; Zheng, G.; Chen, N.; Xu, W.; Li, Y.; Shen, F.; Wang, S.; Cao, G.; Li, J. Occurrence Characteristics and Ecological Risk Assessment of Organophosphorus Compounds in a Wastewater Treatment Plant and Upstream Enterprises. *Water* **2022**, *14*, 3942. [CrossRef]
43. Liu, R.; Mabury, S.A. Synthetic Phenolic Antioxidants: A Review of Environmental Occurrence, Fate, Human Exposure, and Toxicity. *Environ. Sci. Technol.* **2020**, *54*, 11706–11719. [CrossRef]
44. Dương, T.-B.; Dwivedi, R.; Bain, L.J. 2,4-di-tert-butylphenol exposure impairs osteogenic differentiation. *Toxicol. Appl. Pharmacol.* **2023**, *461*, 116386. [CrossRef]
45. Chen, Y.; Chen, Q.; Zhang, Q.; Zuo, C.; Shi, H. An Overview of Chemical Additives on (Micro)Plastic Fibers: Occurrence, Release, and Health Risks. *Rev. Environ. Contam. Toxicol.* **2022**, *260*, 22. [CrossRef]
46. Simoneau, C.; Van Den Eede, L.; Valzacchi, S. Identification and quantification of the migration of chemicals from plastic baby bottles used as substitutes for polycarbonate. *Food Addit. Contam. Part A* **2012**, *29*, 469–480. [CrossRef] [PubMed]
47. Fouyer, K.; Lavastre, O.; Rondeau, D. Direct Monitoring of the Role Played by a Stabilizer in a Solid Sample of Polymer Using Direct Analysis in Real Time Mass Spectrometry: The Case of Irgafos 168 in Polyethylene. *Anal. Chem.* **2012**, *84*, 8642–8649. [CrossRef] [PubMed]
48. Sommers, C.H.; Sheen, S. Inactivation of avirulent Yersinia pestis on food and food contact surfaces by ultraviolet light and freezing. *Food Microbiol.* **2015**, *50*, 1–4. [CrossRef] [PubMed]
49. FDA. Irradiation in the Production, Processing and Handling of Food. *Final Rule, November 2012*. Available online: https://www.researchgate.net/publication/284923753_Irradiation_in_the_production_processing_and_handling_of_food_Final_rule (accessed on 3 March 2022).

50. Yang, Y.P.; Hu, C.; Zhong, H.; Wang, Z.Y.; Zeng, G.F. Degradation of Irgafos 168 and determination of its degra-dation products. *Mod. Food Sci. Technol.* **2016**, *32*, 304–309. [CrossRef]
51. James, B.D.; De Vos, A.; Aluwihare, L.I.; Youngs, S.; Ward, C.P.; Nelson, R.K.; Michel, A.P.M.; Hahn, M.E.; Reddy, C.M. Divergent Forms of Pyroplastic: Lessons Learned from the M/V *X-Press Pearl* Ship Fire. *ACS Environ. Au* **2022**, *2*, 467–479. [CrossRef]
52. Zhao, F.; Wang, P.; Lucardi, R.; Su, Z.; Li, S. Natural Sources and Bioactivities of 2,4-Di-Tert-Butylphenol and Its Analogs. *Toxins* **2020**, *12*, 35. [CrossRef]
53. Shi, J.; Xu, C.; Xiang, L.; Chen, J.; Cai, Z. Tris(2,4-di-*tert*-butylphenyl)phosphate: An Unexpected Abundant Toxic Pollutant Found in $PM_{2.5}$. *Environ. Sci. Technol.* **2020**, *54*, 10570–10576. [CrossRef] [PubMed]
54. Luque-García, J.; de Castro, M.L. Ultrasound: A powerful tool for leaching. *TrAC Trends Anal. Chem.* **2003**, *22*, 41–47. [CrossRef]
55. Lama-Muñoz, A.; Contreras, M.D.M. Extraction Systems and Analytical Techniques for Food Phenolic Compounds: A Review. *Foods* **2022**, *11*, 3671. [CrossRef] [PubMed]
56. Sachon, E.; Matheron, L.; Clodic, G.; Blasco, T.; Bolbach, G. MALDI TOF-TOF characterization of a light stabilizer polymer contaminant from polypropylene or polyethylene plastic test tubes. *J. Mass Spectrom.* **2010**, *45*, 43–50. [CrossRef] [PubMed]
57. Feng, G.; Wang, X.; Zhang, D.; Xiao, X.; Qian, K. Fabrication of bilayer antioxidant microcapsule and evaluation of its efficiency in stabilization of polypropylene. *Mater. Res. Express* **2019**, *6*, 125327. [CrossRef]
58. Farajzadeh, M.A.; Goushjuii, L.; Ranji, A.; Feyz, E. Spectrophotometric determination of Irgafos 168 in polymers after different sample preparation procedures. *Microchim. Acta* **2007**, *159*, 263–268. [CrossRef]
59. Fiorio, R.; D'Hooge, D.R.; Ragaert, K.; Cardon, L. A Statistical Analysis on the Effect of Antioxidants on the Thermal-Oxidative Stability of Commercial Mass- and Emulsion-Polymerized ABS. *Polymers* **2018**, *11*, 25. [CrossRef]
60. Li, B.; Wang, Z.-W.; Lin, Q.-B.; Hu, C.-Y.; Su, Q.-Z.; Wu, Y.-M. Determination of Polymer Additives-Antioxidants, Ultraviolet Stabilizers, Plasticizers and Photoinitiators in Plastic Food Package by Accelerated Solvent Extraction Coupled with High-Performance Liquid Chromatography. *J. Chromatogr. Sci.* **2015**, *53*, 1026–1035. [CrossRef]
61. Rodil, R.; Quintana, J.B.; Basaglia, G.; Pietrogrande, M.C.; Cela, R. Determination of synthetic phenolic antioxidants and their metabolites in water samples by downscaled solid-phase extraction, silylation and gas chromatography–mass spectrometry. *J. Chromatogr. A* **2010**, *1217*, 6428–6435. [CrossRef]
62. Hernández-Fernández, J.; Ortega-Toro, R.; López-Martinez, J. A New Route of Valorization of Petrochemical Wastewater: Recovery of 1,3,5-Tris (4-tert-butyl-3-hydroxy-2,6-dimethyl benzyl)–1,3,5-triazine-2,4,6-(1H,3H,5H)-trione (Cyanox 1790) and Its Subsequent Application in a PP Matrix to Improve Its Thermal Stability. *Molecules* **2023**, *28*, 2003. [CrossRef] [PubMed]
63. Badri, K.; Redwan, A. Molecular Characterization of Synthetic Polymers by Means of Liquid Chromatography. In *Physical Chemistry of Macromolecules: Macro to Nanoscales*; Apple Academic Press: Bratislava, Slovakia, 2014; pp. 237–348. [CrossRef]

MDPI AG
Grosspeteranlage 5
4052 Basel
Switzerland
Tel.: +41 61 683 77 34

Molecules Editorial Office
E-mail: molecules@mdpi.com
www.mdpi.com/journal/molecules

www.ingramcontent.com/pod-product-compliance
Lightning Source LLC
LaVergne TN
LVHW072337090526
838202LV00019B/2434